Esterases, Lipases, and Phospholipases

From Structure to Clinical Significance

NATO ASI Series

Advanced Science Institutes Series

A series presenting the results of activities sponsored by the NATO Science Committee, which aims at the dissemination of advanced scientific and technological knowledge, with a view to strengthening links between scientific communities.

The series is published by an international board of publishers in conjunction with the NATO Scientific Affairs Division

A	**Life Sciences**	Plenum Publishing Corporation
B	**Physics**	New York and London
C	**Mathematical and Physical Sciences**	Kluwer Academic Publishers
D	**Behavioral and Social Sciences**	Dordrecht, Boston, and London
E	**Applied Sciences**	
F	**Computer and Systems Sciences**	Springer-Verlag
G	**Ecological Sciences**	Berlin, Heidelberg, New York, London,
H	**Cell Biology**	Paris, Tokyo, Hong Kong, and Barcelona
I	**Global Environmental Change**	

Recent Volumes in this Series

Volume 264 — Magnetic Resonance Scanning and Epilepsy
edited by S.D. Shorvon, D.R. Fish, F. Andermann, G.M. Bydder, and H. Stefan

Volume 265 — Basic Mechanisms of Physiologic and Aberrant Lymphoproliferation in the Skin
edited by W.Clark Lambert, Benvenuto Giannotti, and Willem A. van Vloten

Volume 266 — Esterases, Lipases, and Phospholipases: From Structure to Clinical Significance
edited by M.I. Mackness and M. Clerc

Volume 267 — Bioelectrochemistry IV: Nerve Muscle Function— Bioelectrochemistry, Mechanisms, Bioenergetics, and Control
edited by Bruno Andrea Melandri, Giulio Milazzo, and Martin Blank

Volume 268 — Advances in Molecular Plant Nematology
edited by F. Lamberti, C. De Giorgi, and D. McK. Bird

Volume 269 — Ascomycete Systematics: Problems and Perspectives in the Nineties
edited by David L. Hawksworth

Volume 270 — Standardization of Epidemiologic Studies of Host Susceptibility
edited by Janice Dorman

Series A: Life Sciences

Esterases, Lipases, and Phospholipases

From Structure to Clinical Significance

Edited by

M. I. Mackness

University of Manchester
Manchester, United Kingdom

and

M. Clerc

University of Bordeaux II
Bordeaux, France

Plenum Press
New York and London
Published in cooperation with NATO Scientific Affairs Division

Proceedings of a NATO Advanced Research Workshop on
Esterases, Lipases, and Phospholipases,
held September 22–24, 1993,
in Bordeaux, France

QP
609
, E8
E88
1994

NATO-PCO-DATA BASE

The electronic index to the NATO ASI Series provides full bibliographical references (with keywords and/or abstracts) to more than 30,000 contributions from international scientists published in all sections of the NATO ASI Series. Access to the NATO-PCO-DATA BASE is possible in two ways:

—via online FILE 128 (NATO-PCO-DATA BASE) hosted by ESRIN, Via Galileo Galilei, I-00044 Frascati, Italy

—via CD-ROM "NATO Science and Technology Disk" with user-friendly retrieval software in English, French, and German (©WTV GmbH and DATAWARE Technologies, Inc. 1989). The CD-ROM also contains the AGARD Aerospace Database.

The CD-ROM can be ordered through any member of the Board of Publishers or through NATO-PCO, Overijse, Belgium.

```
              Library of Congress Cataloging-in-Publication Data


    Esterases, lipases, and phospholipases : from structure to clinical
      significance / edited by M.I. Mackness and M. Clerc.
          p.   cm. -- (NATO ASI series. Series A, Life sciences ; v.
    266)
      "Proceedings of a NATO Advanced Research Workshop on Esterases,
    Lipases, and Phospholipases, held September 22-24, 1993, in
    Bordeaux, France"--T.p. verso.
      "Published in cooperation with NATO Scientific Affairs Division."
      Includes bibliographical references and index.
      ISBN 0-306-44802-5
      1. Esterases--Congresses.  2. Lipase--Congresses.
    3. Phospholipases--Congresses.   I. Mackness, M. I. (Michael I.)
    II. Clerc, M. (Michel)  III. NATO Advanced Research Workshop on
    Esterases, Lipases, and Phospholipases (1993 : Bordeaux, France)
    IV. Series.
      [DNLM: 1. Esterases--congresses.  2. Lipase--congresses.
    3. Phospholipases--congresses.   QU 136 E79 1994]
    QP609.E8E88  1994
    612'.0151--dc20
    DNLM/DLC
    for Library of Congress                              94-36095
                                                              CIP
```

ISBN 0-306-44802-5

©1994 Plenum Press, New York
A Division of Plenum Publishing Corporation
233 Spring Street, New York, N.Y. 10013

PREFACE

The NATO-Advanced Research Workshop "Esterases, Lipases and Phospholipases: From Structure to Clinical Significance" was held at the University of Bordeaux II, France from 22nd-24th September 1993 under the Directorship of Professor Michel Clerc of the University of Bordeaux II. The meeting was organised by Hugues Chap (INSERM U 326, Toulouse, France), Georges Ferard (University of Strasbourg, France), Wolfgang Junge (University of Kiel, Germany) and Michael Mackness (University of Manchester, UK).

In recent years it has become increasingly apparent that hydrolytic enzymes of the esterase, lipase and phospholipase type play central roles in the pathophysiology of many human diseases. The purpose of this NATO-ARW was to bring together experts (both clinical and scientific) in all three interrelated fields to review the current basic and clinical position and discuss future developments particularly with respect to future research aimed at determining the basic biochemical lesion involving hydrolytic enzymes involved in human disease and the use of these enzymes in diagnosis.

As well as formal lectures from established researchers, the meeting also involved a number of lively round-table discussions on future developments and presentations from younger research workers, all of which are recorded in this Proceedings and which contribute to the success of the meeting.

As well as the substantial financial assistance which was received from NATO, the Organising Committee would also like to express their appreciation to the following, without whose assistance (financial and otherwise) the meeting would not have been as successful:-

INSERM, Society of Biochemistry and Molecular Biology (France), Society of Clinical Biology (France), Association of Clinical Biochemists (UK), Abbot, Baxter, Beaufour, Behring, Bioxytech, Boehringer-Mannheim, Corata, French Foreign Office, Fournier-Dijon, Helena France, Iris-Servier, Jonan, Kodak, Lipha Oberval, Pierre Fabre, Radiometer, Sanofi, Sebia, and the Regional and local councils and University of Bordeaux II and the Hospitals of Bordeaux.

CONTENTS

PHOSPHOLIPASES A_2

PHOSPHOLIPASES

LIPASES AND ESTERASES:

REASSESSMENT OF THEIR CLASSIFICATION IN THE LIGHT OF THEIR THREE DIMENSIONAL STRUCTURE

Verger R., Ferrato F., Aoubala M., de Caro A., Ivanova M., de la Fournière L. ,Carrière F., Cudrey C., Rivière C., Rugani N. and Gargouri Y.
Laboratoire de Lipolyse Enzymatique. GDR-1000 - CNRS. 31 chemin Joseph Aiguier. 13402 Marseilles - France

Hjorth A., Wöldike H., Boel E. and Thim L.
Bioscience, Novo-Nordisk A/S. 2880 Bagsvaerd - DK

Lawson D. M. and Dodson G. G.
Department of Chemistry. University of York. YO15DD - UK

van Tilbeurgh H., Egloff M. P. and Cambillau C.
LCCMB du CNRS. Faculté de Médecine Nord. 13336 Marseilles - France

Contributions of the Marseilles, York and Novo-Nordisk teams to the European Communities BRIDGE T-programme on lipases (BIOT-CT91-0274 and BIOT-CT90-0181).

Apart from their general biological significance, lipolytic enzymes play an increasingly important role in biotechnology and medicine.

1. INTERFACIAL ACTIVATION OF THE LIPASE- PROCOLIPASE COMPLEX BY MIXED MICELLES REVEALED BY X-RAY CRYSTALLOGRAPHY

The three dimensional structure of the lipase procolipase complex, cocrystallized either without [1] or with [2] mixed micelles of phosphatidylcholine and bile salt, has been determined at 3 Å resolution by X-ray crystallography. Considerable changes in the secondary and tertiary structure are observed in the region around the active site. Especially the lid, a surface loop covering the catalytic triad, adopts a totally different conformation that allows a phospholipid to bind to the active site, displacing some residues over a distance of 30 Å. The open lid is an integral part of the active site and is intensively interacting with the bound phospholipid. The open lid also interacts with procolipase and together they form the lipid-water interface binding site. This reorganisation of the lid structure provokes a second drastic conformational change in an active site loop, which in turn creates the oxyanion hole. This is an example of "induced fit" triggered by the physical properties of the substrate. Besides this, the variable conformation of the lid illustrates how the same amino acid sequence can adopt different secondary structures, depending on the local physico chemical environment.

The present crystal structure shows that the pancreatic lipase lid is a complicated structural device. Upon lipid binding it reorganizes in an ingenious way. This conformational change serves many purposes : it creates the interfacial lipid binding site, gives shape and access to the active site canyon, stabilizes the oxyanion hole loop, and finally it strengthens the interaction with procolipase. The importance of the lid in mammalian lipases has recently been illustrated by an elegant study on lipoprotein lipase. A chimeric construction consisting of the pancreatic lipase lid and the lipoprotein lipase core yielded a completely inactive enzyme. The pancreatic lipase lid grafted on a lipoprotein lipase core may not encounter the correct stabilizing interactions. It may well be that the lid regulates the different substrate specificity, interfacial behaviour and cofactor dependency of these mammalian lipases.

2. A STRUCTURAL DOMAIN (THE LID) FOUND IN PANCREATIC LIPASES IS ABSENT IN THE GUINEA PIG (PHOSPHO)LIPASE

Pancreatic lipolytic enzymes in the guinea pig have previously been demonstrated to differ significantly from those found in other mammals. Following a partial purification that revealed a high phospholipase activity remaining associated with lipase activity, two cationic lipases with high phospholipase A_1 activity from guinea pig pancreas were purified by Durand *et al* [3] and Fauvel *et al.* [4]. They demonstrated an unusually high phospholipase/lipase activity ratio of 1, higher by 3-5 orders of magnitude compared to lipases from other sources. Futhermore, no evidence suggesting the existence of a classical secretory phospholipase A_2 in guinea pig pancreas was found [5]. The high phospholipase A_1 activity of GPL might thus be of physiological significance for the degradation of dietary phospholipids and detailed study of this enzyme seemed pertinent.

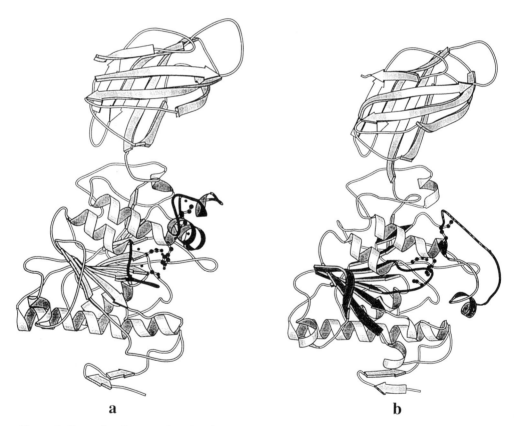

a   b

Figure 1. Comparison between the "closed" (a) and "open" (b) form of pancreatic lipase. Colipase is not represented. The catalytic triad is represented in balls and sticks. From van Tilbeurgh *et al.*[1,2].

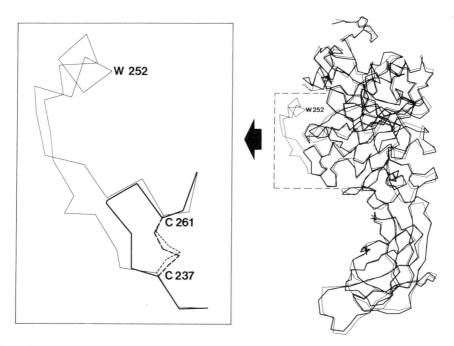

Figure 2. *View of the superposition of GPL model (thick line) and HPL (thin line) Cα backbones.* GPL modelling was based on the HPL structure from Winkler *et al.* [7] and sequence comparison. The close-up view shows the major difference between HPL and GPL: the lid-domain in HPL between Cys 237 and Cys 261, being substituted by the GPL "mini-lid". From Hjorth *et al.* [6].

A cDNA clone encoding a guinea pig pancreatic (phospho)lipase (GPL) has been sequenced and expressed [6]. The amino acid sequence of GPL is highly homologous to that of other known pancreatic lipases, with the exception of a deletion in the so-called lid domain that regulates access to the active centres of other lipases (figure 2). By inspection of the 3-D structure of human pancreatic (HPL) [7] and *Rhizomucor miehei* lipases [8] it was proposed independently that interfacial activation of lipases, probably involved a reorientation of the lid-domain.

The recombinant like the native GPL displayed no interfacial activation using tributyrin as substrate. The catalytic activity of GPL on such a partially soluble triglyceride is thus clearly more typical of an esterase than of pancreatic lipase according to Sarda and Desnuelle [9]. Another interesting kinetic feature of GPL is its insensitivity to the presence of bile salts when using emulsified tributyrin as substrate. This behaviour contrasts sharply with the well known inhibition by bile salts of all other pancreatic lipases tested so far. In the later case, colipase is required to counteract bile salt effects. Finally, GPL hydrolyze monomolecular films of various phospholipids at rates comparable to those observed using 1,2-diglyceride as a substrate.

The question arises whether there might be a direct structure-function relationship between the existence of a "mini-lid" structure in GPL and the unusual kinetic characteristics of this enzyme. In a lipase containing a "mini-lid", the catalytic site is probably freely accessible to monomeric short chain triglycerides (figure 2). The lack of interfacial activation in GPL might be a consequence of a maximal enzymatic activity being displayed on monomeric substrates. In fact, interfacial activation of classical lipases may be viewed as a depressed action on monomeric esters rather than an increased interfacial activity on aggregated substrates.

The lack of bile salt inhibition and the colipase-effect on GPL activity on tributyrin is probably indirectly related to the presence of a "mini-lid". In fact, the maximal enzymatic activity of GPL on monomeric substrates does not allow one to see any effect of bile salts.

3

Even if the enzyme is desorbed from the interface, it will remain active on monomers. Yet, the question remains whether there is a simple structure-function relationship between the presence of a "mini-lid" domain and the unusually high phospholipase activity of GPL. From the GPL modelling based on HPL 3-D structure, the core of the N-terminal domain appeared to be conserved overall, with the exception of the lid-domain (figure 2).

One is therefore tempted to speculate that the lipase lid is indeed involved in substrate selectivity of these enzymes. It is known that porcine pancreatic lipase can hydrolyze negatively charged phospholipids with a very low specific activity [10]. Accordingly, the active centre in pancreatic lipases can catalyze the hydrolysis of ester bonds in both triglycerides and phospholipids. Here again, the lack of high phospholipase activity in classical pancreatic lipases may be viewed as a depressed action on phospholipids due to the presence of the lid-domain, rather than a total absence of catalysis.

Thus GPL is a natural mutant challenging the classical distinction between lipases, esterases and phospholipases.

3. MONOCLONAL ANTIBODIES AGAINST HUMAN GASTRIC LIPASE

In humans, the hydrolysis of dietary triacylglycerols begins in the stomach and is catalyzed by one major enzyme: human gastric lipase (HGL) which is located in the chief cells of fundic mucosa [11]. HGL is a glycoprotein with a molecular mass of around 50 kDa. In vitro studies have shown that HGL hydrolyses both short and long-chain triacylglycerols at comparable rates. The enzyme is highly stable and active under acidic pH conditions. The optimal pH for HGL activity is 5.4 in the case of long-chain and 6 in the case of short-chain triacylglycerols.

The amino acid sequence of HGL, deduced from the cDNA, consists of 379 residues and shows a conservative pentapeptide (Gly-X-Ser-X-Gly) common to all the known lipases as well as serine esterases. The essential serine within this pentapeptide is supposed to be part of the Asp/His/Ser catalytic triad to be found in esterases and lipases.

Two topographically and functionally distinct sites (a lipid or interface binding site and a catalytic site) have been postulated to exist in lipolytic enzymes, based on monolayer experiments. No information about the three dimensional structure of HGL is yet available, although several crystal types have been obtained. In order to better characterize these functional sites of HGL, an immunological approach was developed using monoclonal antibodies (MAbs) against HGL as probes for the spatial identification of these domains.

3.1. Production of Mabs against HGL [12]

Hybridomas prepared by fusing spleen cells with myeloma cells were placed in 576 wells containing the selective medium H.A.T. and mouse peritoneal macrophages as feeder cells. Ten days after the fusion, 422 wells produced one or more clones of which 52 clones had anti-HGL activity in a direct binding ELISA test and five of them (4-3, 25-4, 35-2, 83-15, and 218-13) were selected and cloned using the limiting dilution technique. The five clones selected were then cultured for storage and production of antibodies in mice. MAbs were then purified from ascitic fluid. All these MAbs belong to the IgG1 class with a κ light chain.

3.2. A New Elisa Procedure to Quantitate HGL in Duodenal Contents [13]

The usual method for estimating enzymatic activity is based on the use of insoluble substrates present in an emulsified state. The lipolytic activity measurements in biological fluids, such as in duodenal contents, are sometimes unreliable, due to the presence of lipases of various cellular origins acting upon the same lipidic substrate. For this reason we developed a sensitive and specific double sandwich ELISA for measuring HGL in the duodenal contents, where HGL and HPL are both present. With this new procedure using purified rabbit anti-HGL polyclonal antibodies as captor antibody and biotin-labeled MAb 35-2 as detector antibody, it was possible to detect HGL concentrations as low as 1 ng/ml. Since HGL levels in

Table 1. HGL detection ratio (%) of pure exogeneous HGL added to duodenal contents. The detection ratio was calculated as follows

$$\text{HGL detection ratio} = \frac{\text{measured HGL - endogenous HGL}}{\text{exogeneous HGL}} \times 100$$

Endogeneous HGL in duodenal contents (μg/ml)	Exogeneous HGL (μg/ml)	Total measured HGL (μg/ml)	HGL detection ratio (%)
2.5	1.5	3.8	86.6
2.5	5.0	7.2	94.0
2.5	10.0	12.4	99.0
5.1	1.5	6.6	100.0
5.1	5.0	9.3	84.0
5.1	10.0	16.1	110.0

duodenal contents are of the order of μg/ml the sensitivity of the ELISA test turned out to be sufficient. Moreover, the good correlation between the expected and measured concentrations of duodenal HGL (Table 1) indicates that no other duodenal compound interfered with the immunoassay. Furthermore, the HGL concentrations determined either by the ELISA or an enzymatic test, using standard pH stat titration, shows a good correlation between both assays. With the enzymatic assays used to measure lipase activity in biological fluids (gastric juice, duodenal contents and serum), there is a risk of overestimating lipolysis due to the presence of lipases from various tissular origins such as pancreatic lipase. The double sandwich ELISA developed in this study provides a particularly suitable means of specifically estimating HGL in biological fluids.

3.3. Immunopurification of HGL [13]

Human gastric juice is a biological fluid which is rich in mucus and pepsinogens. An improved procedure was developed for isolating HGL by ion exchange chromatography in combination with immunoaffinity chromatography, using a column of immobilized MAb 35-2. As expected, the elution of HGL under acidic conditions (pH 2.2) from the immunoaffinity column does not alter its enzymatic activity and gives rise to very pure HGL fractions. This procedure offers the advantage of being rapid, reproducible with a better overall yield (58 %) as compared to the previously described methods using conventional chromatographic steps.

3.4. Immunoinactivation of HGL by Mabs [12]

Since the screening of anti-HGL MAbs was performed using a native enzyme, all epitopes recognized by these antibodies are probably located on the surface of the enzyme. In order to investigate the involvement of several epitopes in the catalytic activity of HGL, we studied the effects of each MAb on the lipolytic activity of this enzyme using the following three substrates: tributyrin or soybean oil emulsions and 1,2 didecanoyl *sn*-glycerol (dicaprin) monomolecular films.

3.4.1. HGL Activity on Tributyrin and Soybean Oil Emulsions. Three MAb classes could be defined: class 1 includes MAbs 4-3, 25-4, and 35-2 (Figure.3A); class 2 includes MAb 83-15 (Figure 3B) and class 3 composed of MAb 218-13. The three MAbs of class 1 give very similar patterns, reaching a maximum inhibition level at a MAb/HGL molar ratio of 0.5. Figure 3A illustrates typical MAb 35-2 inhibition curves with tributyrin or soybean oil emulsions as HGL substrates. In the case of class 2 MAb, which contains MAb 83-15 (Figure 3B), no significant effect on HGL lipolytic activity was observed when the substrate was tributyrin, whereas on soybean oil the HGL activity was reduced by 50%. With class 3 MAb (218-13) no inhibitory effect on HGL activity was observed with either tributyrin or soybean oil emulsions.

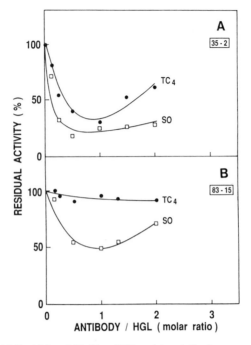

Figure 3. Effect of both MAbs 35-2 and 83-15 on HGL activity. A fixed amount of HGL was incubated with each MAb at various molar ratios. The relative residual activity of HGL was measured on either tributyrin (●) or soybean oil (□) emulsions.

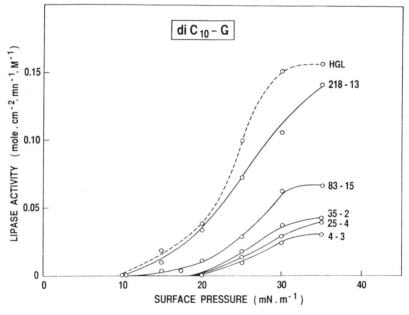

Figure 4. Effects of the five MAbs on HGL activity measured with 1,2-*sn*-didecanoylglycerol films as substrate at variable surface pressures.

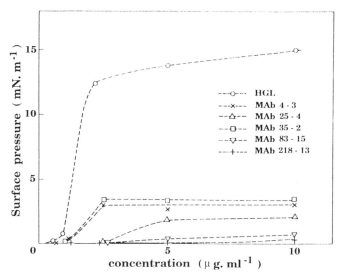

Figure 5. Variation in surface pressure of the air/water interface after injection of HGL and anti-HGL MAbs. Each point coresponds to the maximal value of the surface pressure reached after protein injection into the water subphase.

3.4.2. HGL Activity on 1,2-sn-didecanoyl Glycerol Monolayers. In order to optimize the inhibitory effect of various MAbs, HGL was incubated for 1 h at 37°C with each MAb at a MAb/HGL molar ratio of 0.5 (Figure 4). We checked that the optimal inhibition of HGL activity tested on monolayers was obtained at MAbs/HGL molar ratios of 0.5 as previously observed in the case of the bulk assay systems. The lipolytic activity of HGL was measured at surface pressures ranging from 0 to 35 mN.m^{-1}. As shown in Figure 4, HGL does not significantly hydrolyse dicaprin films at surface pressures below 15 mN.m^{-1}. Above this pressure the HGL activity increases rapidly reaching a maximum value at 35 mN.m^{-1}. When HGL was incubated with each of the five MAbs, the previously obtained classification of the MAbs, based on bulk assay systems, was confirmed.

3.4.3. Effect of Mabs on Hgl Penetration into Lipid Monolayers [14]. In order to further characterize the catalytic and lipid binding domains on the HGL surface, we investigated the lipid monolayer penetration capacities of the five anti-HGL MAbs as well as their respective complexes with HGL.

Figure 5 shows the adsorption patterns at the air/water interface of HGL and the five purified anti-HGL MAbs. Each point represents the maximal surface pressure value reached after protein injection into the aqueous subphase. The HGL and MAbs adsorption behaviours are clearly very different. HGL possesses a much higher surface tensioactivity than any of the MAbs. This is a particularly favorable situation which facilitated the interpretation of the adsorption experiments performed with MAb/HGL complexes.

To identify the lipid binding domain on the HGL surface, the penetration of MAbs/HGL complexes into egg phosphatidylcholine monomolecular films was investigated (Figure 6). MAbs 4-3, 25-4, 35-2 and 83-15, unlike MAb 218-13, decreased the penetration capacity of native HGL. These results show that the inibitory MAbs 4-3, 25-4, 35-2 and 83-15 recognize epitopes overlapping the lipid binding domain of HGL, unlike the epitope reacting with MAb 218-13.

3.5. Epitope Mapping With Mabs [12]

To test whether the MAbs recognized overlapping epitopes on HGL, the ELISA double antibody binding test (ELISA additivity test) developed by Friguet *et al.* [15] was used. The

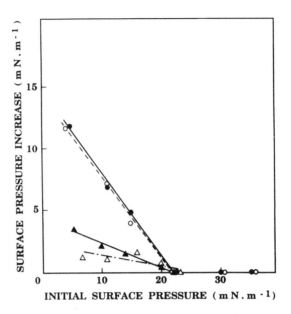

Figure 6. Maximal surface pressure increase after injections of HGL [●], MAb [△] and the MAb/HGL complexes (MAb 4-3/HGL [▲] and MAb 218-13/HGL [○]) under a monomolecular film of egg phosphatidylcholine spread at various initial surface pressures.

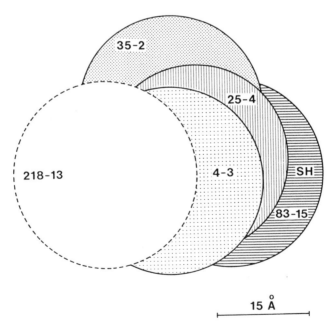

Figure 7. Spatial map of overlapping epitope recognized by the five MAbs against HGL.

titration studies have shown that the five MAbs expressed different binding properties with HGL adsorbed on a PVC plate. Competition between couples of antibodies for the antigen was expressed by means of the additivity index. This index (expressed as an overlapping percentage) makes it possible to evaluate the simultaneous binding of two monoclonal antibodies to the antigen. The five anti-HGL MAbs were studied in all possible pairs and the values obtained ranged between 20.5 % and 75 %. From these results, it emerged that four (4-3, 25-4, 35-2 and 83-15) out of the five anti-HGL MAbs recognize closely overlapping epitopes in the same antigenic region. We have tentatively positioned the epitopes of HGL by means of a schematic spatial map (Fig 7). The additivity indexes between MAbs have been taken to be inversely related to the overlapping surfaces between pairs of epitopes. If one assumes first that an epitope has a relatively flat surface (700 Å^2) with protuberances and depressions formed by the amino-acid side chains of the combining antibody site and secondly that the complex formation between a MAb and its antigen does not induce any substantial conformational changes that might affect the subsequent binding of other MAbs, it is possible to calculate the distances (in Å) between the centers of the five epitopes and to build a spatial model (Fig 7). A general overview of the resulting model confirms that the various overlapping epitopes are all located in the same antigenic region of HGL.

REFERENCES

1. van Tilbeurgh H., Sarda L., Verger R. and Cambillau C. *Nature* **359** (1992) 159-162.
2. van Tilbeurgh H., Egloff M. P., Martinez C., Rugani N., Verger R. and Cambillau C. *Nature* **362** (1993) 814-820.
3. Durand S., Clemente F., Thouvenot J. P., Fauvel-Marmouyet J. & Douste-Blazy L. *Biochimie* **60** (1978) 1215-1217.
4. Fauvel J., Bonnefis M. J., Sarda L., Chap H., Thouvenot J. P. & Douste-Blazy L. *Biochim. Biophys. Acta* **663** (1981) 446-456.
5. Fauvel J., Bonnefis M. J., Chap H., Thouvenot J. P. & Douste-Blazy L. *Biochim. Biophys. Acta* **666** (1981) 72-79.
6. Hjorth A., Carrière F., Cudrey C., Wöldike H., Boel E., Lawson D. M., Ferrato F., Cambillau C., Dodson G.G., Thim L. & Verger R. *Biochemistry* **32** (1993) 4702-4707.
7. Winkler F. K., d'Arcy A. & Hunziker W. *Nature* **343** (1990) 771-774.
8. Brady L., Brzozowski A. M., Derewenda Z. S., Dodson E., Dodson G., Tolley S., Turkenburg J. P., Christiansen L., Huge-Jensen B., Norskov L., Thim L. & Menge U. *Nature* **343** (1990) 767-770.
9. Sarda L. & Desnuelle P. *Biochim. Biophys. Acta* **30** (1958) 513-521
10. Verger R., Rietsch J. & Desnuelle P. *J. Biol. Chem.* **252** (1977) 4319-4325.
11. Gargouri Y., Moreau H. & Verger R. *Biochim. Biophys. Acta* **1006** (1989) 255-271.
12. Aoubala M., Daniel C., De Caro A., Ivanova M., Hirn M., Sarda L. & Verger R. *Eur. J. Biochem.* **211** (1993) 99-104.
13. Aoubala M., Douchet I., Laugier R., Hirn M., Verger R. & De Caro A. *Biochim. Biophys. Acta* **1169** (1993) 183-188.
14. Ivanova M., Aoubala M., De Caro A., Daniel C., Hirn J. & Verger R. *Colloïds and Surfaces B: Biointerfaces* **1** (1993) 17-22.
15. Friguet B., Djavadi-Ohaniance L., Pagès J., Busard A. & Goldberg M.E. *J. Immunol. Methods* **60** (1983) 351-358.

ESTERASES

SERUM CHOLINESTERASE: GENETICS, ENZYMOLOGY, DIAGNOSTIC USE AND THE ASSOCIATION WITH CLINICAL DISORDERS

Ellen and Friedrich Werner Schmidt
Pregelweg 4, D-30916 Isernhagen NB, Germany

Cholinesterase (EC 3.1.1.8, CHE or BChE) also called butyrylcholinesterase or serum cholinesterase, has evoked the interest of geneticists, biochemists, pharmacologists, and clinicians, particularly hepatologists. Recent molecular biological investigations have thrown light on new facets of this plasma enzyme. Some of them will be reported, and a number of still open questions discussed, preferably under clinical aspects, without claiming comprehensiveness.

GENETICS

The genetic polymorphism of BChE has been recognized since 1956, when Kalow observed the familial incidence of low BChE levels in serum associated with prolonged apnea after the muscle relaxant suxamethonium and with Genest, developed a method for the detection of the atypical (A) variant by determining the dibucaine number in serum [1]. By the use of other inhibitors of BChE activity, pedigree analyses and, recently, molecular biological methods, the fluoride resistant F variants [2], three silent forms S [3,4], and the less severely deficient quantitative variants K [5], J [6] and H [7] with 30, 60, and 90 % reduced catalytic activity respectively, have been identified. Many combinations of two or more inhibitors of BChE with various substrates have been proposed. In our experience, with butyrylthiocholine as substrate and assay temperature 37°C, dibucaine, propranolol and Ro 02-0683 discriminate the commoner phenotypes UU, UA, AK, AA, AF and UF better than other combinations [8].

Mostly due to the endeavors of the working groups of Oksana Lockridge and Bert La Du, but also of researchers in France, Great Britain and Israel, the BCHE gene has been localized, its structure elucidated and the point mutations which are responsible for the hereditary variants, identified [9]. The site of the BCHE gene is on the long arm of chromosome 3, at 3q26.1-3 [10,11] The gene is at least 72 kB long and contains 4 exons. Exon 1 contains two potential transcription initiation sites. Exon 2 codes for 83% of the mature protein, including the N-terminus and a third, probably functional, transcription initiation site. Exon 3 is rather short. Exon 4 is longer and codes for the C-terminus of BChE [9,10]. Table 1 shows the nucleotide alterations in exons 2 - 4, the ensuing changes in the amino acid sequence of the respective genetic variants and the functional consequences.

In the atypical variant the change of aspartate 70 to glycine causes a reduced Km for positively charged substrates and inhibitors. This suggests that aspartate 70 is an important component of the anionic site [12]. Two different point mutations in exon 2, associated with the fluoride resistant variants, have similar consequences [13]. In the silent variants two frame shifts in exon 2 and a new stop codon in exon 3 lead to the synthesis of a defective protein or no protein at all [4]. The three less severe quantitative variants, H, J

Table 1. Nucleotide and amino acid alterations in human BChE genetic variants

Common name	Afflicted exon	DNA alteration	Amino acid alteration	Phenotypic description
Usual	-	none	none	normal
Silent-2	2	nt 16 (ATT to TT)	6 Ile to frame shift	no catalytic activity
Atypical	2	nt 209 (GAT to GGT)	70 Asp to Gly	Dibucaine resistant
Silent-1	2	nt 351 (GGT to GGAG)	117 Gly to frame shift	no catalytic activity
H variant	2	nt 424 (GTG to ATG)	142 Val to Met	90% lower activity
Fluoride-1	2	nt 728 (ACG to ATG)	243 Thr to Met	Fluoride resistant
Fluoride-2	2	nt 1169 (GGT to GTT)	390 Gly to Val	Fluoride resistant
J variant	3	nt 1490 (GAA to GTA)	497 Glu to Val	66% lower activity
Silent -3	3	nt 1500 (TAT to TAA)	500 Tyr to stop	no catalytic activity
K variant	4	nt 1615 (GCA to ACA)	539 Ala to Thr	33% lower activity

[from C.F. Bartels et al., Am. J. Hum. Genet. 50, 1086-1103, 1992, abridged]

and K, are due to point mutations in exons 2, 3 and 4. The H variant is conspicuous by its very low activity, and difficult to detect [14].   The J variant is also rare and is associated with a reduction of BChE catalytic activity of approximately 66%. The amount of protein is also reduced, whether by impaired synthesis or by instability of the enzyme with rapid degradation, is not yet clear [9,15,16]. The K type is the most common hereditary BChE variant. In Caucasian populations it occurs in homozygous form in 1.7%, and in heterozygous form in nearly 23%. It is found in about 90% combined with the atypical allel. The variant has a 33% lower than normal catalytic activity, concomitant with a reduction of the amount of enzyme protein as determined by immunological quantitation [4,17,18].

In earlier studies on the geographic and ethnic distribution of the genetically determined BChE variants, the omission of the not yet identified K variant introduced a considerable bias.. A high frequency of the atypical allele occurs in Caucasians, while its prevalence in African and Oriental populations is much lower , even 0. Within the large racial groups there is a considerable variation. Exceptional and extreme frequencies are observed in communities with a high rate of intermarriage and not much migration [19-21]. In mixed populations, e.g. in USA or Brazil, the white admixture to the negroid population can be calculated from the prevalence of the atypical allele [22]. The data on the ethnic distribution of the fluoride resistant variant are not so extensive, and the data on the quantitative variants still more fragmentary.

The clinical significance of the genetically controlled variants of BChE lies in their increased sensitivity to suxamethonium with the risk of prolonged apnea. The patients homozygous for deficient genes are markedly sensitive, and those with the phenotypes FS, AS, KS, AJ and AF are moderately so. The majority of patients with the common phenotype AK are not sensitive, but exceptions, due to its genetic heterogeneity, are not as rare as in the individuals, with the phenotypes UA, UF and US [4,20,23,24]. Generally, the variants with reduced activity interfere with the diagnosis of acquired depressions of BChE in serum.

It could be argued that this is not the place to discuss the so-called E2 or CHE2 variants, because there is convincing evidence that a BCHE2 gene does not exist [25]. The C5+

variant with up to 30% more catalytic activity was detected in 1963 by Harris et al.[26] as an additional slow moving band in gel electrophoresis. It turned out that this molecular form of BChE is a conjugate of tetrameric BChE with a protein of 60 kD, resulting in a relative molecular mass of the whole compound of 400 kD. The nature of the associated protein is still unknown: It is not albumin, fibronectin, an immunoglobulin or collagen fragment and contains no phospholipids [27]. It must be assumed that the so-called second BCHE locus codes for this protein. The localization of this gene is still under discussion: It has been postulated to be on chromosome 16 linked to the haptoglobin site [28,29] or on chromosome 2, linked to the gamma-crystalline cluster [30]. The inheritance of C5+ is autosomal dominant. However, it is not always mendelian, and there are some C5+ individuals turned transiently negative [27]. The frequency of the C5+ variant varies between 0 and 50%[31]. In Caucasians its prevalence lies around 10%, while in negroid and oriental populations it is around 5% and 3% respectively [20,32,33].

The clinical significance of the C5+ variant consists only in its usually undetected occurrence in healthy people. This leads to a blurred upper limit of the reference interval of BChE in serum and complicates the diagnosis of disorders with elevated BChE activity in plasma [33,34].

STRUCTURE AND ENZYMOLOGY

Serum cholinesterase is synthesized in the liver and secreted in trans-Golgi exocytic vesicles into the plasma [35]. Its half-life in the circulation has been estimated as 10 (3.4-12) days [36]. BChE is a glycoprotein with 24% carbohydrate, arranged in 9 oligosaccharide chains per subunit (G1) with 574 amino acids. More than 90% of the enzyme in serum is in the tetrameric G4 form (C4 in electrophoresis) with a relative molecular mass of 340 kD, which consists of two disulfide-linked dimers, which in turn are linked together by non-covalent bonds [37].

In electrophoresis at least three faster moving molecular forms of BChE can be identified: C1,which corresponds to the monomer G1, C3, which is the dimer G2 and C2, which represents an albumin-G1 conjugate [27]. In addition, there are slow moving molecular forms. The most important one is the C5 band, which gave the name to the respective variant. A stronger slow band is found in the sera of individuals with the rare hereditary variant E Cynthiana, who exhibit a BChE with several times higher activity than the usual enzyme [38-40]. Even slower migrating bands have been found in malignant diseases or are caused by storage of the serum in the cold., at the expense of C4. Therefore, they are not associated with elevated catalytic activity [41].

The amino acid sequence of the usual form of BChE has been determined [42], and the structure of the active site elucidated [43,44]. Each subunit carries in its active site gorge one "esteratic" site around serine 198 with the amino acid sequence Phe-Gly-Glu-Ser-Ala-Gly-Ala, and one "anionic" site around aspartate 70 [45]. Both sites, as well as large portions of the primary and secondary structure of the enzyme are conserved in evolution [44]. There is more than 50% homology in vertebrates and more than 30% homology between human BChE and Acetylcholinesterase (EC 3.1.1.7, AChE) from Drosophila [46].

The esteratic site binds to the carbonyl group of the choline esters and is responsible for their hydrolysis. The anionic site binds to choline and regulates the substrate affinity, the latter function being the basis for the functional sequelae of the replacement of negatively charged aspartate 70 by neutral glycine in the atypical variant of BChE [47]. BChE hydrolyzes various choline esters. Substrates positively charged at pH 7.4, like quaternary or tertiary amines, have lower Michaelis constants for the usual enzyme than neutral and negatively charged compounds. Although generally the kM values are higher for the atypical variant, neutral substrate kMs are very similar in both forms. Likewise, the turnover numbers of the usual and the atypical enzyme are similar.

One might presume that an enzyme, found in a concentration of 3 - 6 mg/l in adult human plasma, and highly conserved in the phylogenetic tree, must have an important metabolic function. However, despite extensive investigations, which included also the embryonic and tissular forms of BChE, its physiological role is still unclear. The great variation between the regulation of BChE in different species precludes the direct transferability to humans of findings in animals [48-50]. The proposed functions include the regulation of the levels of free choline and of the choline/acetylcholine ratio in plasma and tissues [20,51], actions as esterase, peptidase and amidase toward neurotransmitters and hormones [52-54], effects on the permeability of cell membranes, on transport processes, growth, thermoregulation and respiration [55-57] and participation in the metabolism of fatty acids and lipoproteins in liver, reticulo-endothelial system and plasma [20,51]. Whichever of these proposed functions of BChE finally turns out to be the real physiological role of the enzyme, it must be compatible with the fact that homozygotes for one of the silent BCHE genes have no BChE in liver or plasma and still display no metabolic disturbance.

DEVELOPMENT AND HORMONAL REGULATION

There are indications that very early in the evolution of vertebrates a gene duplication occurred, which gave rise to the two cholinesterase genes ACHE and BCHE [58], now located at different chromosomal sites - in humans at 7q22 [59,60] and at 3q26, respectively [10,11]. Both enzymes exist as polymers of catalytic subunits, both occur as water-soluble forms, secreted into body fluids, both are found as immobilized forms in tissues. However, in the latter AChE prevails, while BChE is abundant in plasma, where in adults only traces of AChE are found, G1 and G2 forms probably derived from the erythrocyte membranes, G4 forms from the central nervous system. [46,61].

There is evidence that in fetal muscle and nervous tissues of chicken and rats BChe is frequently present transiently before the occurrence of AChE or independent of the latter [62]. Observations that in embryonic brain BChE is expressed before and during mitosis, while AChE appears 11 hours after mitosis, suggest an involvement of BChE in cell proliferation and of AChE in cell differentiation [57]. This conception is supported by reports on amplification of cholinesterases in various tumor cells [63,64]. There are, however, many discrepancies between studies on the regulation of cholinesterases, which are conceivable in view of the striking differences between species and even between strains [62].

It appears that the regulation of AChE is tissue-specific and that of BChE is homeiostatic throughout the body [46]. Thus, hormonal effects offer themselves as regulatory factors for BChE. We found that adult female Wistar rats possess 5- to 8-fold higher BChE levels in plasma than adult male rats [49]. Many years ago, others demonstrated a similar difference in liver tissue and its disappearance after castration [65,66].Later, it was shown that the sex hormone effects are reversible, require mediation by growth hormone from the pituitary in-situ and are not reproducible in vitro[67-69]. In rats, estrogens increase and androgens decrease BChE activity in plasma. In humans, sex steroids influence it in the opposite direction: With all substrates, slightly higher activities in male adults than in female adults have been reported. The depression by female sex steroids becomes even more marked in women taking oral contraceptives and during pregnancy [62,70-72]. The reactions of BChE to oral contraceptives are dose-dependent, and in only 20 - 25% of women a decrease >10% is observed [73]. These effects lead to a blurred lower limit of the reference interval of BChE in serum, which is , therefore, too low in many references given [71-73].

Glucocorticoids reduce BChE activity in plasma in several species by inhibition of its synthesis in the liver, probably during their first pass, when given perorally [74-77]. In patients undergoing glucocorticoid therapy, the reversible drop in enzyme activity was less with dexamethasone (14-57%) than with prednisone (23-69%) as compared with the individual control values [76], whereas in rats dexamethasone had the most pronounced effect [75].

PHARMACOLOGIC ASPECTS

A great variety of drugs lower BChE activity in plasma not only by reducing the synthesis, like the glucocorticoids, but by direct reaction with the enzyme . All types of inhibiton occur due to the structure of the molecule [78]. The inhibition can be reversible or irreversible. The latter group includes cytotoxic substances, which are used in chemotherapy of malignancies. The former group includes muscle relaxants, tranquillizers, anesthetics and analgesics. The list of drugs can never be complete, as new drugs, which may inhibit BChE activiy, are continually introduced [79-85].

The degree of inhibition varies between BChE and AChE [83,86] and among the genetic variants, so that some of the drugs are used for their discrimination [1,8,87]. The degree of inhibition may also vary due to slight molecular modifications within a chemical class of drugs, e.g. the benzodiazepines, where alkalytion at nitrogen 1 as well as double halogenation increase the inhibitory effect, at least in vitro [88].

Clinically, these drug effects may be of little or no significance if the inhibitory concentration far exceeds the therapeutic concentration. Drug effects become important, however, in cases of overdose, addiction or intoxication, or with particularly susceptible persons [89]. Moreover, in view of the common polypragmatism, drug interactions may affect BChE levels in plasma in unpredictable ways , thereby blurring the diagnostic pattern.

Cholinesterase inhibitors, carbamates and organophosphates, are widely used as pesticides. Although AChE is the target enzyme of these toxins, the easier determination of BChE is used for the monitoring of accidental and chronic intoxications in workers, occupied in pesticide production and agriculture, particularly in greenhouses during the spraying season [90], as well as in patients following suicide attempts. A recent retrospective cohort study from California, comprising 103 worker-years, reports that 24% of the workers were temporarily removed from spraying because their BChE levels in serum had been below 60% of baseline. Only 5 of them had mild symptoms, no accident was recorded [91]. Similar preventive programs are reported from other countries [92-95]. The question if chronic dietary intake of anticholinesterase agents depresses BChE activity, has been investigated in Denmark and found to be measurable, although not necessarily hazardous [96]. However, one might speculate that observed seasonal variations [97] were related to the intake of sprayed vegetables and fruits.

An observation made in Israel by the group of Soreq and Zakut should not go unmentionned: In two generations of a farming family, expressing the silent BChE phenotype and chronically exposed to parathion, a 100-fold amplification of DNA at the site of the BCHE gene, similar to that occurring in leukemic and other malignant cells, was found [98,99].

METHODOLOGY

Besides the various usual spectrometric assays, fluorometric, chromatographic, electrometric and immunological methods have been proposed, which are, in part, adapted to automated devices [100-104]. The mere number of procedures precludes a comprehensive discussion or even citation in this paper, although it suggests, as such, a certain uneasiness with the present state of performance. The main reason for the dissatisfaction is the high catalytic activity of BChE in serum. Particularly at 37°C this requires either a predilution of the sample, interfering with automated procedures or very small sample volumes: both, necessarily, increase the unavoidable imprecision [105]. At present, there is neither an acknowledged reference method, nor, apart from the manual in the former GDR [106] and a European method, [107]apparently without much acceptance in clinical medicine, standardization of routine assays. Recently, the German Society for Clinical Chemistry recommendeded a standard method at 37°C with butyrylthiocholine as substrate and hexacyanoferrate(III) as indicator substance [108], which avoids the above-mentionned disadvantages and is also suitable for phenotyping (own unpublished results).

Method standardization combined with certified reference materials and quality control surveys [23,109] would enhance reliability and comparability of results, thus, saving unnecessary work and leading to generally accepted reference intervals and inhibition numbers. On such a basis, the true value of BChE determinations in plasma in clinical medicine beyond toxicology, might be better appreciated.

PHYSIOLOGICAL VARIATIONS, EFFECTS OF NUTRITION AND THERAPY

Qualified interpretation of variations of BChE in clinical medicine implies the awareness of its biological variability and of possible jatrogenic modifications of activity. Besides the genetically determined variants and the sex differences, which include also the significant decrease during pregnancy and early puerperium [110], there are also variations with age, from low levels at birth to higher adult concentrations, to a depression between the 25th and the 40th year, and to a marked fall after transient elevation around the age of 55, which are more pronounced in women than in men [72]. For given age classes, BChE activity in serum correlates well with body height and overweight, more significantly in men than in women [97,111]. In comparison to the inter-individual variation of immunoreactive protein, corresponding to 22% in a healthy population, the intra-individual variation is low, only 6.4%. If the specific catalytic activity (kU/mg) of BChE is measured, there is no more effect of sex, height, body weight and age. Only the phenotype remains as a modifying factor, and the intra-individual variation becomes negligible [112].

The interactions with many prescribed drugs have already been mentioned. Obviously, during treatment with anticholinesterase agents, e.g. for acute episodes of myasthenia gravis, BChE activity should be monitored to prevent cholinergic crises [113]. Furthermore, attention has been directed to the effect on BChE of repeated plasmapheresis, especially in pregnancy or in malignancies, when the starting level is low anyway. BChE activity concentrations lowered to 30% or less can lead to a risky situation for subsequent surgery [114]. BChE is remarkably stable in vitro [115-117]. In heparin or ACD blood, stored for weeks, and in fresh frozen plasma or conserved human serum the initial BChE activity is maintained. In contrast, in plasma protein preparations, which have been treated by heating, low pH or elimination of higher molecular weight proteins, BChE activity is markedly reduced or even nil [115,118, own unpublished results].

The clinical problem is that blood, protein solutions or synthetic substitutes are often lavishly applied in emergency situations or intra- and post-operatively, and their BChE activity is not balanced. So, they can either dilute the patient's own enzyme or restore his reduced levels: in any case, this blurs the true picture [119]. As a rule, there is a drop in BChE in serum post-operatively [120], due to post-surgical catabolism and aggravated by the usual fasting before and after the event [121,122]. Fasting or undernutrition alone must be more prolonged to cause the same degree of BChE reduction as surgical interventions [123-125]. Reports on the role of anesthetic drugs in the latter are contradictory [126].

ASSOCIATION WITH DISEASES AND DIAGNOSTIC USE

A higher or lower than usual frequency of some of the genetic variants of BChE has been associated with various diseases: The atypical gene has been found to be more frequent in leprosy and less frequent in Down's syndrome, the fluoride resistant alleles have been noticed with increased frequency in patients with mental illnesses, Huntington 's Chorea and malignant hyperthermia, the C5+ variant has also been more frequently observed in Huntington's Chorea and in Down's Syndrome, as well as in myeloma patients [20]. Since 1986, numerous publications have been dedicated to the pathobiology of Alzheimer's disease, and to its therapy with cholinesterase inhibitors, which are meant to improve the cholinergic deficiency, due to reduced activity of choline acetyltransferase. The clinical results are judged contradictory [127-129], and the disease is still poorly understood. The altered AChE/BChE ratios in distinct brain regions [130], the more peptidase-like properties of probably embryonic forms of AChE and BChE in the plaques and tangles [131,132] and the decrease of both cholinesterases in cerebrospinal fluid, but not in plasma, of patients with advanced disease stages may rather be secondary to the primary disturbance and are not suited for diagnosis or therapeutic control [133,134]. Determination of the AChE/BChE ratio in colonic tissue for enhanced diagnosis of Hirschsprung's disease [135] and in amniotic fluid for the distinction of open neural tube defects from abdominal wall defects have recently been proposed [136].

In obesity, type II diabetes, hyperlipoproteinemia and fatty liver BChE in plasma is frequently above normal. It has been shown that increased BChE results in the elevation of ß- and pre-ß-lipoproteins and a decrease of alpha-lipoproteins and vice versa, and that there is frequently an inverse correlation between BChE and HDL in patients with type II and IV hyperlipoproteinemia [51]. Kutty et al. proposed the ratio of BChE/HDL as a complementary risk factor for atherosclerosis and reported a 20% enhancement of the prediction of coronary heart disease [137]. We found in adult patients with nephrotic syndrome a significant correlation between BChE and the concentrations of cholesterol, triglycerides and VLDL in the collective and in the individual patients under longitudinal observation.

The depression of BChE levels found in the acute stages of chronic inflammatory bowel diseases does not so much reflect the patients' nutritional state or their actual weight loss or gain [138], but the overall clinical picture, as expressed by the so-called activity indexes. Although recently disputed for single measurements [139], BChE determinations can be used for monitoring therapeutic effects [140,141]. Likewise, they may be taken as a liver function test and a prognostic factor in extrahepatic infectious, toxic, malignant and other systemic diseases with liver involvement, from mild fatty infiltration or cholestasis to inflammation and necrosis [142]. Secondary liver disorders are more common in the average population than primary ones. It is their high prevalence, which accounts for the apparent "false-positive" BChE values in the diagnosis of liver diseases.

Nevertheless, the most important reason for the determination of BChE in serum is detection, differential diagnosis and prognosis of liver diseases. Although the diagnostic sensitivity of BChE is lower than that of alanine aminotransferase and gamma-glutamyltransferase [143,144], the different pathogenetic mechanisms, which lead to the increase of cell enzymes and the variations of secreted enzymes, like BChE, in plasma and isolated changes of the latter justify the inclusion of BChE into the screening enzyme profile: In liver cirrhosis, when the activities of cell enzymes in plasma tend to become normal, BChE falls to its most pathological levels [145,146]. The triad of screening enzymes also allows cautious presumptions as to differential diagnosis: Low BChE activity indicating either an advanced stage of chronic liver disease [147], or malignancy, primary or secondary to the liver, or toxic liver damage with inhibition of export protein synthesis. Elevated BChE levels, on the contrary, are highly suggestive of a nutritional fatty liver [146,148] . Normal BChE signifies in cases of obstructive jaundice a benign cause, like gallstones, in chronic hepatitis of viral or toxic origin an early stage or a persistent or scarcely active process, in acute viral hepatitis an uncomplicated course. In acute toxic liver damage it indicates an undisturbed protein synthesis and, thereby, rules out a number of drugs and toxins [146]. Whether observations on the absence of the C2 band in PAGE of the sera from patients with liver cirrhosis [149], which is presumably due to low albumin levels, enhances differential diagnosis, remains to be confirmed. The prognostic significance of decreased and further decreasing BChE activity in serum in fulminant liver failure [150], in liver cirrhosis [151], and after liver transplantation [152] has stood the test for years and has been confirmed by models for computer-supported decision-making [153].

REFERENCES

1. Kalow, K. and K. Genest, A method for the detection of atypical forms of human cholinesterase, determination of dibucaine numbers. Can. J. Biochem. 35:1957; 339-346

2. Harris, H. and M. Whittaker, Differential inhibition of human serum cholinesterase with fluoride: recognition of two new phenotypes. Nature191:1961; 496-498

3. Liddell, J., H. Lehmann and E. Silk, A "silent" pseudocholinesterase gene. Nature 193:1962;561-562

4. Bartels, C.F., F.S. Jensen, O. Lockridge, A.F.L. van der Spek, H.M. Rubinstein, T. Lubrano and B.N. La Du, DNA mutation associated with the human butyrylcholinesterase K-variant and its linkage to the atypical variant mutation and other polymorphic sites. Am. J. Hum. Genet. 50:1992; 1086-1103

5. Rubinstein, A.A. Dietz and T. Lubrano, E1k, another quantitative variant at cholinesterase locus 1. J. Med. Genet. 15:1978; 27-29

6. Garry, P.J., A.A. Dietz, T. Lubrano, P.C. Ford, K. James and H.M. Rubinstein, J. med. Genet. 13:1976; 38-42

7. Whittaker, M. and J.J. Britten, E1h, a new allele at cholinesterase locus 1. Hum. Hered. 37:1987; 54-58

8. Schmidt, E., R. Klauke and F.W. Schmidt, Proposed routine method for phenotyping of cholinesterase (CHE) locus E1 variants with butyrylcholine as substrate at 37°C, Fresenius Z. Anal. Chem. 330:1988; 364

9. Lockridge, O. and B. N. La Du, Structure of the human butyrylcholinesterase gene and expression in mammalian cells, in: Cholinesterases, J. Massoulié et al. eds., Am. Chem. Soc. Washington, DC, 1991, 168-171

10. Arpagaus, M., M. Kott, K.P. Vatsis, C.F. Bartels, B.N. La Du and O. Lockridge, Structure of the gene for human butyrylcholinesterase. Evidence for a single copy. Biochemistry 29:1990; 124-131

11. Gaughan, G., H. Park, J. Priddle, I. Craig and S. Craig, Refinement of the localization of human butyrylcholinesterase to chromosome 3q26.1-q26.2 using a PCR-derived probe. Genomics 11:1991; 455-458

12. McGuire, M.C., C.P. Nogueira, C.F. Bartels, H. Lightstone, A. Hajra, A.F. van der Spek, O. Lockridge and B.N. La Du, Identification of the structural mutation responsible for the dibucaine-resistant (atypical) variant form of human serum cholinesterase. Proc. Nat. Acad. Sci. U.S.A. 86:1989; 953-957

13. Nogueira, C.P., C.F. Bartels, M.C. McGuire, S. Adkins, T. Lubrano, H.M. Rubinstein, H. Lightstone, A.F. van der Spek, O. Lockridge and B.N. La Du, Identification of two different point mutations associated with the fluoride-resistant phenotype for human butyrylcholinesterase. Am. J. Hum. Genet. 51:1992; 821-828

14. Jensen, F.S. C.F. Bartels and B.N. La Du, A DNA point mutation associated with the H-variant of human butyrylcholinesterase, in: Cholinesterases, J. Massoulié et al. eds. Am. Chem. Soc. Washington DC ,1991, 189

15. Whittaker, M. and J.J. Britten, Segregation of the sub(1)(j) gene for plasma cholinesterase in family studies. Hum. Her. 39:1989; 1-6

16. Bartels, C.F., K. James, B.N. La Du, DNA mutations associated with the human butyrylcholinesterase J-variant. Am. J. Hum. Genet. 50:1992; 1104- 1114

17. Evans and J Wardell, On the identification and frequency of the J and K cholinesterase phenotypes in a Caucasian population. J. Med. genet. 21:1984; 99-102

18. Bartels, C.F., O. Lockridge and B.N. La Du, DNA coding for the K polymorphism in linkage disequilibrium with atypical human butyrylcholinesterase complicates phenotyping, in: Cholinesterases, J. Massoulié et al. eds.Am. Chem. Soc., Washington, DC, 1991, 191

19. Szeinberg, A., S. Pipano, M. Assa, J.H. Medalie and H.N. Neufeld, High frequency of atypical pseudocholinesterase gene among Iraqi and Iranian Jews. Clin. Genet. 3:1972; 123-127

20. Whittaker, M., Cholinesterase, Karger, Basel, 1986

21. Arnaud, J., H. Brun, R. Llobera and J. Constans, Serum cholinesterase polymorphism in France: an epidemiological survey of the deficient alleles detected by an automated micro-method. Ann. Hum. Biol. 18:1991; 1- 8

22. Chautard-Freire-Maia, E.A. S.L. Primo-Parmo, M.A. Canever de Lourenço and L. Culpi, Frequencies of atypical serum cholinesterase among Caucasians and Negroes from Southern Brazil. Hum. Hered. 34:1984; 388-392

23. Evans, R.T., A. Walker and K.M. Bowness, Improved accuracy of cholinesterase phenotyping after participation in a proficiency survey. Clin. Chem. 33:1987; 823-825

24. Rosalki, S.B., Genetic influences on diagnostic enzymes in plasma. Enzyme 39:1988; 95-109

25. Masson, P., A. Chatonnet and O. Lockridge, Evidence for a single butyrylcholinesterase gene in individuals carrying the C5 plasma cholinesterase variant (CHE2). FEBS 262:1990; 115-118

26. Harris, H., D.A. Hopkinson, E.B. Robson and M. Whittaker, Genetical studies on a new variant of serum cholinesterase detected by electrophoresis. Ann. Hum. Genet. 26:1963; 359-382

27. Masson, P., Molecular heterogeneity of human plasma cholinesterase, in: Cholinesterases, J. Massoulié et al. eds. Am. Chem. Soc. Washington, DC , 1991, 42-46

28. Soreq, H., R. Zamir, D. Zevin-Sonkin and H. Zakut, Human cholinesterase genes localized by hybridization to chromosomes 3 and 16. Hum. genet. 77:1987; 325-328

29. Marazita, M.L., B.J.B. Keats, M.A. Spence, R.S. Sparkes, L.L. Field, M.C. Sparkes and M. Crist, Mapping studies of the serum cholinesterase-2-locus (CHE2). Hum. Genet. 83:1989; 139-144

30. Eiberg, H., L.S. Nielsen, J. Klausen, M. Dahlen, M. Kristensen, M.L. Bisgaard, N. Moller and J. Mohr, Linkage between serum cholinesterase 2 (CHE2) and gamma-crystalline gene cluster (CRYG): assignment to chromosome 2. Clin. Genet. 35:1989; 313-321

31. Gueirrero, J.F., S.E. Santos and G.F. Aguiar, Serum cholinesterase polymorphism (CHE1 and CHE2 loci) among several Indian groups from Amazon region of Brazil, and segregation of the C5 variant in families. Gene Geography 3:1989; 11-20

32. Steegmüller, H., On the geographical distribution of pseudocholinesterase variants. Humangenetik 26:1975; 167- 185

33. Oimomi, M., J. Ohkawa, S. Saeki and S. Baba, A familial study of C5+cholinesterase and its frequency in the normal population. Gastroent. japon. 23:1988; 680-683

34. Schmidt, E., F.W. Schmidt, A. Delbrück and E. Henkel, Variants of cholinesterase. Adv. Clin. Enzymol. 2:1982; 55-66

35. Lockridge, O., H.W. Eckerson and B.N. La Du, Interchain disulphide bonds and subunit organization in human serum cholinesterase. J. biol. Chem. 254:1979; 8324-8330

36. Fishman, J.B. and R.E. Fine, A trans Golgi-derived exocytic coated vesicle can contain both newly synthesized cholinesterase and internalized transferrin. Cell 48:1987; 157-164

37. Ostergaard, D., J. Viby-Mogensen, H.K. Hanel and L.T. Skovogaard, Half-life of plasma cholinesterase. Acta Anaesthesiol. Scand. 32:1988; 266-269

38. Neitlich, H.W., Increased plasma cholinesterase activity and succinylcholine resistance: a genetic variant. J. clin. Invest. 45:1966; 380-387

39. Yoshida, A. and A.G. Motulski, A pseudocholinesterase variant (E Cynthiana) associated with elevated plasma enzyme activity. Am. J. Hum. genet. 21:1969; 486-498

40. Delbrück, A. and E. Henkel, A rare genetically determined variant of pseudocholinesterase in two German families with high plasma enzyme activity. Eur. J. Biochem. 99:1979; 65-69

41. Brock, A., Additional electrophoretic components of cholinesterase in plasma: a phenomenon of no importance to the total plasma cholinesterase activity. J. Clin. Chem.Clin. Biochem. 27:1989; 429-431

42. Lockridge, O., C.F. Bartels, T.A. Vaughan, C.K. Wong, S.E. Norton and L.L. Johnson, Complete amino acid sequence of human serum cholinesterase. J. biol. Chem. 262:1987; 549-557

43. Yamato, K., I. Huang, H. Muensch, A. Yoshida, H.-W. Goedde and D.P. Agarwal, Amino acid sequence of the active site of human pseudocholinesterase. Biochem. Genet. 21:1983; 135-145

44. Arpagaus, M., A. Chatonnet, P. Masson, M. Newton, T.A. Vaughan, C.F. Bartels, C.P. Nogueira, B.N. La Du and O. Lockridge, Use of the polymerase chain reaction for homology probing of butyrylcholinesterases from several vertebrates. J. biol. Chem. 266:1991; 6966-6974

45. Neville, L.F., A. Gnatt, R. Padan, B. Seidman and H. Soreq, Anionic site interactions in human butyrylcholinesterase disrupted bytwo single point mutations. J. biol. Chem. 265:1990; 20735-20738

46. Chatonnet, A. and O. Lockridge, Comparison of butyrylcholinesterase and actylcholinesterase. Biochem. J. 260:1989; 625-634

47. Lockridge, O. and B.N. La Du, Comparison of atypical and usual human serum cholinesterase. J. biol. Chem. 253:1978; 361-366

48. Augustinsson, K.B., Cholinesterases: a study in comparative enzymology. Acta Physiol. Scand. 15:1948, Suppl. 52; 1-182

49. Schmidt, E., F.W. Schmidt, Sex differences of plasma cholinesterases in the rat. Enzyme 23:1978; 52-55

50. Wright, P.G., V. de Vos, E. Marcus, M. Ganhao and J. Hattingh, Species variation in plasma cholinesterase activity. Comp. Biochem. Physiol. 70:1981; 289-291

51. Kutty, K.M., Biochemical functions of cholinesterase. Clin. Biochem. 13:1980; 239-243

52. Lockridge, O., Substance P hydrolysis by human serum cholinesterase. J. Neurochem. 39:1982; 106-110

53. Myers, C., O. Lockridge, B.N. La Du, Hydrolysis of methylprednisolone acetate by human serum cholinesterase. Drug Metab. Dispos. 10:1982; 279-280

54. Rao, R.V., A.S. Balasubramanian, The peptidase activity of human serum butyrylcholinesterase:studies using monoclonal antibodies and characterization of the peptidase.J. prot. chem. 12:1993; 103-110

55. Ram, Z., M. Molcho, Y.L. Danon, S. Almong, J. Baniel, A. Karni and J. Shemer, The effect of pyridostigmine on respiratory function in healthy and asthmatic volunteers. Israel J. med. Sci. 27:1991; 664-668

56. Gordon, C.J. and L. Fogelson, Relationship between serum cholinesterase activity and the change in body temperature and motor activity in the rat: A dose-response study of diisopropyl fluorophosphate. Neurotox. Teratol. 15:1993; 21-25

57. Layer, P.G., Expression and possible functions of cholinesterases during chicken neurogenesis, in: Cholinesterases, J. Massoulié et al. eds. Am. Chem. Soc., Washington, DC, 1991, 350-357

58. Massoulié, J., S. Bon and M. Vigny, The polymorphism of cholinesterase in vertebrates. Neurochem. int. 2:1980; 161-184

59. Ehrlich, G., E. Viegas-Péquinot, D. Ginzberg, L. Sindel, H. Soreq and H. Zakut, Mapping the human acetylcholinesterase gene to chromosome 7q22 by fluorescent in situ hybridization coupled with selective PCR amplification from a somatic hybrid cell panel and chromosome-sorted DNA libraries. Genomics 13:1992; 1192-1197

60. Getman, D.K., J.H. Eubanks, S. Camp, G.A. Evans and P. Taylor, The human gene encoding acetylcholinesterases located on the long arm of chromosome 7. Am. J. Hum. Genet. 51:1992; 170-177

61. Sorensen, K., U. Brodbeck, A.G. Rasmussen and B. Norgaard-Pedersen, Normal human serum contains two forms of acetylcholinesterase. Clin. Chim. Acta 158:1986; 1-6

62. Edwards, J.A. and S. Brimijoin, Divergent regulation of acetylcholinesterase and butyrylcholinesterase in tissues of the rat. J. Neurochem. 38:1982; 1393-1403

63. Lapidot-Lifson, Y., C.A. Prody, D. Ginzberg, D. Meytes, H. Zakut and H. Soreq, Coamplification of human acetylcholinesterase and butyrylcholinesterase genes in blood cells: correlation with various leukemias and abnormal megakaryocytopoiesis. Proc. Nat. Acad. Sci. U.S.A. 86:1989; 4715-4719

64. Soreq, H., Y. Lapidot-Lifson and H. Zakut, A role for cholinesterase in tumorigenesis? Cancer Cells 3:1991; 511-516

65. Birkhauser, H. and E.A. Zeller, Cholinesterase und Sexualhormone. Helv.chim.Acta 23:1940; 1460-1464

66. Everett, H.W. and C.H. Sawyer, Effects of castration and treatment with sex steroids on synthesis of serum cholinesterase in rat. Endocrinology 39:1946; 323-343

67. Illsley, N.P. and C.A. Lamartinière, Endocrine regulation of rat serum cholinesterase activity. Endocrinology 108:1981; 1737-1743

68. Lamartinière, C.A.,Growth hormone modulates serum cholinesterase. Endocrinology 118:1986; 1252-1254

69. Kambam, J.R., R.J. Naukam, W. Parris, J.J. Franks, S.M. Perry, B.V.R. Sastry and B.E. Smith, Effects of progesterone, estriol, and prostaglandins on pseudocholinesterase activity. Anesthesiology 71:1989; A 883

70. Sidell, F.R. and A. Kaminskis, Influence of age, sex and oral contraceptives on human blood cholinesterase activity. Clin. Chem. 21:1975; 1393-1395

71. Blaauwen, D.H. den, W.A. Poppe and W. Tritschler, Cholinesterase (EC 3.1.1.8) mit Butyryl-thiocholin-jodid als Substrat: Referenzwerte in Abhängigkeit von Alter und Geschlecht unter Berücksichtigung hormoneller Einflüsse und Schwangerschaft. J. Clin. Chem. Clin. Biochem. 21:1983; 381-386

72. Lepage, L., Cholinesterase, in: Interpretation of Clinical Laboratory Tests. G. Siest, J. Henny, F. Schiele and D.S. Young, eds. Biochemical Publications, Forster City, Ca, U.S.A., 1985, 209-219

73. Pritchard, J.A., Plasma cholinesterase activity in normal pregnancy and in eclamptogenic toxemias. Am. J. Obstet. Gynecol. 70:1955; 1083-1086

74. Foldes, F.F., T. Arai, H.H. Gentsch and Y. Zarda, The influence of glucocorticoids on plasma cholinesterase. Proc. Soc. exp. Biol. 146:1974; 918-920

75. Bradamante, V. and E. Kunec-Vaji´c, Effect of glucocorticoids on plasma cholinesterase in the rat. Biomed. Biochim. Acta 46:1987; 439-443

76. Bradamante, V., E. Kunec-Vaji´c, M. Lisic, I. Dobric and I. Beus, Plasma cholinesterase activity in patients during therapy with dexamethasone or prednisone. Eur. J. clin. Pharmacol. 36:1989; 253-257

77. Tiefenbach, B., L. Jordanov and G. Henninghausen, The glucocorticoid-induced inhibition of cholinesterase activity and the importance for drug interactions. Arch. Pharmacol. 345:1992; Suppl. 1, R101

78. Seto, Y. and T. Shinohara, Structure-activity relationship of reversible cholinesterase inhibitors including paraquat. Arch. Toxicol. 62:1988; 37-40

79. Hansen, W.E. and K. Nehammer, Inhibition of cholinesterase by oxmetidine. Gastroenterology 86:1984; 1107

80. Buccafusco, J.J. and M.D. Smith, In vivo and in vitro cholinesterase inhibitor property of the antitumor agent caracemide. Res. Comm. Chem. Pathol. Pharmacol. 67:1990; 219-227

81. Aden-Abdi, Y., T. Villen, O. Ericsson, L.L. Gustavsson and M.L. Dahl-Puustinen, Metrifonate in healthy volunteers: interrelationship between pharmacokinetic properties, cholinesterase inhibition and side-effects. Bull. WHO 68:1990; 731-736

82. Bang, U., J. Viby-Mogensen and J.E. Wiren, The effect of bambuterol on plasma cholinesterase activity and suxamethonium-induced neuromuscular blockade in subjects heterozygous for abnormal plasma cholinesterase. Acta Anaesthesiol. Scand. 34:1990; 600-604

83. Sharma, M. and L.-Å. Svensson, Bambuterol, a selective inhibitor of human plasma butyrylcholinesterase, in: Cholinesterases, J. Massoulié et al. eds., Am. Chem. Soc. Washington, DC, U.S.A., 1991, 345

84. Berman, H.A. and K. Leonard, Interaction of tetrahydroaminoacridine with acetylcholinesterase and butyrylcholinesterase. Mol. Pharmacol. 41:1992; 412-418

85. Laine-Cessac, P., A. Turcant, A. Premel-Cabic, J. Boyer and P. Allain, Inhibition of cholinesterases by histamine 2 receptor antagonist drugs. Res. Comm. Chem. Path. Pharm. 79:1993; 185-193

86. Kao, Y.J., J. Tellez and D.R. Turner, Dose-dependent effect of metoclopramide on cholinesterases and suxamethonium metabolism. Br. J. Anaesth. 65:1990; 220-224

87. Whittaker, M., J.J. Britten, R.J. Wicks, Inhibition of the plasma cholinesterase variants by propranolol. Br. J. Anaesth. 53:1981; 511-516

88. Holmes, J.H., P. Kaufer and H. Zwarenstein, Effect of benzodiazepine derivatives on human blood cholinesterase in vitro. Res. Comm. Chem. Path. Pharm. 21:1978; 367-370

89. Hoffman, R.S., G.C. Henry, M.A. Howland, R.S. Weisman, L. Weil and L..R. Goldfrank, Association between life-threatening cocaine toxicity and plasma cholinesterase activity. Ann. Emerg. Med. 21:1992; 247-253

90. Lander, F., E. Pike, K. Hinke, A. Brock and J.B. Nielsen, Anti-cholinesterase agents uptake during cultivation of greenhouse flowers. Arch. Envir. Contam. Toxicol. 22:1992; 159-162

91. Filimore, C.M. and J.E. Lessenger, A cholinesterase testing program for pesticide applicators. J. occup. Med. 35:1993; 61-70

92. Wu, Y.Q., J.D. Wang, J.S. Cheng, S.C. Chung and S.Y. Hwang, Occupational risk of decreased plasma cholinesterase among pesticide production workers in Taiwan. Am. J. Industr. Med. 16:1989; 659-666

93. Kamal, A.A., N.T. Elgarthy, F. Maklady, M.A. Mostafa and A. Massoud, Serum cholinesterase and liver function among a group of organophosphorus pesticides sprayers in Egypt. J. toxicol. clin. exper. 10:1990; 427-435

94. Faustini, A., F. Arpala, P. Pagliarella, F. Forastiere, P. Papini and C.A. Perucci, Monitoraggio delle colinesterasi in lavoratori agricoli e commercianti esposito ad esteri fosforici e carbammati. Med. Lavoro 83:1992; 135-145

95. Rama, D.B. and K. Jaga, Pesticide exposure and cholinesterase levels among farm workers in the Republic of South Africa. Sci. Total Envir. 122:1992; 315-319

96. Lander, F., A. Brock, E. Pike, K. Hinke, Chronic subclinical intake of dietary anticholinesterase agents during the spraying season. Food Chem. Toxicol. 30:1992; 37-40

97. Brock, A. and V. Brock, Plasma cholinesterase activity in a healthy population group with no occupational exposure to known cholinesterase inhibitors: relative influence of some factors related to normal inter- and intra-individual variations. Scand. J. Clin. Lab. Invest. 50:1990; 401-408

98. Prody, C.A., P. Dreyfus, R. Zamir, H. Zakut and H. Soreq, De novo amplification within a "silent" human cholinesterase gene in a family subjected to prolonged exposure to organophosphorous insecticides.
Proc. Nat. Acad. Sci. US.A. 86:1989; 690-694

99. Soreq, H. and H. Zakut, Amplification of butyrylcholinesterase and acetylcholinesterase genes in normal and tumor tissues: putative relationship to organophosphorous poisoning. Pharma. Res. 7:1990; 1-7

100. Panteghini, M. and R. Bonora, Evaluation of a new continuous colorimetric method for determination of serum pseudocholinesterase catalytic activity and its application to a centrifugal fast analyser. J. Clin. Chem. Clin. Biochem. 22:1984; 671-676

101. Rostron, P. and T. Higgins, Serum pseudocholinesterase and dibucaine numbers as measured with the Technicon RA-1000 analyzer. Clin. Chem. 34:1988; 1924-1925

102. Hasselberg, S., L. Mauck and D Nealon, Development of a Kodak Ektachem thin-film assay for serum cholinesterase. Clin. Chem. 35:1989; 1120

103. Takeuchi, T., Y. Kabasawa, R. Horikawa and T. Tanimura. Mechanized assay of serum cholinesterase by specific colorimetric detection of released acid. Clin. Chim. Acta 205:1992; 117-126

104. Krull, N.B., J. Kropf and A.M. Gressner, Influence of reagent composition of atypical pseudocholinesterase activity measurement: comparison of a manual and an automated method and implications for routine. Eur. J. Clin. Chem. Clin. Biochem. 30:1992;545-546

105. Thomsen, T., H. Kewitz and O. Pleul. Estimation of cholinesterase activity (EC 3.1.1.7; 3.1.1.8) in undiluted plasma and erythrocytes as a tool for measuring in vivo effects of reversible inhibitors. J. Clin. Chem. Clin. Biochem. 26:1988; 469-475

106. Vorschlag zum Arzneibuch der DDR, 2. Ausg., Bestimmung der Aktivität der Cholinesterase in Serum. Zbl. Pharm. 119:1980; 1293-1297

107. Alcini, D., M. Maroni, A. Colombi, D. Xaiz and V. Foa, Evaluation of a standardized European method for the determination of cholinesterase activity in plasma and erythrocytes. Med. Lavoro 79:1988; 42-53

108. Deutsche Gesellschaft für Klinische Chemie, Proposal for standard methods for the determination of enzyme catalytic concentrations in serum and plasma at 37°C, II. Cholinesterase (acylcholine acylhydrolase, EC 3.1.1.8) Eur. J. Clin. Chem. Clin. Biochem. 30:1992; 163-170

109. Rama, D.B.K. and M. Deneys, Quality control of red blood cell cholinesterase estimations. South Afr. Med. J. 81:1992; 530

110. Whittaker, M., J.W. Jones and J. Braven, Immunological studies of plasma cholinesterase during pregnancy and the puerperium. Clin. Chim. Acta 199:1991; 223-230

111. Lepage, L. F. Schiele, R. Gueguen and G. Siest, Total cholinesterase in plasma: Biological variations and reference limits. Clin. Chem. 31:1985; 546-550

112. Brock, A., Immunoreactive plasma cholinesterase (EC 3.1.1.8) substance concentration compared with cholinesterase activity concentration and albumin: Inter- and intra-individual variations in a healthy population group. J. Clin. Chem. Clin. Biochem. 28:1990; 851-856

113. Pourrat, O., R. Robert, J.P. Neau, P. Deleplanque and D. Alcalay, Crises cholinergiques chez un myasthenique traité par échanges plasmatiques et anticholinesterasiques. Ann. Med. Int. 139:1988; 51-52

114. Evans, R.T. and A. Robinson, The combined effects of pregnancy and repeated plasma exchange on serum cholinesterase activity. Acta anaesthesiol. Scand. 28:1984; 44-46

115. Schuh, F.T., Pseudocholinesterase activity of human whole blood, bank blood and blood protein solutions. Anaesthesist 24:1975; 103-106

116. Huizenga, J.R., K. van der Belt and C.H. Gips, The effect of storage at different temperatures on cholinesterase activity in human serum. J. Clin. Chem. Clin. Biochem. 23:1985; 283-285

117. Balland, M., M. Vincent-Viry and J. Henny, Effect of long-term storage on human plasma cholinesterase activity. Clin. Chim. Acta 211:1992; 129-131

118. Braun, B.-E., R. Goes, M. Tryba, D. Hüppe, H.-D. Kuntz and M. Krieg, Anstieg der Cholinesterase-Aktivität bei Patienten mit dekompensierter Leberzirrhose nach Gabe von gerinnungsaktivem Frischplasma (FFP). Lab. med. 15:1991; 485-489

119. Lovely, M.J., S.K. Patteson, F.J. Beuerlein and J.T. Chesney, Perioperative blood transfusion may conceal atypical pseudocholinesterase. Anest. Analg. 70:1990; 326-327

120. Puche, E., E. Gomez-Valverde, M- Garcia-Morillas, F. Jorde F. Fajardo and J.M. Garcia Gil, Postoperative decline in plasma aspirin-esterase and cholinesterase activity in surgical patients. Acta anaesthesiol. Scand. 37:1993; 20-22

121. Burnett, W. and Y. Conen: Liver function after surgery: A study of 50 cases with particular reference to serum cholinesterase. Br. J. Anaesth. 27:1955; 66-71

122. Ryan, D.W., Postoperative serum cholinesterase activity following successful renal transplantation. Br. J. Anaesth. 51:1979; 881-884

123. Waterlow, J., Liver cholinesterase in malnourished infants. Lancet 1:1950; 908-909

124. Barclay, G.P.T., Pseudocholinesterase activity as a guide to prognosis in malnutrition. Am. J. Clin. Path. 59:1973; 712-716

125. Venkataraman, B.V., J. Thangam, P.S. Shetty and P.M. Stephen, Cholinesterase in starvation. Ind. J. Physiol. Pharmacol. 26:1982; 137-140

126. Foldes, F.F., Enzymes in Anesthesiology. Springer, New York , 1978, 107-131

127. Jenike, M.A., M. Albert, L. Baer and J. Gunther, Oral physostigmine treatment for primary dementia: a double-blind placebo-controlled inpatient trial. J. Geriat. Psychiat. Neurol. 3:1990; 13-16

128. Volger, B.W., Alternatives in the treatment of memory loss in patients with Alzheimer's disease. Clin. Pharm. 1o:1991; 447-456

129. Davis, K.L., L.J. Thai, E.R. Gamzu, C.S. Davis, R.F. Woolson, S.I. Gracon, D.A. Drachman, L.S. Schneider, P.J. Whitehouse, T.m. Hoover and the Tacrine Collaborative Study Group, A double-blind, placebo-controlled multicenter study of tacrine for Alzheimer's disease. New Engl. J. Med. 327:1992; 1253-1259

130. Mesulam, M.M. and C Goula, Shifting patterns of cortical cholinesterase in Alzheimer's disease: implications for treatment, diagnosis, and pathogenesis. Adv. Neurol. 51:1990; 235-240

131. Arendt, T., M.K. Brueckner, M. Lange and V. Bigl, Changes in acetylcholinesterase and butyrylcholinesterase in Alzheimer's disease resemble embryonic development: a study of molecular forms. Neurochem. Int. 21:1992; 381-396

132. Wright, C.I., C. Guela and M.M. Mesulam, Protease inhibitors and indoleamines selectively inhibit cholinesterases in the histopathologic structures of Alzheimer disease. Proc. Nat. Acad. Sci. U.S.A. 90:1993; 683-686

133. Atack, J.R., C. May, J.A. Kaye, A.D. Kay and S.I. Rapoport, Cerebrospinal fluid cholinesterases in aging and in dementia of the Alzheimer type. Ann. Neurol. 23:1988; 161-167

134. Sirvioe, J. and P.J. Riekkinen, Brain and cerebrospinal fluid cholinesterases in Alzheimer's disease, Parkinson's disease and aging. A critical review of clinical and experimental studies. J. Neur.Transmiss. 4:1992; 337-358

135. Bonham, J.R., G. Dale, D.J. Scott and J. Waggett, Diagnostic value of acetylcholinesterase / butyrylcholinesterase ratio in Hirschsprung's disease. Am. J. Clin. Path. 90:1988; 520-521

136. Rasmussen Loft, A.G., Determination of amniotic fluid acetylcholinesterase activity in the antenatal diagnosis of foetal malformations: The first ten years. J. Clin. Chem. Clin. Biochem. 28:1990; 893-911

137. Kutty, K.M., R. Jain, S.N. Huang and K. Kean, Serum pseudocholinesterase:high density lipoprotein cholesterol ratio as an index of risk for cardiovascular disease. Clin. Chim. Acta 115:1981; 55-61

138. Ollenschläger, G., M. Schrappe-Bächer, M. Steffen, B. Bürger and B. Allolio, Erhebung des Ernährungszustandes - ein Bestandteil der klinischen Routine-Diagnostik: Cholinesterase-Aktivität als Ernährungsindikator. Klin. Wschr. 67:1989; 1101-1107

139. Novacek, G., H. Vogelsang, B. Schmidt and H. Lochs, Are single measurements of pseudocholinesterase and albumin markers for inflammatory activity or nutritional status in Crohn's disease? Wien. Klin. Wschr. 105:1993; 111-115

140. Wellmann, W., R. Kubale, T.G.A. Nyman, J. Oestmann, E. Schmidt and F.W. Schmidt, Enzyme screening in inflammatory bowel disease: A preliminary report, in: Progress in Clinical Enzymology. D.M. Goldberg, M. Werner, eds. Masson, New York , 1980, 145-149

141. Tromm, A., D. Hüppe, I. Thau, U. Schwegler, H.D. Kuntz, M. Krieg and B. May, Die Serumcholinesterase als Aktivitätsparameter bei chronisch entzündlichen Darmerkrankungen. Z. Gastroenterol. 30:1992; 449-453

142. Schmidt, E. and F.W. Schmidt, Mitreaktionen der Leber bei systemischen Erkrankungen unter besonderer Berücksichtigung der Infektionen, in: Die Leber bei extrahepatischen Erkrankungen und Stoffwechselleiden. W. Tittor, G. Schwalbach, eds. Demeter, Gräfelfing 1984, 41-76

143. Guder, W.G., Anwendung von Bewertungsverfahren: Modell Lebererkrankungen, in: Validität klinisch-chemischer Befunde, H. Lang, W. Rick, H. Büttner, eds., Springer, Berlin, 1980, 84-91

144. Schmidt, E. and F.W. Schmidt, Strategie-Probleme bei der Diagnostik von Lebererkrankungen, in: Strategien für den Einsatz klinisch-chemischer Untersuchungen, H. Lang, W. Rick, H. Büttner, eds. Springer, Berlin, 1982, 152-169

145. Lautz, H.U., E. Schmidt, F.W. Schmidt and M. Barthels, Korrelationen zwischen Einschränkung der Sekretionsleistung der Leber und dem Austritt von Zellenzymen. Z. Gastroenterol. 171979; :99-105

146. Schmidt, E. and F.W. Schmidt, Enzyme diagnosis in diseases of the liver and the biliary system. Adv. Clin. Enzymol. 1:1979; 239-292

147. Tinè, F. and L. Pagliaro, Cirrhosis and its recognition in asymptomatic subjects with aminotransferase elevation. Hepatology 11:1990; 516-517

148. Nomura, F., K. Ohnishi, H. Koen, Y. Hiyama, T. Nakayama, Y. Itoh, K. Shirai, Y. Saitoh and K. Okuda, Serum cholinesterase in patients with fatty liver. J. Clin. Gastroenterol. 8:1986; 599-602

149. Hada, T., T. Ohue, H. Imanishi, H. Nakaoka, M. Fujikura, T. Yamamoto, Y. Amuro and K. Higashino, Alteration of serum cholinesterase isozyme in patients with liver cirrhosis. Clin. Chim. Acta 178:1988; 111-112

150. Schmidt, E. and F.W. Schmidt, Enzyme patterns in liver failure, in: Artificial Liver Support. G. Brunner, F.W. Schmidt, eds. Springer, Berlin, 1987, 8-17

151. Burghardt, M., A. Henze, U. Russmann, J. Lobers, E. Schmidt, F.W. Schmidt, On the prognosis of liver cirrhosis, in: Experimental and Clinical Hepatology. C.E. Broelsch, O. Zelder, eds. MTP Press, Lancaster, 1986, 80-88

152. Schmidt, E., F.W. Schmidt S. Ohlendorf, R. Raupach, T. Wittig, C.E. Broelsch and R. Pichlmayr, Enzyme patterns in serum after liver transplantation, in: A. Burlina, L. Galzigna, eds. Clinical Enzymology Symposia 5. Piccin, Padua, 1986, 143-156

153. Irrgang, B., N. Adam, A. Schwab and F. Przybylski, Computer-supported decision making in liver diagnosis based on a graphical presentation of laboratory data. J. Clin. Chem. Clin. Biochem. 28:1990; 733

SERUM CHOLINESTERASE IN DIABETES

K.M. Kutty

Dr. Charles A. Janeway Child Health Centre and Faculty of Medicine, Memorial University of Newfoundland, St. John's, Newfoundland, Canada

The history of cholinesterase in regard to its discovery, proposed biological functions and involvement in lipoprotein metabolism is provided as a preamble to explain the rationale behind the relationship between serum cholinesterase and diabetes mellitus.

Cholinesterase (ChE; acylcholine acyl hydrolase E.C.3.1.1.8) was first discovered in the horse serum in 1932 [1]. The analog of this enzyme is acetylcholinesterase (AChE; E.C.3.1.1.7) which has a very important function in the cholinergic nervous system; hence it is called true cholinesterase. Since serum ChE has no such function it is called pseudocholinesterase. However, this classification may be an oversimplification. For example, the cholinesterase in the red cell membrane has no known cholinergic function yet it is clearly an AChE. Although both serum ChE and neuromuscular AChE hydrolyse acetylcholine and are inhibited by physostigmine, they are different proteins conditioned by different genes. Serum ChE is a tetramer with a molecular weight of about 350,000 daltons [2]. Serum ChE is a typical blood protein in that liver is its primary source in humans, mice and rats. The human serum ChE gene is located on chromosome 3 [3].

ChE is also found in adipose tissue, small intestine, smooth muscle cells, white matter of the brain and other organs and tissues. Only liver ChE is secreted into the blood whereas in other locations the enzyme appears to be associated with the plasma membrane. In spite of high ChE activity in many tissues and its ubiquitous presence in mammalian species, no true biological function has been discovered as yet. Hypotheses have been put forward and a few of these are cited below.

ChE and choline ester metabolism

In humans, ChE is important in the hydrolytic detoxification of succinylcholine, a xenobiotic muscle relaxant used during anaesthesia [4]. Some individuals undergoing surgery who respond atypically to this drug by excessively prolonged apnea are homo-zygous for a low-

Esterases, Lipases and Phospholipases, Edited by M.I. Mackness and M. Clerc, Plenum Press, New York, 1994

activity ChE. Despite this important pharmaco-logical property of ChE, succinylcholine is not a true biological substrate and its hydrolysis by ChE may have no direct relation-ship to its biological function. In 1964, Clitherow et al [5] proposed that ChE is involved in fatty acid metabolism. Their contention was that during fatty acid metabolism, butyryl-CoA formed as an intermediate reacts with choline to form butyryl-choline. As butyrylcholine is toxic, the function of ChE is to detoxify this metabolite by hydrolysis. As the increased energy requirement during accelerated cell growth is derived from the enhanced rate of fatty acid oxidation, Ballantyne and Burwell [6] suggested that ChE has a function in sustaining cell growth by preventing intracellular butyrylcholine accumulation. As choline is an important precursor for the synthesis of acetylcholine, lecithin and sphingomyelin, Funnell and Olivier [7] proposed that ChE is involved in choline homeostasis by controlling the balance between choline esters and free choline. The common theme in all these proposals is that ChE being a choline ester hydrolase by definition, its biological function must be as a cholinesterase. In order to test this assumption, we treated mice for one month with tetraisopropylpyrophosphoramide (iso-OMPA). Serum ChE activity was reduced by about 80% (unpublished results) but we were unable to demonstrate any increase in levels of serum butyryl-, propioyl-, or acetylcholine esters. The study was repeated in rats by treating them with iso-OMPA for two weeks, but again we failed to detect accumulation of choline esters in the liver, serum or urine of these animals [8]. This led us to suggest that the true biological function of ChE is unrelated to choline ester hydrolysis. What then is the function or functions of ChE?

ChE and lipid metabolism

We have proposed [9] that serum ChE can stabilize LDL by interacting with the phosphorylcholine site of lecithin. The basis of the suggestion is that the phosphorylcholine site of lecithin has a very similar structure to acetylcholine, but with a phosphate group instead of acetate. Rats and guinea pigs treated with anti-ChE showed reduced incorporation of H3 lysine into LDL and increased incorporation into HDL [10]. As further evidence, we reported [11] that a boy who had accidentally ingested an organophosphate anti-ChE insecticide showed considerable reduction in serum LDL and an increase in HDL. As the patient recovered, serum ChE levels increased concurrently with increased LDL and decreased HDL levels.

We have observed increased serum ChE activity in patients with hyperlipoproteinemia. In an early study directed at determining the types of hyperlipoproteinemia most commonly associated with raised serum ChE activity [12], we found that out of a total of 159 serum specimens from patients whose lipid profile was investigated, 49% were classified as type IIb or IV hyperlipoproteinemia, 5% as type IIa, and the rest as normal.

Only patients labelled as type IIb or IV hyperlipoproteinemia showed a significant increase in serum ChE activity. Similar observations were also made in another group of 400 patients [13]. In these patients with type IIb and IV hyperlipoprotein-emia the increase in serum VLDL and triacylglycerols has been suggested to be due to hepatic over-production [14]. Obesity and diabetes are usually associated with increased levels of serum triacylglycerol and VLDL. Increased serum ChE has been found to be characteristic of obese patients, especially when they are also hyperlipoproteinemic [12].

To better understand these observations in patients, we studied ChE changes in the liver and serum of genetically obese (ob/ob) mice, diabetic (db/db) mice and Zucker fat rats. These animals are hyperinsulinemic, hyperphagic, and have an increased rate of lipogenesis and VLDL production in the liver. In many respects these animal models mimic non-insulin dependent diabetes mellitus (NIDDM) in humans. Both obese (ob/ob) and diabetic (db/db) mice showed a marked increase in liver and serum ChE activity [15]. Control of diet in diabetic mice prevented the elevation of serum ChE activity and body weight [16]. Electron microscopic studies showed that ChE increase in the obese mice is due to its over-production in the liver [17]. Zucker fat rats also showed increased ChE activity in the liver and serum and over-produce VLDL and triacylglycerols [18]. As these rodents resemble humans in exhibiting concurrently increased synthesis of VLDL and ChE, it appears that there is a common event leading to the increased synthesis of VLDL and ChE in pathologic states involving hepatic triacylglycerol over-production.

ChE in diabetes

Diabetic patients were also reported to have increased serum ChE activity [19]. Diabetes mellitus is known to be associated with abnormal lipid metabolism; pre-dominantly elevated serum triacylglycerols and VLDL [20]. We have studied serum ChE levels in diabetic patients ranging in age from 12 to 65 years. Patients below 18 years of age were classified as juvenile diabetics and above 18 years as adult diabetics. Controls were age-matched individuals with no evidence of diabetes, hyperlipo-proteinemia, obesity or heart disease. Since the analytes measured in the controls showed no significant difference between the two age groups the data from the diabetic patients were compared against the pooled control values (Table 1).

Our results demonstrate that diabetic patients showed a significant increase in serum glucose consistent with their diabetic status. Juvenile diabetic patients showed considerably reduced C-peptide levels which is an indication of low insulin secretory activity. These younger patients were dependent on insulin administration for the control of their diabetes (IDDM). Adult diabetic patients, on the other hand, showed a significant increase in C-peptide and insulin which is characteristic of insulin resistance in non-insulin dependent diabetes mellitus (NIDDM). Haemoglobin A_{1c} showed

Table 1. A comparison of the different analytes between diabetic patients (both IDDM and NIDDM) and the controls

ANALYTES	CONTROLS (62)	IDDM (110)	NIDDM (41)
Cholinesterase (u/l)	3125 ± 70.7	4336 ± 70.5	3836 ± 126
Glucose (mmol/l)	4.74 ± 0.09	14.6 ± 0.6	10.45 ± 0.52
C-Peptide (pmol/l)	448.8 ± 58.4	160 ± 11.8	869 ± 82.4
HbA$_{1C}$ (%)	5.5 ± 0.08	8.57 ± 0.23	8.44 ± 0.25
Triacylglycerols (mmol/l)	1.26 ± 0.08	2.15 ± 0.07	2.58 ± 0.20
Cholesterol (mmol/l)	4.53 ± 0.13	4.67 ± 0.07	6.02 ± 0.21

The results are presented as means ± S.E.M. with the number of samples in parentheses. All the analytes are significantly higher ($P > 0.01$) in the diabetic patients except for cholesterol in IDDM (Statistical analysis by unpaired Student's t-test). Methods of analysis of cholinesterase, glucose, triacylglycerols and cholesterol are described in reference [21]. HbA$_{1C}$ was measured using Bio-Rad Mini Column Test and C-peptide by Diagnostic Products Corporation, Double Antibody method.

a significant increase in both types of diabetic patients indicating chronic hyperglycemic state. Mean serum ChE activities and triacylglycerol levels were found to be significantly elevated in both categories of diabetes. Serum cholesterol was found to be elevated only in NIDDM. Lipoprotein electrophoresis showed prominent pre-beta lipoprotein band in the diabetic patients indicating elevated VLDL levels as the reason for the increased serum triacylglycerols.

These observations in diabetic patients led us to find out more about the relationship between serum ChE increase in diabetes using an experimental animal model. Streptozotocin induced diabetic rats were used for this purpose [21]. Serum ChE activity increased with the induction of diabetes concomitant with an increase in serum triacylglycerol, glycerol and ApoB containing lipoproteins. In contrast HDL cholesterol decreased. When diabetes was controlled with insulin treatment, serum ChE activity declined in concert with the reduction in serum lipids. Heparin treatment decreased serum triacylglycerols in the diabetic rats without a concurrent reduction in serum ChE activity. The removal of triacylglycerols appears to be due to the activation of lipoprotein lipase. That serum triacyl-glycerols can be reduced without influencing ChE activity confirms that increased ChE in hypertriacylglycerolemia is not simply due to enhanced activation of the enzyme in the presence of high levels of VLDL but that there is enhanced secretion of ChE from the liver. Increased serum glycerol in diabetic animals indicates increased adipose tissue lipolysis. Increased fatty acid mobilization and uptake by the liver results in the over-production of triacylglycerols, as has also been suggested [22] to explain the increased serum lipids in human diabetes.

It is apparent from the evidence presented above that serum ChE and VLDL levels are tightly linked. To further investigate this phenomenon, we treated diabetic rats with iso-OMPA, a specific serum ChE inhibitor [21]. Treatment with iso-OMPA resulted in a marked decrease in serum and liver triacylglycerols and VLDL. These results are an indication that ChE may have a function in either VLDL assembly or secretion. Chronic inhibition of ChE also decreased serum glycerol concentration. This suggests that adipose tissue ChE may have an effect on hormone mediated lipolysis. Further work is required to better understand the reasons for serum ChE increase in hyperlipo-proteinemia in general and diabetes mellitus in particular.

REFERENCES

1. Alles, G.A. and Hawes, R.C. Cholinesterase in the blood of man. J. Biol. Chem. 133, (1940) 375-390.

2. Lockridge, O. and LaDu, B.N. Loss of interchain disulfide peptide and dissociation of the tetramer following limited proteolysis of native human serum cholin-esterase. J. Biol. Chem. 257, (1982) 12012-12018.

3. Lockridge, O. and LaDu, B.N. Structure of human butyryl-cholinesterase gene and expression in mammalian cells, in: "Cholinesterases", J. Massoulie, ed., Amer. Chem. Soc., Washington (1991).

4. Svensmark, O. Molecular properties of cholinesterase. Acta Physiol. Scand. 64, Suppl. 245, (1965) 1-74.

5. Clitherow, J.W., Mitchard, M. and Harper, N.J. The possible biological function of pseudocholinesterase. Nature 199, (1963) 1000-1001.

6. Ballantyne, B. and Burwell, R.G. Distribution of cholin-esterase in normal lymph nodes and its possible relation to regulation of tissue size. Nature 206, (1965) 1123-1125.

7. Funnell, H.S. and Oliver, W.T. Proposed physiological function for plasma cholinesterase. Nature 208, (1965) 689-690.

8. Kutty, K.M., Annapurna, V. and Prabhakaran, V. Pseudocholin-esterase: a protein with functions unrelated to its name. Biochem Soc Trans 17, (1989) 555-556.

9. Kutty, K.M., Rowden, G. and Cox, A.R. Interrelationship between serum beta-lipoprotein and cholinesterase. Can. J. Biochem. 51, (1973) 883-887.

10. Kutty, K.M., Redheendran, R. and Murphy, D. Serum cholin-esterase, function in lipoprotein metabolism. Experientia 33, (1977) 420-421.

11. Kutty, K.M., Jacob, J.C., Hutton, C.J., Davis, P.J. and Peterson, S.C. Serum beta lipoprotein: Studies in a patient and in guinea pigs after the ingestion of organophosphorus compounds. Clin. Biochem. 8, (1975) 379-383.

12. Chu, M.I., Fontaine, P., Kutty, K.M., Murphy, D. and Redheendran, R. Cholinesterase in serum and low density lipoprotein of hyperlipidemic patients. Clin. Chim. Acta 85, (1978) 55-59.

13. Jain, R., Kutty, K.M., Huang, S.N. and Kean, K. Pseudo-cholinesterase/high-density lipoprotein cholesterol ratio in serum of normal persons and of hyperlipo-proteinemics. Clin. Chem. 29, (1983) 1031-1033.

14. Levy, R.I and Glueck, C.J. Hypertriglyceridemia, diabetes mellitus and coronary vessel disease. Arch. Intern. Med. 123, (1969) 220-228.

15. Kutty, K.M., Huang, S.N. and Kean, K. Pseudocholinesterase in obesity hypercaloric diet induced changes in experimental obese mice. Experientia 37, (1981) 1141-1142.

16. Kutty, K.M., Kean, K.T., Jain, R. and Huang, S.N. Plasma pseudocholinesterase: A potential marker for early detection of obesity. Nutr. Res. 3, (1983) 211-216.

17. Kean, K.T., Kutty, K.M., Huang, S.N. and Jain, R. A study of pseudocholinesterase induction in experimental obesity. Jour. Amer. Coll. of Nutr. 5, (1986) 253-261.

18. Kutty, K.M., Jain, R., Kean, K. and Peper, C. Cholinesterase activity in the serum and liver of Zucker fat rats and controls. Nutr. Res. 4, (1984) 99-104.

19. Antopol, W., Tuchman, T. and Schifrin, W. Cholinesterase activity in human serum with special reference to hyperthyroidism. Proc. Soc. Exp. Biol. Med. 36, (1937)46-49.

20. Howard, B.V. Lipoprotein metabolism in diabetes mellitus. J. Lip. Res. 28, (1987) 613-628.

21. Annapurna, V., Senciall, I., Davis, A.J. and Kutty, K.M. Relationship between serum pseudocholinesterase and triglycerides in experimentally induced diabetes mellitus in rats. Diabetologia 34, (1991) 320-324.

22. Nikkila, E.A and Kekki, M. Plasma triglyceride transport kinetics in diabetes mellitus. Metabolism 22, (1973) 1-22.

LIPID FEEDING AND SERUM ESTERASES

Hein A. Van Lith, Bert F.M. Van Zutphen and Anton C. Beynen

Department of Laboratory Animal Science, Faculty of Veterinary
Medicine, Utrecht University, P.O. Box 80.166, NL-3508 TD
Utrecht, The Netherlands

I. SUMMARY

A brief review is given of the effects of the amount and type of dietary lipid on serum
esterase activities in rabbits and selected rodent species. The feeding of a cholesterol-rich
diet causes an increased activity of serum total esterase in rats, rabbits and mice. Cholesterol loading of rats, but not rabbits, lowers serum butyrylcholinesterase activity. The activity
or level of a fast anodal serum esterase zone in rats (ES-1) and mice (ES-2) is raised after
cholesterol feeding. The concentration and type of dietary fat influence ES-1 and total
esterase activity in the serum of rats. Increased fat intakes markedly elevate serum ES-1
activity and slightly increase serum total esterase activity. The magnitude of this effect
depends on the type of fat. In contrast to rats, the activity of an anodal fast moving serum
esterase zone is decreased in gerbils after feeding a high-fat diet. Serum butyrylcholinesterase activities of rats and rabbits are slightly increased with increasing fat intakes. In
gerbils, such an effect does not occur. Dietary fish oil when compared with either coconut
fat, corn oil or olive oil produces increased activities of serum butyrylcholinesterase in rats.

II. INTRODUCTION

The serum of vertebrate animals contains various enzymes, so-called esterases, that
can catalyze the hydrolysis of water soluble artificial fatty acid esters, such as α-naphthylacetate and p-nitrophenylacetate [1]. The serum esterase isozymes differ in pH optimum,
substrate specificity and inhibitor sensitivity [2].

Because of the high degree of polymorphism, serum esterases can be used for the
genetic characterization and quality control of inbred strains of laboratory animals [3]. The
esterases are involved in the detoxification or activation of a wide range of xenobiotic
carboxylesters and aromatic amides, including various drugs and anaesthetics [4]. Within
and between species variation in serum esterase isozyme activities [5,6] may thus influence
the efficacy of ester or amide-type compounds.

The physiological function of serum esterases is still obscure, but they are probably
essential because their genetic codes have been preserved throughout evolution. Genetic

Esterases, Lipases and Phospholipases, Edited by M.I. Mackness
and M. Clerc, Plenum Press, New York, 1994

homology of serum esterase isozymes among various animal species has been described [3]. There is evidence that serum esterases are involved in lipid metabolism. *In-vitro* studies have shown that serum esterases can hydrolyze various classes of lipids, including mono- and triacylglycerols [7,8]. The administration of fat by stomach tube to rats, produced an increase in the activity of a serum esterase with a high electrophoretic mobility (esterase-1 , ES-1) [9]. In mice, a raised activity of esterase-2 (ES-2) in serum was seen after infusion of fat into the duodenum [10]. Here, we describe that the amount and/or type of lipid in the diet affects serum esterase activities in rabbits and selected rodent species.

Table 1. Effect of dietary cholesterol on serum esterase activities of IIIVO/JU and AX/JU inbred rabbits[1]

Measure	Diet without added cholesterol		Diet with added (0.3%, w/w) cholesterol		Significance[2]
	IIIVO/JU	AX/JU	IIIVO/JU	AX/JU	
Serum total esterase activity (μmol/min/ml)[4]					
initial	1.59$\pm$0.18	0.74$\pm$0.06	1.43$\pm$0.06	0.76$\pm$0.04	S[3]
final	2.01$\pm$0.27	0.79$\pm$0.07	3.06$\pm$0.42	0.93$\pm$0.07	S,C[3]
change	+0.42$\pm$0.29	+0.05$\pm$0.03	+1.63$\pm$0.37	+0.17$\pm$0.06	S,C[3]
Serum arylesterase activity (μmol/min/ml)					
initial	418$\pm$31	179$\pm$22	424$\pm$16	191$\pm$14	S
final	441$\pm$25	182$\pm$21	416$\pm$17	167$\pm$10	S
change	+23$\pm$21	+3$\pm$ 4	-8$\pm$14	-24$\pm$ 7	C

[1] Results expressed as means $\pm$ SEM for six (commercial diet without added cholesterol; two males and four females) or nine (commercial diet with added cholesterol; two males and seven females) animals per group. Restricted amounts of each diet (80 g/rabbit/day) were fed for 21 days.

[2] Significance (P<0.05) was calculated by two-way analysis of variance. S, effect of rabbit strain; C, effect of dietary cholesterol.

[3] Analysis of variance after logarithmic transformation of the data.

[4] Total esterase activity was measured with *p*-nitrophenylbutyrate as substrate.

III. EFFECT OF HIGH-CHOLESTEROL DIETS

A. Studies with rabbits

In two inbred strains of rabbits the amount of cholesterol in the diet affected total esterase and arylesterase activity in serum (Table 1). Total esterase and arylesterase activities were assayed using *p*-nitrophenylbutyrate and phenylacetate, respectively, as substrate. Baseline esterase activities were higher in IIIVO/JU than AX/JU rabbits. Cholesterol feeding increased serum total esterase and decreased arylesterase activities in both strains.

Total esterase activities were also measured with ß-naphthylpropionate as substrate (Table 2). In the AX/JU, but not in IIIVO/JU and New Zealand White rabbits, dietary cholesterol caused a significant increase in serum total esterase activity. IIIVO/JU rabbits had the highest serum total esterase activities.

Cholesterol feeding did not significantly influence serum butyrylcholinesterase activity in rabbits (Table 3). New Zealand White rabbits had much higher activities than the two inbred strains.

Table 2. Effect of dietary cholesterol on serum total esterase activity in random-bred New Zealand White and inbred IIIVO/JU and AX/JU female rabbits [Based on data taken from references 11 and 12][1]

	Experiment 1	Experiment 2	
	New Zealand White	IIIVO/JU.	AX/JU
Serum total esterase activity (µmol/min/ml)[2]			
initial	15.60±1.65	21.86±1.66	14.05±0.83[*]
final	15.66±1.42	23.45±1.54	16.10±0.39[*]
change	+0.06±0.62	+1.59±2.26	+2.05±0.44[#]

[1] Results expressed as means ± SEM for three (Experiment 2) or nine (Experiment 1) animals per group. Values are given before (initial) and 28 days after (final) cholesterol feeding. The New Zealand White rabbits were fed *ad libitum* a cholesterol-rich (0.2%, w/w) semipurified diet. The IIIVO/JU and AX/JU rabbits received on restricted basis (90 g/rabbit/day) a commerical diet fortified with 0.5% (w/w) cholesterol.

[2] Total esterase activity was measured with ß-naphthylpropionate as substrate.

[*] Significantly different from the IIIVO/JU rabbits (P<0.05; two-tailed Student's *t*-test).

[#] Significantly different from zero (P<0.05; two-tailed paired Student's *t*-test).

Table 3. Effect of dietary cholesterol on serum butyrylcholinesterase activity in random-bred New Zealand White and inbred IIIVO/JU and AX/JU rabbits [Partly based on data taken from reference 13][1]

	Experiment 1		Experiment 2			
	Diet without added cholesterol	Diet with added (0.1%, w/w) cholesterol	Diet without added cholesterol		Diet with added (0.3%, w/w) cholesterol	
	New Zealand White	New Zealand White	IIIVO/JU	AX/JU	IIIVO/JU	AX/JU
Serum butyrylcholinesterase activity (nmol/min/ml)						
initial	671±138	559±166	94±5	85±4	93±3	78±4
final	604±123	592±178	108±6	105±3	106±4	125±7

[1] Results expressed as means ± SEM for six (Experiment 2; diet without added cholesterol, two males and four females), eight (Experiment 1, eight males) or nine (Experiment 2, diet with added cholesterol, two males and seven females) animals per group. Restricted amounts of each diet (Experiment 1, 75 g/rabbit/day; Experiment 2, 80 g/rabbit/day) were fed for 21 (Experiment 2) or 42 (Experiment 1) days. The IIIVO/JU and AX/JU rabbits received commericial diets (with or without added cholesterol), whereas the New Zealand White rabbits were fed purified diets (with or without added cholesterol).

B. Studies with rats

Random-bred and inbred rats were fed a natural-ingredient, commercial diet without or with added cholesterol. ß-Naphthylpropionate hydrolyzing activity in serum was higher in the SHR/Cpb than in the random-bred Wistar and SD/Cpb rats (Table 4). Cholesterol feeding increased esterase activity in each strain, the increase varying from 30 (Wistar) to 73% (SD/Cpb).

The addition of cholesterol to the diet of Wistar rats significantly decreased serum butyrylcholinesterase activity, but raised that of serum ES-1, a fast anodal esterase zone after polyacrylamide gel electrophoresis (Table 5).

Table 4. Effect of dietary cholesterol on serum total esterase activity in male random-bred Wistar (Cpb:WU) and inbred SHR/Cpb and SD/Cpb rats [Based on data taken from reference 11][1]

Strain	Diet	Serum total esterase activity (μmol/min/ml)[2]
	(Experiment 1)[3]	
Wistar (Cpb:WU)	Diet without added cholesterol	3.56±0.27
	Diet with 2% (w/w) added cholesterol	4.63±0.25
	(Experiment 2)[4]	
SHR/Cpb	Diet without added cholesterol	5.91±0.24
	Diet with 2% (w/w) added cholesterol	9.23±0.16
SD/Cpb	Diet without added cholesterol	3.44±0.17
	Diet with 2% (w/w) added cholesterol	5.96±0.14

[1] Results expressed as means ± SEM for five (Experiment 1) or six (Experiment 2) animals per group. The commercial diets were fed for 19 (Experiment 1) or 23 (Experiment 2) days.

[2] Total esterase activity was measured with ß-naphthylpropionate as substrate.

[3] The data of Experiment 1 were analyzed with a two-tailed Student's t-test. The group fed the diet with added cholesterol was significantly different (P<0.05) from the group fed the diet without added cholesterol.

[4] The data of Experiment 2 were analyzed by two-way analysis of variance. There were significant effects (P<0.05) of rat strain (S), dietary cholesterol (C) and interaction (SxC).

Table 5. Serum ES-1 and butyrylcholinesterase activities of female random-bred Wistar rats (Cpb:WU) fed either a cholesterol-free (control) or a cholesterol-rich (1%; w/w) purified diet[1]

	Days on diets			
Diet	7	14	21	Significance[2]
	Serum ES-1 activity			
	(% of standard)			
Control	194±21	155±18	134±12	
Cholesterol-rich	259±27	193±14	238±23	C,T
	Serum butyrylcholinesterase activity			
	(nmol/min/ml)			
Control	104± 9	138±14	194±11	
Cholesterol-rich	102±11	109± 9	132±11	C,T,CxT

[1] Results expressed as means ± SEM for six animals per group.

[2] Significance (P<0.05) was calculated by two-way analysis of variance. C, effect of dietary cholesterol; T, effect of time; CxT, effect of interaction.

C. Studies with mice

Mice of seven inbred strains were either fed a commercial diet or a high-fat, high-cholesterol diet. Baseline serum total esterase activity, as measured with ß-naphthyl-propionate as substrate, varied between strains (Table 6). Feeding the high-cholesterol diet caused a significant increase in serum total esterase activity in five of the seven strains, the mean increase varying from 2 (V/U) to 56% (C57BL/U). Table 6 also shows the levels of serum ES-1 and ES-2 as determined by rocket immunoelectrophoresis. There were marked strain differences in the levels of both esterases. In all strains there was a dramatic increase in ES-2, but not in ES-1, after cholesterol loading.

Table 6. Effect of a high-fat, high-cholesterol diet on total esterase activity and levels of ES-1 and ES-2 in serum of male inbred mice [Based on data taken from reference 14][1]

	Serum total esterase activity (μmol/min/ml)[2]		Serum ES-1 level (μg isozyme/ml)		Serum ES-2 level (μg isozyme/ml)	
	Diet without added cholesterol	Diet with added (2%, w/w) cholesterol	Diet without added cholesterol	Diet with added (2%, w/w) cholesterol	Diet without added cholesterol	Diet with added (2%, w/w) cholesterol
C57BL/U	34±2	53±3[*]	465± 93	536± 98	0.60±0.20	2.74±0.17[*]
R/U	45±4	55±2[*]	393± 83	270± 53	0.09±0.01	1.68±0.33[*]
V/U	62±2	63±4	495± 35	452± 22	0.17±0.02	0.76±0.13[*]
C3H/U	48±3	67±4[*]	603± 52	608± 51	0.22±0.04	1.79±0.25[*]
H/U	60±2	74±3[*]	491± 37	511± 40	0.71±0.13	2.29±0.23[*]
S/U	60±3	72±5	713± 98	760±115	0.25±0.03	2.56±0.56[*]
FT/U	57±2	78±4[*]	711± 27	812± 35	0.85±0.19	1.63±0.24[*]

[1] Results expressed as means ± SEM for six animals per group. The diets were fed for 28 days.
[2] Total esterase activity was measured with ß-naphthylpropionate as substrate.
[*] Significantly different from the group fed the diet without added cholesterol (P<0.05; two-tailed Student's t-test).

IV. EFFECT OF AMOUNT AND TYPE OF DIETARY FAT

A. Studies with rabbits

The consumption of extra corn oil, which was added to the diets at the expense of isoenergetic amounts of carbohydrates (corn starch plus dextrose; 1:1, w/w), slightly raised serum butyrylcholinesterase activities in rabbits (Table 7). There was a positive correlation (Pearson's r=0.999, P<0.01, n=4) between dietary corn oil concentration and group mean serum butyrylcholinesterase activities.

Serum butyrylcholinesterase activity was measured in inbred rabbits fed either corn oil or coconut fat. Dietary fat type did not significantly influence butyrylcholinesterase (Table 8).

Table 8. Effect of dietary coconut fat versus corn oil on serum butyryl-cholinesterase activity in inbred IIIVO/JU rabbits[1]

Time point	Dietary fat	
	Corn oil	Coconut fat
	Serum butyrylcholinesterase activity (nmol/min/ml)	
Day 0	109± 5	99±15
Day 7	100±11	97±14
Day 14	108±15	104±18
Day 21	98±15	97±15
Day 28	105±11	97±12
Day 35	101±10	94±14
Day 42	105±12	95±12

[1] Results expressed as means ± SEM for seven animals (three males and four females) per group. The purified diets were fed *ad libitum* for 42 days.

Table 7. Effect of dietary corn oil: carbohydrate ratio on serum butyrylcholinesterase activity in male New Zealand White rabbits [Based on data taken from reference 15][1]

Dietary composition (% of energy)		Serum butyrylcholinesterase activity (nmol/min/ml)
Corn oil	Carbohydrate	
5.5	74.9	450±157
11.0	69.4	489±118
22.0	58.4	543±112
44.0	36.4	675±158

[1] Results expressed as means ± SEM for eight animals per group. Restricted, isoenergetic amounts of each purified diet were fed for 56 days.

B. Studies with gerbils

Increasing the amount of dietary high-cis oleic acid, sunflower oil from 17.8 to 37.8% of energy at the expense of an isoenergetic amount of carbohydrates (corn starch plus dextrose; 1:1, w/w) did not affect serum butyrylcholinesterase and p-nitrophenylbutyrate hydrolyzing activities in Mongolian gerbils (Table 9). However, this dietary change decreased the activity of a fast moving anodal esterase zone in the serum zymogram (Table 9). There was no effect of high cis-oleic acid versus isomerized sunflower oil containing half of the oleic molecules in the trans configuration.

Table 9. Serum esterase activities in male Mongolian gerbils fed diets containing either high cis-oleic acid or isomerized sunflower oil[1]

	Dietary fat (% of energy)		
	Isomerized sunflower oil	Sunflower oil	
Measure	37.8	37.8	17.8
Serum total esterase activity (μmol/min/ml)[2]			
initial	4.11±0.20	4.34±0.24	4.05±0.12
final	5.32±0.28	5.27±0.18	5.20±0.22
change	+1.21±0.31	+0.93±0.26	+1.15±0.23
Serum butyrylcholinesterase activity (nmol/min/ml)			
initial	5598±242	5988±302	5648±139
final	5619±174	5685±171	5803±215
change	-21±219	-303±310	+155±264
Serum ES-FA activity (% of standard)[3]			
initial	75±16	66±14	86±18
final	121±38	80±26	251±46*
change	+46±28	+14±14	+165±35*

[1] Results expressed as means ± SEM for 14 to 18 animals per group. The purified diets were fed *ad libitum* for 28 days.
[2] Total esterase activity was measured with p-nitrophenylbutyrate as substrate.
[3] The fast moving anodal serum esterase zone is provisionally indicated as ES-FA.
* Significantly different from group fed the 37.8% sunflower oil diet (P<0.05; two-tailed Student's t-test).

C. Studies with rats

Increased intakes of corn oil at the expense of isoenergetic amounts of carbohydrates (corn starch plus dextrose; 1:1, w/w) resulted in a marked increase in serum ES-1 activity in random-bred Wistar rats (Table 10 and Fig. 1). The feeding of extra olive oil or coconut fat produced a lesser increase. An increase in fat content of the diets also significantly raised serum total esterase and ES-2 activities, but there was no clear effect of fat type. Fat feeding produced a slight, but significant increase in serum butyrylcholinesterase activity.

Upon the isoenergetic replacement of macronutrients, omission theoretically determines the resulting change in serum esterase activities as much as addition. One approach to determine specific effects is to add the fat under study to a restricted base diet so that total energy intake is still below that consumed under *ad libitum* conditions. The observed effect of the added fat, after correction for the effect of the extra energy by using body weight gain as covariate, can be considered to reflect the intrinsic fat effect. In two studies with rats this principle was followed to assess the influence of the amount and type of dietary fat on total esterase, ES-1 and butyrylcholinesterase activity in serum.

Table 10. Effect of dietary fat:carbohydrate ratio on serum esterase activities in male Wistar rats (Cpb:WU)[Partly based on data taken from reference 16][1]

| | | Dietary composition (% of energy) | | |
| | Fat: | 5.5 | 44.0 | |
Serum esterase activity	Carbohydrates:	74.9	36.4	Significance[2]
Esterase-1 (ES-1)	Coconut fat	19 ± 2	147 ± 12	
(% of standard)	Corn oil	20 ± 3	223 ± 12	A,T,AxT
	Olive oil	20 ± 4	179 ± 9	
Esterase-2 (ES-2)	Coconut fat	95 ± 11	136 ± 21	
(% of standard)	Corn oil	74 ± 8	110 ± 8	A
	Olive oil	99 ± 12	124 ± 14	
Butyrylcholinesterase	Coconut fat	58 ± 2	68 ± 10	
(nmol/min/ml)	Corn oil	61 ± 9	92 ± 9	A
	Olive oil	59 ± 5	82 ± 11	
Total esterase[3]	Coconut fat	2.63 ± 0.13	3.29 ± 0.18	
(µmol/min/ml)	Corn oil	2.41 ± 0.22	3.09 ± 0.18	A
	Olive oil	2.68 ± 0.10	2.98 ± 0.14	

[1] Results expressed as means $\pm$ SEM for six animals per group. The purified diets were fed for 62 days.

[2] Significance ($P<0.05$) was based on two-way analysis of variance. A, effect of amount of dietary fat; T, effect of type of dietary fat; AxT, interaction.

[3] Total esterase activity was measured with p-nitrophenylacetate as substrate.

PLASMA

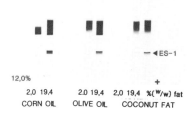

Fig. 1. Gradient PAGE of esterases in serum from male Wistar rats (Cpb:WU) fed low or high-fat diets for 62 days. α-Naphthylacetate was used as substrate for visualizing enzyme activity.

The *ad libitum* consumption of the high-coconut fat versus control diet significantly raised serum total esterase and ES-1 activity, but did not affect butyrylcholinesterase activity (Table 11). Fortification of the restricted control diet with coconut fat, but not with either casein or glucose, produced significantly higher activities of serum ES-1 and tended to produce higher activities of serum total esterase. Butyrylcholinesterase activity in serum was not significantly influenced by the source of energy added to the restricted control diet.

Enrichment of the restricted control diet with fish oil produced an increase in total esterase activity (Table 12). Olive oil tended to raise total esterase activity but no effects were seen with coconut fat or corn oil. Group mean ES-1 activity was raised by all four fat supplements, the increment being greatest with coconut fat and fish oil. Serum butyrylcholinesterase activity was significantly higher with the fish oil supplement than with the other fats.

Table 11. Effects of a high-coconut fat diet with *ad libitum* access and isoenergetic supplements of either glucose, coconut fat or casein on serum esterase activities of male Wistar rats (Cpb:WU) [Based on data taken from reference 17][1]

| | Control diet (*ad libitum*) | High coconut fat diet (*ad libitum*) | Control diet (restricted) (8.0 g/d) | Addition to restricted Control diet | | | |
				Glucose (11.9 g/d)	Coconut fat (9.7 g/d)	Casein (11.9 g/d)	Sign.[2]
Serum total esterase activity (μmol/min/ml)[3]							
initial	3.79±0.27	4.13±0.29	4.03±0.12	3.85±0.31	4.28±0.25	3.90±0.18	
change	-0.71±0.34	+0.49±0.29[a,4]	-0.50±0.15*	-0.46±0.22	+0.22±0.16	-0.90±0.16*	S
Serum butyrylcholinesterase activity (μmol/min/ml)							
initial	128±5	117±6	121±5	128±6	129±8	126±4	
change	-31±4*	-20±4*	-23±4*	-27±3*	-30±5*	-25±4*	
Serum ES-1 activity (% of standard)							
initial	54± 6	42± 4	44± 5	33± 5	31± 5	46± 7	
change	-24± 4*	+260±30[a,*]	-19± 4*	-12± 3*	+105±20[b,*,5]	-2±19	S[5]

[1] Results expressed as means ± SEM for ten animals per group. The purified diets were fed for 14 days.
[2] Analysis of variance significance (P<0.05) for the three groups with additions to the (restricted) control diet. Body weight gain was used as covariate. S, effect of energy source.
[3] Total esterase activity was measured with *p*-nitrophenylbutyrate as substrate.
[4] Contrast significance (analysis of variance with body weight gain as covariate and dietary group as factor): a, high coconut fat (*ad libitum*) vs. control (*ad libitum*), P<0.05; b, coconut fat addition vs. (restricted) control diet, P<0.02.
[5] Significance after logarithmic transformation of the data.
* Change significantly different from zero (P<0.05; two-tailed paired Student's *t*-test).

Table 12. Effects of a high-coconut fat diet given *ad libitum* access and isoenergetic supplements of either olive oil, coconut fat, corn oil or fish oil on serum esterase activities of male Wistar rats (Cpb:WU) [Based on data taken from reference 17][1]

| | Control diet (*ad libitum*) | High coconut fat diet (*ad libitum*) | Control diet (restricted) (8.0 g/d) | Addition to (restricted) control diet | | | | |
				Olive oil (9.7 g/d)	Coconut fat (9.7 g/d)	Corn oil (9.7 g/d)	Fish oil (9.7 g/d)	Sign.[2]
Serum total esterase activity (μmol/min/ml)[3]								
initial	3.35±0.19	4.10±0.27	3.83±0.36	3.55±0.34	4.37±0.43	4.16±0.29	4.23±0.17	
change	+1.34±0.43*	+2.20±0.46*	+0.57±0.47	+1.72±0.39*	+1.15±0.39*	+0.77±0.31*	+2.74±0.46[d,*]	T
Serum butyrylcholinesterase (nmol/min/ml)								
initial	86±4	93±4	93±4	86±4	91±4	91±2	96±6	
change	-35±4*	-38±3*	-42±4*	-34±4*	-35±4*	-33±2*	-8±6*	T
Serum ES-1 (% of standard)								
initial	56± 9	66± 8	58± 5	52± 7	57± 8	53± 6	49± 7	
change	-20± 6*	+340±28[a,*,4,5]	-29± 3*	+49± 7[b,*,5]	+101±11[c,*,5]	+59±16*	+114±16[d,*,5]	T

[1] Results expressed as means ± SEM for ten animals per group. The purified diets were fed for 14 days.
[2] Analysis of variance significance (P<0.05) for the four groups with additions to the (restricted) control diet. Body weight gain was used as covariate. T, effect of dietary fat.
[3] Total esterase activity was measured with *p*-nitrophenylbutyrate as substrate.
[4] Contrast significance (analysis of variance with body weight gain as covariate and dietary group as factor): a, high coconut fat (*ad libitum*) vs. control (*ad libitum*), P<0.05; b, olive oil addition vs. (restricted) control diet, P<0.01; c, coconut fat addition vs. (restricted) control diet, P<0.01; d, fish oil addition vs. (restricted) control diet, P<0.01.
[5] Significance after logarithmic transformation of the data.
* Change significantly different from zero (P<0.05; two-tailed paired Student's *t*-test)

IV. EFFECT OF INDIVIDUAL FATTY ACIDS

For the experiment in Table 12, Pearson correlation coefficients were computed between group mean change in serum total esterase or ES-1 activity and fatty acid intakes (Table 13). There was a significant positive association between serum total esterase activity and the intake of palmitic acid [16:0]. The change in serum ES-1 activity was highly correlated with total saturated fatty acid intake and with the intakes of octanoic [8:0], decanoic [10:0], lauric [12:0], myristic [14:0], palmitic [16:0] and stearic acid [18:0].

Table 13. Pearson correlation coefficients for group mean change in serum total esterase or ES-1 activity and fatty acid intake [Based on data taken from reference 17 and Table 12].

Fatty acid[1,2] intake (mg; d 0 to d 14)	Change in serum total esterase activity (μmol/min/ml)	Change in serum ES-1 activity (% of standard)
	Pearson correlation coefficients[3]	
8:0	0.30	0.89**
10:0	0.29	0.88**
12:0	0.30	0.89**
14:0	0.44	0.93***
15:0	0.71	0.09
16:0	0.79*	0.90**
16:1(n-9)	0.72	0.08
17:0	0.65	0.03
18:0	0.63	0.97****
18:1(n-9)	0.18	0.07
18:2(n-6)	-0.33	0.03
18:3(n-6)	0.71	0.09
18:3(n-3)	0.37	0.02
18:4(n-3)	0.71	0.09
20:0	0.08	0.14
20:1(n-9)	0.64	-0.00
20:2(n-6)	0.71	0.09
20:3(n-6)	0.71	0.09
20:3(n-3)	0.71	0.09
20:4(n-6)	0.71	0.09
20:5(n-3)	0.71	0.09
22:0	0.17	-0.34
22:1(n-9)	0.71	-0.09
22:4(n-6)	0.71	0.09
22:5(n-3)	0.71	0.09
22:6(n-3)	0.71	0.09
24:0	0.75	-0.23
24:1(n-15)	0.73	-0.03
Saturated, total	0.41	0.94***
Monounsaturated, total	0.32	0.08
Polyunsaturated, total	0.12	0.09

[1] Fatty acid intake was calculated on the basis of feed intake, dietary fat content and composition, and the assumption that 100 g fat contains 96 g fatty acids.

[2] Fatty acid notation: numbers before and after the colon represent the number of carbon atoms and of double bonds, respectively.

[3] Significance was calculated by a two-tailed test; n=7.
* $0.01 \leq P \leq 0.05$; ** $0.005 \leq P \leq 0.01$; *** $0.001 \leq P \leq 0.005$; **** $P \leq 0.001$.

V. CONCLUSIONS

There are marked species and strain differences in serum esterase activities as measured with ß-naphthylpropionate as substrate. Rats have the lowest and mice the highest activities. With p-nitrophenylbutyrate as substrate, the inbred rabbits have the lowest and gerbils the highest serum total esterase activities; rats and mice show intermediate activities. When compared with rats, serum butyrylcholinesterase activity is higher in New Zealand White rabbits and much higher in gerbils. The rabbit data demonstrate strain differences in serum butyrylcholinesterase activities. In mice there are strain differences in serum ES-1 and ES-2 levels.

In rabbits, mice and rats serum total esterase activity can be modified by the type and/or amount of lipid in the diet. The feeding of a cholesterol-rich or high-fat diet cause an increased activity of ES-1 in rats and ES-2 in mice. However, in gerbils increased fat intake results in a diminished activity of the fast anodal serum esterase zone. Nutritional studies with rats have revealed that serum butyrylcholinesterase activities can by modified by the amount of cholesterol and fat in the diet and by dietary fat type. Rabbit serum butyrylcholinesterase is also influenced by dietary corn oil amount. In gerbils more work is necessary to identify nutritional determinants of this enzyme, if any. Since serum esterases play an important role in the activation of prodrugs and metabolism of xenobiotics, animal species and strain and also dietary composition should be considered when testing ester or amide-type foreign compounds.

VI. REFERENCES

1. Holmes, R.S., Masters, C.J. and Webb, E.C., A comparative study of vertebrate esterase multiplicity, *Comp. Biochem. Physiol.* **26**, (1968), 837-852.
2. Van Zutphen, L.F.M., Revision of the genetic nomenclature of esterase loci in the rat (Rattus norvegicus), *Transplant. Proc.* **5**, (1983), 1687-1688.
3. Van Zutphen, L.F.M. and Den Bieman, M.G.C.W., Gene mapping and linkage homology, *in*: "New Developments in Biosciences: Their Implications for Laboratory Animal Science", Beynen, A.C. and Solleveld, H.A., eds., Martinus Nijhoff Publishers, Dordrecht, (1988), 197-200.
4. Williams, F.M., Clinical significance of esterases in man, *Clin. Pharmacokinet.* **10**, (1985), 392-403.
5. Reidenberg, M.M., The procaine esterase activity of serum from different mammalian species, *Proc. Soc. Exp. Biol. Med.* **140**, (1972), 1059-1061.
6. Van Zutphen, L.F.M. and Den Bieman, M.G.C.W., Genetic variation in hydrolysis of clofibrate, cyclaine and piperocaine in rabbit serum, *IRCS Med. Sci.* **6**, (1978), 460.
7. Shirai, K., Ohsawa, J., Saito, Y. and Yoshida, S., Effects of phospholipids on hydrolysis of trioleoylglycerol by human serum carboxylesterase, *Biochim. Biophys. Acta* **962**, (1988), 377-383.
8. Tsujita, T. and Okuda, H., Carboxylesterases in rat and human sera and their relationship to serum aryl acylamidases and cholinesterases, *Eur. J. Biochem.* **133**, (1983), 215-220.
9. Lewis, A.A.M. and Hunter, R.L., The effect of fat ingestion on the esterase isozymes of intestine, intestinal lymph and serum, *J. Histochem. Cytochem.* **14**, (1966), 33-39.
10. Wassmer, B., Augenstein, U., Ronai, A., De Looze, S. and Von Deimling, O., Lymph esterases of the house mouse (Mus musculus)-II. The role of esterase-2 in fat resorption, *Comp. Biochem. Physiol.* **91B**, (1988), 179-185.
11. Beynen, A.C., Weinans, G.J.B. and Katan, M.B., Arylesterase activities in the plasma of rats, rabbits and humans on low- and high-cholesterol diets, *Comp. Biochem. Physiol.* **78B**, (1984), 669-673.
12. Beynen, A.C., Katan, M.B. and Van Zutphen, L.F.M. Plasma lipoprotein profiles and arylesterase activities in two inbred strains of rabbits with high or low response of plasma cholesterol to dietary cholesterol, *Comp. Biochem. Physiol.* **79B**, (1984), 401-406.

13. Van Lith, H.A., Meijer, G.W., Haller, M. and Beynen, A.C., Serum pseudocholinesterase activity in rabbits fed simvastatin, *Biochem. Pharm.* **41**, (1991), 460-461.
14. Beynen, A.C., Lemmens, A.G., De Bruijne, J.J., Ronai, A., Wassmer, B., Von Deimling, O, Katan, M.B. and Van Zutphen, L.F.M., Esterases in inbred strains of mice with differential cholesterolemic responses to a high-cholesterol diet, *Atherosclerosis* **63**, (1987), 239-249.
15. Van Lith, H.A., Van Zutphen, L.F.M. and Beynen, A.C., Butyrylcholinesterase activity in plasma of rats and rabbits fed high-fat diets, *Comp. Biochem. Physiol.* **98A**, (1991), 339-342.
16. Van Lith, H.A., Meijer, G.W., Van Der Wouw, M.J.A., Den Bieman, M., Van Tintelen, G., Van Zutphen, L.F.M. and Beynen, A.C., Influence of amount of dietary fat and protein on esterase-1 (ES-1) activities of plasma and small intestine in rats, *Br. J. Nutr.* **67**, (1992), 379-390.
17. Van Lith, H.A., Haller, M., Van Tintelen, G., Van Zutphen, L.F.M. and Beynen, A.C., Plasma esterase-1 (ES-1) activity in rats is influenced by the amount and type of dietary fat, and butyryl cholinesterase activity by the type of dietary fat, *J. Nutr.* **122**, (1992), 2109-2120.

LIVER CARBOXYLESTERASES: TOPOGENESIS OF INTRACELLULAR AND SECRETED FORMS

Mariette Robbi* and Henri Beaufay

Laboratoire de Chimie Physiologique
Université de Louvain and
International Institute of Cellular and Molecular
Pathology, Brussels

INTRODUCTION

In the liver carboxylesterase activity is shared between several isoenzymes which hydrolyse a wide variety of substrates, including therapeutic drugs and xenobiotics. Carboxylesterases act also, in vitro, on endogenous lipid substrates, but the physiological meaning of their action on these compounds is still unclear [1, 2]. Most liver carboxylesterases reside in the lumen of the endoplasmic reticulum (ER) where they are either free, or loosely bound to the inner surface of the membrane [3].

Several major proteins of the ER lumen (GRP 78, GRP 94 and protein disulphide isomerase) have the C-terminal sequence -KDEL, and this tetrapeptide is responsible for their retention in the ER [4]. However, the C-termini originally reported for rat ES-4 and ES-10 were -EVX (X being S, N or Q) and -AVL, respectively [5]. Other proteins are retained in the ER indirectly, by binding to another ER resident protein: a substantial proportion of beta-glucuronidase is bound to egasyn [6, 7] and the immunoglobulin heavy chains are transiently maintained in the ER by BiP [8]. Therefore the precise mechanism by which microsomal esterases are retrieved from the traffic of secretory proteins and reside in the ER was still to be established.

Abbreviations used: ER, endoplasmic reticulum; endo-H, endo-beta-N-acetylglucosaminidase H; endo-F, endoglycosidase F/N-glycosidase F; PAGE: polyacrylamide gel electrophoresis.

*To whom correspondence should be addressed.

Esterases, Lipases and Phospholipases, Edited by M.I. Mackness
and M. Clerc, Plenum Press, New York, 1994

RESULTS AND DISCUSSION

The Esterases in Rat Liver Extracts

Esterases are easily extracted from microsomes by disruption of the membranes with a detergent. The isoenzymes are best separated by isoelectric focusing (Fig. 1A). Four major peaks are seen, whether esterase activity is assayed with alpha-naphthylacetate or o-nitrophenylacetate, but the relative height of the peaks varies according to the substrate [9, 3].

The enzymes can also be resolved by nondenaturing PAGE and histochemical staining with alpha-naphthylacetate (Fig. 1B). Esterases focusing at pH 5.0, 5.5 and 6.4 are monomeric proteins of about 60 kDa which separate on the basis of their charge. pI 6.1 esterase does not migrate between pI 5.5 and 6.4 esterases, because it is a homotrimer made of 60 kDa subunits. The isoenzymes demonstrated in Fig. 1 have been genetically identified by comparison to esterases extracted from the liver of inbred animals. Initially, these enzymes were named by their pI value; the genetic nomenclature [10, 11] indicated in Fig. 1B is now currently used.

Behaviour of ES-10 and ES-10 Variants with Different C-Termini in Transfected COS Cells

When the sequence of ES-10 became known [12], we compared its C-terminal end (-HVEL) with that of three other carboxylesterases: the amino acid sequences of purified rabbit carboxylesterase forms 1 and 2 end in -HIEL [13] and

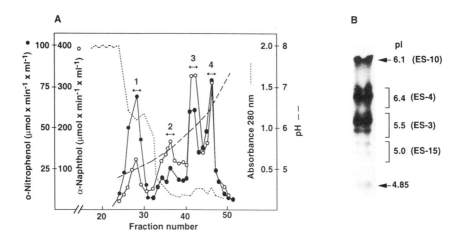

Figure 1. Rat liver esterase isoenzymes. Proteins were extracted from rat (Wistar strain) liver microsomes by 0.2 % Triton X-100. A: the extract was resolved by isoelectric focusing in a pH 4-7 gradient and fractions were assayed for esterase activity with o-nitrophenylacetate, and alpha-naphthylacetate. B: PAGE and histochemical staining with alpha-naphthylacetate and fast blue RR; pI and genetic nomenclature (in parenthesis) are indicated on the right side. From [3].

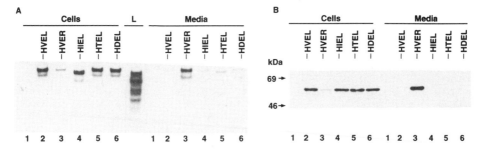

Figure 2. Expression of ES-10-variants in COS cells. The cells were transfected with the void plasmid vector (lanes 1), with the ES-10 cDNA (lanes 2), or with ES-10 cDNAs mutated in the coding region to alter the C-terminus as indicated (lanes 3-6). A: equivalent samples of the cell extracts (0.2 % Triton X-100) and of the culture media were developed by PAGE and stained with alpha-naphthylacetate; lane L was loaded with an extract from rat liver microsomes (arrowhead points to ES-10). B: cultured cells were labeled overnight with ^{35}S-methionine; immunoprecipitates derived from the cell extracts and the culture media were developed by SDS-PAGE and fluorography. Mr standards are indicated in the left margin. From [19].

-HTEL [14], respectively, and the sequence predicted for a rat microsomal carboxylesterase ends in -HTEHT [15, 16]. In spite of some homology, three of the compared C-termini differ more from the -KDEL signal, which functions in mammalian cells, than from the -HDEL signal, which operates in yeast cells. Puzzling enough, a chimeric lysozyme engineered to end in -HDEL was secreted by transfected COS cells, which led to the conclusion that the yeast cell retention signal cannot be recognized by the retrieving machinery of animal cells [17].

To shed some light on the structural characteristics responsible for ER retention of esterases, the cDNA of ES-10 was minimally altered in various ways to produce several ES-10 variants: (1) the terminal Leu was replaced by Arg, for it has been shown to be essential for ER retention of other proteins [18]; (2) forms ending in -HIEL and -HTEL were constructed to mimic the C-termini of the rabbit liver enzymes; (3) a variant ending in -HDEL, the yeast cell retention signal, was constructed to reexamine its effect in animal cells. These ES-10 variants differ from the wild-type molecule by a single amino acid substitution, which seems the most appropriate approach to decide which residue is needed for intracellular retention.

COS cells were transfected with the cDNA encoding ES-10, or the ES-10 variants, and the expression products were searched for in the cell lysates and culture media. Histochemical staining of the proteins resolved in native gels led to the observations shown in Fig. 2A. COS cells transfected with the void vector (lanes 1, cells and medium) do not show any detectable esterase activity. In contrast, cells transfected with the ES-10 cDNA (lanes 2), or any of

its mutated forms (lanes 3-5), express an enzyme which can
be identified with ES-10 by comparison with the rat liver
microsomal enzyme (lane L). An additional band which corresponds to the ES-10 monomer and appears in aged extracts
migrates faster. The enzyme activity of the wild-type enzyme
(lanes 2) and of the -HIEL, -HTEL and HDEL variants is high
in the cell lysate, and barely detected in the conditioned
culture medium; in contrast the -HVER variant of ES-10
(lanes 3) is found predominantly in the culture medium. The
conclusions are: (1) the natural C-terminus of ES-10 contains topogenic information of which the terminal Leu is an
essential part; (2) the terminal tetrapeptides (-HIEL and
-HTEL) of rabbit esterase forms 1 and 2 are most likely
responsible for their retention in liver cells; (3) the
-HDEL signal can be decoded by the sorting machinery of
mammalian cells.

These conclusions were confirmed when the transfected
COS cells were metabolically labeled with ^{35}S-methionine
(Fig. 2B). ES-10-related products were extracted by immuno-
adsorption and resolved by SDS-PAGE. A single labeled
product is seen in transfected cells, migrating at about 60
kDa. In all cases this polypeptide is present in the cell
extracts (lanes 2 and 4-6), except for the variant ending in
-HVER (lanes 3). The discriminative efficiency of the signal
is illustrated in Table 1: 94 % or more of the label was
either in the cells, or in the medium, depending on whether
the polypeptide chain ends in Leu, or Arg.

Table 1. Efficiency of different COOH-terminal tetrapeptides
in retaining ES-10 in COS cells

COOH terminus of the transgenic gene product	Radioactivity[a]		Amount secreted
	Cells	Media	
	dpm x 10^{-3}		%
HVEL	177	7	4
HVER	24	368	94
HIEL	201	2	1
HTEL	173	2	1
HDEL	163	4	2

[a]Strips were excised from the dried gel shown in Fig. 2B and counted.
From [19].

The sequence of ES-10 has been recently confirmed by
others [20] who also succeeded in cloning PCR fragments that
encode four different C-termini occurring in rat carboxylesterase isoenzymes. This revealed that the liver also
expresses an enzyme ending in -HNEL. They then manipulated

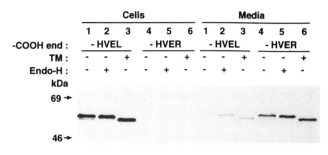

	Cells						Media					
	1	2	3	4	5	6	1	2	3	4	5	6
-COOH end :	- HVEL			- HVER			- HVEL			- HVER		
TM :	-	-	+	-	-	+	-	-	+	-	-	+
Endo-H :	-	+	-	-	+	-	-	+	-	-	+	-

Figure 3. Glycosylation of the cellular and secreted forms of ES-10. After transfection with cDNAs encoding ES-10 variants ending as indicated the COS cells were labeled with ^{35}S-methionine, in the presence of tunicamycin (TM) when indicated (lanes 3 and 6). Immunoprecipitates were obtained directly (lanes 1, 3, 4 and 6), or after digestion with endo-H (lanes 2 and 5), from the cells extracts and the culture media; they were subsequently resolved by SDS-PAGE and fluorography. From [19].

the cDNA of ES-10 to replace a stretch of about 13 amino acid residues at the C-terminus by that present in other isoenzymes. Their results agree completely with our conclusions.

As ES-10 is a glycoprotein of the high mannose type, it was of interest to examine whether the microsomal and the secreted forms have different oligosaccharide chains (Fig. 3). The cellular wild-type enzymes (lanes 1 and 2) and the secreted variant ending in -HVER (lanes 4 and 5) are similarly sensitive to endo-H; thus the oligosaccharide chains of the -HVER variant have not been processed during the transit through the Golgi complex. In addition, the nonglycosylated forms generated in COS cells incubated in the presence of tunicamycin (lanes 3 and 6) are retained in the cells or secreted as the corresponding glycosylated forms (lanes 1 and 4). Thus,glycosylation is in no way a condition for retention, or secretion of these proteins.

A growing number of reports now establish that the signal for retention of esterases in the ER lumen is of the -HXEL type. Ten carboxylesterases from different animal species conform to the rule (Table 2). The three residues H, E and L are conserved, but considerable variation is tolerated at the antepenultimate position: it may be either I, T, V, N or A. Studies on other natural or engineered proteins retained in the ER have similarly shown that some variation around the -KDEL and -HDEL sequences is tolerated, but that the last two residues -EL are the most stringently required [21, 18].

A Rat Carboxylesterase Ending in -HTEHT is a Secreted Isoenzyme

Not all esterase sequences end in -HXEL. A rat isoenzyme which was presumed to be a microsomal protein ends in -HTEHT [15, 16]. Does this mean that an alternative mechanism functions to retain some proteins in the ER lumen? Another example is the mouse esterase Es-N, which ends in -HTEHK [28].

Table 2. Mammalian carboxylesterases ending in -HXEL

X	Origin	Reference
I	Rabbit liver (form 1)	[13]
	Human liver	[22]
		[23]
	Human alveolar macrophages	[24]
	Human monocyte macrophages	[25]
T	Rabbit liver (form 2)	[14]
	Mouse liver (egasyn, ES-22)	[26]
	Rat liver (ES-3)	Robbi (unpublished)
V	Rat liver (ES-10)	[12]
N	Rat liver (ES-4)	Robbi (unpublished)
A	Pig liver	[27]

We have shown that the rat isoenzyme is a secreted form of esterase [29]. To this aim the cDNA was retrieved from our rat liver library, by screening with an antiserum to ES-4, and expressed in COS cells (Fig. 4). The active product, which awaiting further identification was designated Ese-HTEHT, is mostly present in the culture medium, not in the cells (Fig. 4A, lanes 3 and 4). The lack of -EL at the C-terminus of Ese-HTEHT is responsible for its secretion because esterase activity was again retained in the cells when the cDNA was mutated to express the variant Ese-HTEL (Fig. 4A, lanes 5 and 6). Expression and secretion

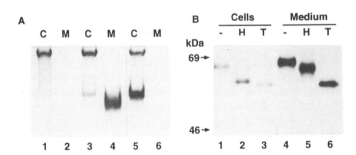

Figure 4. Secretion of the wild-type isoenzyme Ese-HTEHT and retention of the Ese-HTEL variant. A: COS cells were transfected with the void plasmid vector (lanes 1 and 2), with the cloned Ese-HTEHT cDNA (lanes 3 and 4), or with the cDNA engineered to express the mutant Ese-HTEL. Cell extracts (C) and culture media (M) were resolved by PAGE and histochemical staining, as in Fig. 2. B: COS cells transfected with the Ese-HTEHT cDNA were labeled overnight with ^{35}S-methionine in the presence of tunicamycin (T) when indicated (lanes 3 and 6); immunoprecipitates were obtained directly (lanes 1, 3, 4 and 6), or after digestion with endo-H (lanes 2 and 5). From [29].

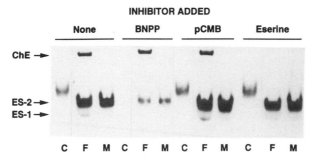

Figure 5. Comparison of Ese-HTEHT with rat serum esterases. The culture medium of COS cells expressing Ese-HTEHT (lanes C) and the serum of adult female (F) and male (M) rats were resolved by PAGE. The gels were soaked for 30 min in 50 mM Tris-HCl buffer, pH 8.0, containing, as indicated, 1 mM bis-p-nitrophenylphosphate (BNPP), p-chloromercuri-benzoate (pCMB), or eserine, and then stained for esterase activity as in Fig. 2. From [29].

of Ese-HTEHT by COS cells were also demonstrated by metabolic labeling with ^{35}S-methionine, immunoprecipitation and analysis of the precipitated material by SDS-PAGE (Fig. 4B, lanes 1 and 4). The half-time of secretion estimated in pulse-labeling experiments is around 1 h (results not shown). These observations have been confirmed by Murakami et al. [30].

The intracellular precursor and the secreted protein are both highly glycosylated, as evidenced by the large shift in Mr when glycosylation is inhibited with tunicamycin (Fig. 4B, lanes 3 and 6). The secreted form is only partially sensitive to endo-H (lane 5), indicating that part but not all of the oligosaccharide chains are processed during transit in the Golgi complex.

Recently, it has been shown that the mouse Es-N isoenzyme is also secreted by transfected COS cells [31].

Identification of Ese-HTEHT as Serum Esterase ES-2

Histochemical staining with alpha-naphthylacetate after non-denaturing PAGE demonstrates two carboxylesterases, ES-1 and ES-2, and a cholinesterase in the rat serum. In Fig. 5, some of their properties are compared to those of Ese-HTEHT, the product secreted by the transfected COS cells. Ese-HTEHT displays the same sensitivity to bis-p-nitrophenylphosphate (BNPP) as the two serum carboxylesterases. In immunoblots of rat serum the rabbit antiserum that precipitates Ese-HTEHT from the transfected COS cells reacts exclusively with ES-2 (results not shown). Ese-HTEHT cannot be identified as ES-1, which is not glycosylated [32], nor to cholinesterase which is inhibited by eserine (Fig. 5). However, Ese-HTEHT does not comigrate with any of the serum esterases (Fig. 5). As Ese-HTEHT remains partially sensitive to endo-H (Fig. 4B), while ES-2 is fully resistant (results not shown), we interpret their different behaviour in PAGE as the consequence of uncomplete processing of the carbohydrate chains of Ese-HTEHT in the Golgi complex of COS cells.

Taken together, the presently available data identify Ese-HTEHT as the serum isoenzyme ES-2. This adds further evidence to our conclusion that the C-terminus of the polypeptide chain commits each esterase isoenzyme to secretion, or to retention in the cell. As for the other proteins vectorially discharged in the ER lumen during their elongation, esterases follow the secretory traffic, unless they are tagged for retrieval and retention in the cell. In the case of carboxylesterases, the sequence -HXEL-COOH is the consensus motif of the tag. Owing the high degree of homology existing among the esterase isoenzymes, an evolutionary process must have led to sequestration of secreted forms in the cells, or, alternatively, to the release of cellular forms in the extracellular space.

REFERENCES

1. MENTLEIN, R. & HEYMANN, E., Hydrolysis of ester- and amide-type drugs by the purified isoenzymes of nonspecific carboxylesterase from rat liver, Biochem. Pharmacol., 33:1984; 1243-1248.
2. MENTLEIN, R., SUTTORP, M. & HEYMANN, E., Specificity of purified monoacylglycerol lipase, palmitoyl-CoA hydrolase, palmitoyl-carnitine hydrolase, and nonspecific carboxylesterase from rat liver microsomes, Arch. Biochem. Biophys., 228:1984a; 230-246.
3. ROBBI, M. & BEAUFAY, H., Purification and characterization of various esterases from rat liver, Eur. J. Biochem., 137:1983; 293-301.
4. MUNRO, S. & PELHAM, H.R.B., A C-terminal signal prevents secretion of luminal ER proteins, Cell, 48:1987; 899-907.
5. MENTLEIN, R., SCHUMANN, M. & HEYMANN, E., Comparative chemical and immunological characterization of five lipolytic enzymes (carboxylesterases) from rat liver microsomes, Arch. Biochem. Biophys., 234:1984b; 612-621.
6. LUSIS, A.J., TOMINO, S. & PAIGEN, K., Isolation, characterization, and radioimmunoassay of murine egasyn, a protein stabilizing glucuronidase membrane binding, J. Biol. Chem., 251:1976; 7753-7760.
7. MEDDA, S., TAKEUCHI, K., DEVORE-CARTER, D. von DEIMLING, O., HEYMANN, E. & SWANK, R.T., An accessory protein identical to mouse egasyn is complexed with rat microsomal beta-glucuronidase and is identical to rat esterase-3, J. Biol. Chem., 262:1987; 7248-7253.
8. BOLE, D.G., HENDERSHOT, L.M. & KEARNEY, J.K., Posttranslational association of immunoglobulin heavy chain binding protein with nascent heavy chains in nonsecreting and secreting hybridomas. J. Cell Biol., 102:1986; 1558-1566.
9. MENTLEIN, R., HEILAND, S. & HEYMANN, E., Simultaneous purification and comparative characterization of six serine hydrolases from rat liver microsomes, Arch. Biochem. Biophys., 200:1980; 547-559.
10. VAN ZUTPHEN, L.F.M., Revision of the genetic nomenclature of esterase loci in the rat, Transplant. Proc., 15:1983; 1687-1688.
11. MENTLEIN, R., RONAI, A., ROBBI, M., HEYMANN, E. & von DEIMLING, O., Genetic identification of rat liver carboxylesterases isolated in different laboratories, Biochim. Biophys. Acta, 913:1987; 27-38.
12. ROBBI, M., BEAUFAY, H. & OCTAVE, J.-N., Nucleotide sequence of cDNA coding for rat liver pI 6.1 esterase (ES-10), a carboxylesterase located in the lumen of the endoplasmic reticulum, Biochem. J., 269:1990; 451-458.

13. KORZA, G. & OZOLS, J., Complete covalent structure of 60-kDa esterase isolated from 2,3,7,8-tetrachlorodibenzo-p-dioxin-induced rabbit liver microsomes, J. Biol. Chem., 263:1988; 3486-3495.
14. OZOLS, J., Isolation, properties, and the complete amino acid sequence of a second form of 60-kDa glycoprotein esterase. Orientation of the 60-kDa proteins in the microsomal membrane, J. Biol. Chem., 264:1989; 12533-12545.
15. TAKAGI, Y., MOROHASHI, K.-i., KAWABATA, S.-i., GO, M. & OMURA, T., Molecular cloning and nucleotide sequence of cDNA of microsomal carboxylesterase E1 of rat liver, J. Biochem., 104:1988; 801-806.
16. LONG, R.M., SATOH, H., MARTIN, B.M., KIMURA, S., GONZALES, F.J. & POHL, L.R., Rat liver carboxylesterase: cDNA cloning, sequencing, and evidence for a multigene family, Biochem. Biophys. Res. Commun., 156:1988; 866-873.
17. PELHAM, H.R.B., HARDWICK, K.G. & LEWIS, M.J., Sorting of soluble ER proteins in yeast, EMBO J., 7:1988; 1757-1762.
18. ANDRES, D.A., DICKERSON, I.M. & DIXON, J.E., Variants of the carboxyl-terminal KDEL sequence direct intracellular retention, J. Biol. Chem., 265:1990; 5952-5955.
19. ROBBI, M. & BEAUFAY, H., The COOH terminus of several liver carboxylesterases targets these enzymes to the lumen of the endoplasmic reticulum, J. Biol. Chem., 266:1991; 20498-20503.
20. MEDDA, S. & PROIA, R.L., The carboxylesterase family exhibits C-terminal sequence diversity reflecting the presence or absence of endoplasmic-reticulum-retention sequences, Eur. J. Biochem., 206:1992; 801-806.
21. PELHAM, H.R.B., The retention signal for soluble proteins of the endoplasmic reticulum, Trends Biochem. Sci., 15:1990; 483-486.
22. LONG, R.M., CALABRESE, M.R., MARTIN, B.M. & POHL, L.R., Cloning and sequencing of a human liver carboxylesterase isoenzyme, Life Sci., 48:1991; PL-43-PL-49.
23. RIDDLES, P.W., RICHARDS, L.J., BOWLES, M.R. & POND, S.M., Cloning and analysis of a cDNA encoding a human liver carboxylesterase, Gene, 108:1991; 289-292.
24. MUNGER, J.S., SHI, G.-P., MARK, E.A., CHIN, D.T., GERARD, C. & CHAPMAN, H.A., A serine esterase released by human alveolar macrophages is closely related to liver microsomal carboxylesterases, J. Biol. Chem., 266:1991; 18832-18838.
25. ZSCHUNKE, F., SALMASSI, A., KREIPE, H., BUCK, F., PARWARESCH, M.R. & RADZUN, H.J., cDNA cloning and characterization of human monocyte/macrophage serine esterase-1, Blood, 78:1991; 506-512.
26. OVNIC, M., SWANK, R.T., FLETCHER, C., ZHEN, L., NOVAK, E.K., BAUMANN, H., HEINTZ, N. & GANSCHOW, R.E., Characterization and functional expression of a cDNA encoding egasyn (esterase-22): the endoplasmic reticulum-targeting protein of beta-glucuronidase, Genomics, 11:1991b; 956-967.
27. MATSUSHIMA, M., INOUE, H., ICHINOSE, M., TSUKADA, S., MIKI, K., KUROKAWA, K., TAKAHASHI, T. & TAKAHASHI, K., The nucleotide and deduced amino acid sequences of porcine liver proline-beta-naphthylamidase, FEBS Letters, 293:1991; 37-41.
28. OVNIC, M., TEPPERMAN, K., MEDDA, S., ELLIOTT, R.W., STEPHENSON, D.A., GRANT, S.G. & GANSCHOW, R.E., Characterization of a murine cDNA encoding a member of the carboxylesterase multigene family, Genomics, 9:1991a; 344-354.
29. ROBBI, M. & BEAUFAY, H., Topogenesis of carboxylesterases : a rat liver isoenzyme ending in -HTEHT-COOH is a secreted protein, Biochem. Biophys. Res. Commun., 183:1992; 836-841.

30. MURAKAMI, K., TAKAGI, Y., MIHARA, K. & OMURA, T., An isozyme of microsomal carboxyesterases, carboxyesterase Sec, is secreted from rat liver into the blood, J. Biochem., 113:1993; 61-66.
31. ZHEN, L., BAUMANN, H., NOVAK, E.K. & SWANK, R.T., The signal for retention of the egasyn-glucuronidase complex within the endoplasmic reticulum, Arch. Biochem. Biophys., 304:1993; 402-414.
32. VAN LITH, H.A., HALLER, M., VAN HOOF, I.J.M., VAN DER WOUW, M.J.A., VAN ZUTPHEN, B.F.M. & BEYNEN, A.C., Characterization of rat plasma esterase ES-1A concerning its molecular and catalytic properties, Arch. Biochem. Biophys., 301:1993; 265-274.

CATALYTIC PROPERTIES AND CLASSIFICATION OF PHOSPHORIC TRIESTER HYDROLASES: EDTA-INSENSITIVE AND EDTA-SENSITIVE PARAOXONASES IN SERA OF NON-DISEASED AND DISEASED POPULATION GROUPS

Elsa Reiner and Vera Simeon

Institute for Medical Research and Occupational Health
University of Zagreb
Ksaverska cesta 2
P.O.B. 291
41001 Zagreb
Croatia

INTRODUCTION

Phosphoric triester hydrolases (EC 3.1.8) are defined as enzymes acting on organophosphorus compounds, including esters of phosphonic and phosphinic acids, and on phosphorus anhydride bonds; the enzymes are subdivided into paraoxonase (EC 3.1.8.1) and DFP-ase (EC 3.1.8.2) [1].

In this paper the EDTA-insensitive and EDTA-sensitive paraoxonases were analysed in serum samples from population groups of different age and health status. A statistical evaluation and comparison of the activity distribution profiles are presented.

SERUM SAMPLES AND ENZYME ANALYSIS

Paraoxonase activities were measured in serum samples from the following population groups: (A) non-diseased individuals (N = 232; 129 male and 103 female) of median age 33 yrs (range: 16 to 59), (B) non-diseased individuals (N = 45; 16 male and 29 female) of median age 67 yrs (range: 60 to 82), (C) patients with

Esterases, Lipases and Phospholipases, Edited by M.I. Mackness
and M. Clerc, Plenum Press, New York, 1994

hyperlipidaemia (N = 159; 105 male and 54 female) of median age 51 yrs (range: 17 to 81) and (D) patients with dementia (N = 74; 14 male and 60 female) of median age 78 yrs (range: 57 to 96). All serum samples were collected in Zagreb, Croatia.

The paraoxonase activities were measured spectrophotometrically in Tris/HCl buffer (0.1 M, pH 7.4, 37 °C) with paraoxon (0,0-diethyl-4-nitrophenol phosphate; 5.0 mM) as substrate in the absence of EDTA (total activity, v_{tot}) and in the presence of 1.0 mM EDTA (EDTA-insensitive activity, v_{ins}) [2]. The difference v_{tot} - v_{ins} amounted to the EDTA-sensitive activity (v_{sen}). The assay volume was 1100 μL and contained 100 μL serum. The increase in absorbance of 4-nitrophenol was measured at 405 nm (ϵ = 1600 $M^{-1}cm^{-1}$).

Differences between median activities were tested by the Wilcoxon-Mann-Whitney rank sum test and differences between activity distribution profiles by the Kolmogorov-Smirnov two-sample test, which is based upon distances between cumulative distribution functions [3]. The critical significance level was chosen to be P = 0.05. The distribution profiles were compared after shifting the origin of the activity scale to the respective median, i.e. the activities were expressed relative to the median. The asymmetry of the distribution curves was quantified in terms of skewness coefficients [3].

RESULTS AND DISCUSSION

The obtained v_{tot} and v_{ins} paraoxonase activities are given in Table 1. Included into the table are also our previously published data: for group A from [2], for v_{tot} activities of group C from [4] and for group D from [5].

Serum donors in groups A and B came from a non-diseased population whose age ranged between 16 and 82 yrs. The v_{tot} activities in sera from donors above 60 yrs (group B) were lower than those below 60 yrs (group A). The age range in patients with hyperlipidaemia (group C) was equally broad, but v_{tot} activities were not different in sera from donors above and below 60 yrs. The v_{tot} activity of group C was not different from that of group B. Patients with dementia (group D) were all elderly, and their v_{tot} activities were not different from those of groups B and C. In all groups, ranges of v_{tot} activities overlapped. An analysis of v_{sen} activities revealed equal characteristics. This was expected, because under the applied experimental conditions v_{sen} amounted to the bulk of the v_{tot} activities.

Published data on the effect of age show no correlation between age and paraoxonase activities, when analysed in terms of correlation coefficients [2,6]. However, Zech and Zürcher [7] analysed average activities in two age groups, and found lower activities in people above 50 yrs than those below 50 yrs. On the other

Table 1. Median activities and activity ranges for v_{tot} and v_{ins} paraoxonases in the listed population groups

Groups	Median (Range) μmol min^{-1} mL^{-1}	
	v_{tot}	v_{ins}
Non-diseased: A	0.126*	0.00848**
	(0.0385-0.750)	(0.00070-0.0235)
B	0.080	0.00258**
	(0.0275-0.660)	(0.00134-0.0093)
Hyperlipidaemia: C	0.065	0.00433**
	(0.0069-0.561)	(0.00089-0.0189)
Dementia: D	0.069	0.00171**
	(0.0118-0.320)	(0.00064-0.0354)

* Significantly higher than in the other three groups.
**Significantly different from one another.

Table 2. Comparison between distribution profiles of v_{tot} and v_{ins} paraoxonases in the listed population groups. Significance levels <0.05 were taken to indicate significant differences

Groups	Significance level	
	v_{tot}	v_{ins}
A *vs* B	0.05	< 0.001
A *vs* C	0.002	< 0.001
A *vs* D	0.001	< 0.001
B *vs* C	> 0.99	0.004
B *vs* D	> 0.99	> 0.99
C *vs* D	0.3	< 0.001

hand Playfer et al. [8] found no significant difference in activities between a group above 70 yrs and one below 48 yrs. The existing discrepancies require clarification.

Under the given experimental conditions, the EDTA-insensitive activities (v_{ins}) represented a small proportion of the total paraoxonases (v_{tot}); v_{ins} activities were different in all groups studied, with group A having the highest median. In all groups, the ranges of v_{ins} activities overlapped.

Many studies have been conducted on the distribution profiles of human serum paraoxonases concerning different ethnic populations and concerning diseased populations [cf.6,9,10] . The v_{tot} and v_{sen} activities have been shown to be polymodally distributed, while the distribution of v_{ins} has been described as unimodal.

The low and high activity modes of v_{tot} and v_{sen} studied in this paper were not well separated, particularly in group C where the polymodal distribution was blurred (Fig. 1). The distribution of the v_{ins} activities appeared unimodal (Fig. 2). In group A the mode was very broad, in group C somewhat narrow, while groups B and D had a very narrow profile.

The asymmetry of the distribution profiles was analysed in terms of skewness coefficients, which represents a measure of asymmetry, with zero indicating a symmetric curve. The calculated coefficients for v_{tot} and v_{sen} activities were found to be equal, and amounted to 1.2, 2.0, 1.8 and 1.1 for groups A, B, C and D, respectively. The corresponding coefficients for v_{ins} activities were 0.4, 2.4, 1.4 and 2.2, respectively. The obtained coefficients indicate a pronounced asymmetry, even for the v_{ins} activity curves. The least asymmetric were v_{ins} activities in group A.

Differences between distribution profiles were analysed by the Kolmogorov-Smirnov test. In order to evaluate only the profiles, and not the activities, we expressed activities relative to their median. Differences between profiles were expressed in terms of significance levels (Table 2); a graphic presentation of the test is shown in Fig. 3. Significance levels for v_{sen} were the same as for v_{tot}. It follows from Table 2 that the four groups differed more in the v_{ins} profiles than in the v_{tot} profiles; the v_{ins} profiles were equal only in groups B and D.

In summary: Paraoxonase activities measured in serum samples from population groups differing in age and health status (non-diseased, hyperlipidaemia, dementia), showed that both, median activities and distribution profiles were different in the groups studied. The statistically evaluated differences were more pronounced for the EDTA-insensitive activity (v_{ins}) than the total or EDTA-sensitive (v_{tot}, v_{sen}) activities.

In the Enzyme Nomenclature [1] it is stated that paraoxonase and DFP-ase require divalent cations for activity and are inhibited by alkylating agents. The existence of EDTA-insensitive enzymes suggests that a broader definition, concerning the effect of alkylating agents, might be required.

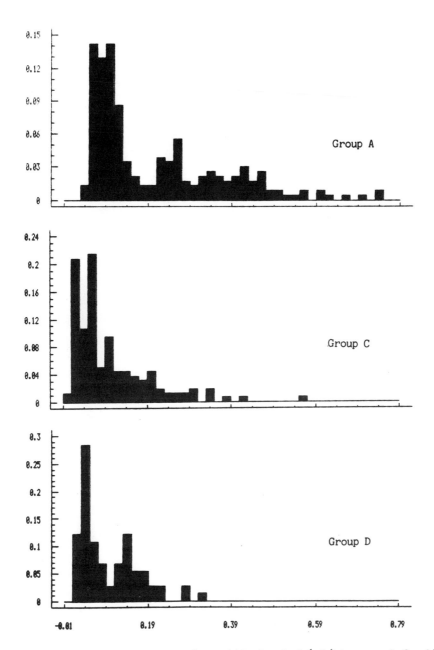

Figure 1. Relative frequency histograms of v_{tot} activities (μmol min^{-1}mL^{-1}) in groups A, C and D.

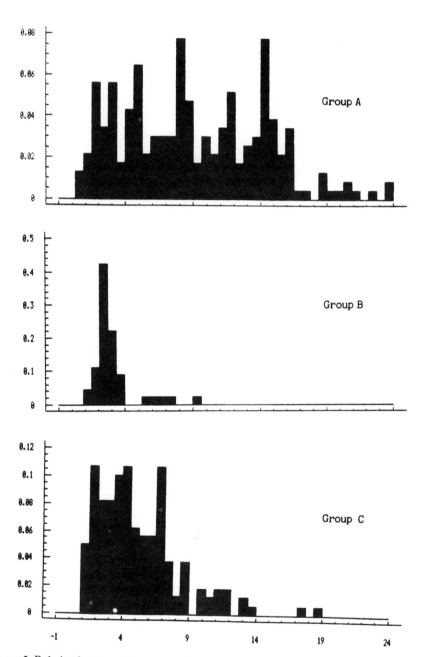

Figure 2. Relative frequency histograms of v_{ins} activities (μmol min^{-1}mL^{-1}) in groups A, B and C.

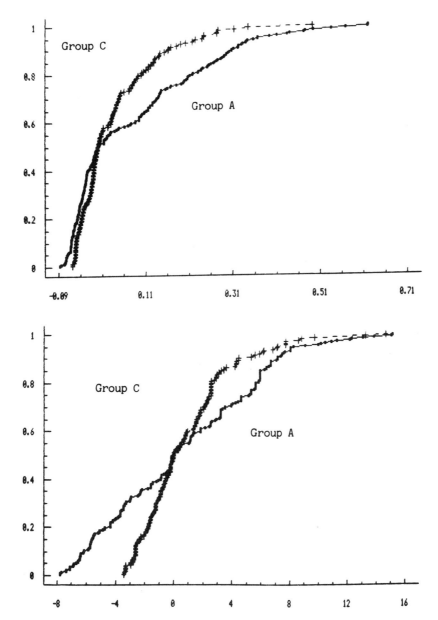

Figure 3. Comparison between distribution profiles. Cumulative frequencies as a function of relative v_{tot} activities (upper figure) and v_{ins} activities (lower figure) in groups A (•) and C (+).

REFERENCES

1. Enzyme Nomenclature, Recommendations (1992) of the Nomenclature Committee of the International Union of Biochemistry and Molecular Biology. Academic Press Inc., San Diego (1992).

2. E. Reiner, Z. Radić, and V. Simeon. Hydrolysis of paraoxon and pheylacetate by human serum esterases, *in:* "Enzymes Hydrolysing Organophosphorus Compounds", E. Reiner, W.N. Aldridge, and F.C.G. Hoskin, eds., Ellis Horwood Ltd., Chichester (1989) 30-40.

3. P. Sprent, "Applied Nonparametric Statistical Methods", Chapman and Hall, London (1989).

4. E. Pavković, V. Simeon, E. Reiner, M. Sučić, and V. Lipovac. Serum paraoxonase and cholinesterase activities in individuals with lipid and glucose metabolism disorders. *Chem.-Biol. Interactions*, 87 (1993) 179-182.

5. E. Reiner, I. Gruić, V. Simeon, Z. Pecotić, V. Dürrigl, I. Peko-Čović, J. Brkljačić, and D. Hodoba. Paraoxonase activities in sera of non-diseased persons and those with dementia (in Croatian). First Croatian Congress of Clinical Biochemistry. Zagreb (1993), Book of Abstracts 05-P17/13.

6. M. Geldmacher-von Mallincrodt, and T.L. Diepgen. The human serum paraoxonase-polymorphism and specificity. *Toxicol. Environ. Chem.* 18 (1988) 79-196.

7. R. Zech, and K. Zürcher. Organophosphate splitting serum enzymes in different mammals. *Comp. Biochem. Physiol.* 48 B (1974) 427-433.

8. J.R. Playfer, C. Powell, and D.A.P. Evans. Plasma paraoxonase activity in old age. *Age and Ageing* 6 (1977) 89-95.

9. M.J. Mackness, D. Harty, D. Bhatnagar, P.H. Winocour, Sh. Arrol, M. Ishola, and P.N. Durrington. Serum paraoxonase activity in familial hypercholesterolaemia and insulin-dependent diabetes mellitus. *Atherosclerosis* 86 (1991) 193-199.

10. B.N. LaDu. Human serum paraoxonase/arylesterase, *in*: "Pharmacogenetics of Drug Metabolism", W. Kalow, ed., Pergamon Press, New York (1991) 51-91.

THE HUMAN SERUM PARAOXONASE POLYMORPHISM AND ATHEROSCLEROSIS

Michael I. Mackness

University Department of Medicine
Manchester Royal Infirmary
Manchester M13 9WL U.K.

SUMMARY

Human serum paraoxonase is associated with HDL and has a polymorphic distribution in Europid populations with 2 allozymes present, one with low activity and one with high activity. Paraoxonase prevents lipid peroxide generation during the Cu^{2+} catalysed oxidation of LDL *in vitro* and may therefore contribute to the *in vivo* protection by HDL against the development of atherosclerosis. The presence of different allozymes of paraoxonase in the population may contribute to this process if they have different efficiencies in preventing LDL oxidation.

INTRODUCTION

Aryldialkylphosphatases (EC.3.1.8.1.) which hydrolyse organosphosphate anticholinesterases are widely distributed in animal tissues with liver and blood generally having the highest activities [1]. Several forms are present in mammalian sera and are responsible for the selective toxicity of organophosphates towards mammals compared to birds which lack this activity in their serum [2]. One form of aryldialkylphosphatase present in mammalian serum which has received widespread attention is the Ca^{2+} -dependent form commonly known as paraoxonase because of its ability to hydrolyse the organophosphate paraoxon (Diethyl-p-nitrophenyl phosphate).

During the 1970's and early 1980's many investigations were carried out on human serum paraoxonase because of its polymorphic distribution and because of the different distribution of the enzyme in different ethnic groups. More recently, evidence has emerged linking human serum paraoxonase to atherosclerosis. This manuscript will review this evidence in relation to the polymorphic distribution of the enzyme in human serum.

Esterases, Lipases and Phospholipases, Edited by M.I. Mackness
and M. Clerc, Plenum Press, New York, 1994

The Human Serum Paraoxonase Polymorphism

Human serum paraoxonase has a bimodal (and possibly trimodal) distribution in white Caucasian populations throughout the world [3]. Figure 1 illustrates this distribution in a population based in the U.K. Approximately 50% of individuals have a low activity phenotype (homozygous low) and 50% have a high activity phenotype (heterozygous and homozygous high). Genetic studies using a double substrate method to determine individual phenotypes [4] indicate the presence of 3 phenotypes AA (homozygous low) AB, (heterozygous) and BB (homozygous high) which are in Hardy-Weinberg equilibrium.

Molecular studies have indicated that the basis of the polymorphism is the presence of two allozymes of paraoxonase, a low activity allozyme (A) and a high activity allozyme (B) which differ in one amino acid in position 191 [5,6]. Glutamine being present in the A allozyme and arginine in the B. However, investigations in our laboratory, in collaboration with Dr R. James, University of Geneva have indicated that a further factor or factors also contribute to the polymorphism by determining the amount of circulating paraoxonase protein [7], AA phenotypes have lower serum paraoxonase protein than AB or BB phenotypes.

The polymorphic distribution of human serum paraoxonase is, however, restricted to Europid populations. The further from Europe a population originates, the less low activity individuals are present and thus some African Negro populations, Chinese and Aboriginals and have no low activity individuals [8] and a unimodal distribution. The reasons for the different distribution of paraoxonase in different ethnic populations are at present unknown.

Paraoxonase Activity in Disease

Human serum paraoxonase is attached to high-density lipoprotein (HDL) presumably via its highly hydrophobic N-terminal end [9], there is a strong positive relationship between the serum concentrations of paraoxonase protein and HDL. Paraoxonase appears to be associated with a specific HDL particle also containing apo A1 and apo J [10]. In the analphalipoproteinaemias 'Fish-eye' and Tangier disease,

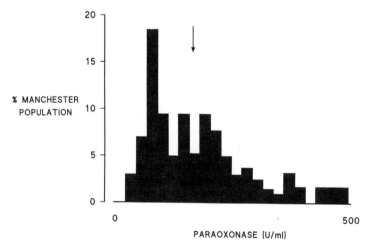

Figure 1. Distribution of paraoxonase in a population from Manchester U.K.

where plasma HDL is greatly reduced or absent, serum paraoxonase activity is also reduced or absent [11,12]. This may be because of the lack of HDL as a carrier in the circulation.

Serum paraoxonase activity is also reduced in diabetes mellitus (with no change in the genotypic distribution) and heterozygous familial hypercholesterolaemia [13] two diseases with a greatly increased prevalence of atherosclerosis and in patients who have suffered a myocardial infarction [14]. These findings indicate the possibility of a relationship between paraoxonase and atherosclerosis.

The Initiation of Atherosclerosis

Atherosclerosis is the largest single cause of premature mortality in countries with a northern European diet. High plasma low-density lipoprotein concentration is a major determinant of susceptability to develop atherosclerosis [15]. One current theory to explain the events which initiate atherosclerosis is the oxidation of low-density lipoprotein (LDL) [16]. Unsaturated fatty acids of the phospholipid outer coat of LDL present in the sub-intimal space are peroxidised by a mechanism which has yet to be elucidated (although arterial endothelial cells, smooth muscle cells and macrophages have all been shown to oxidatively modify LDL *in vitro*). This peroxidation sets in motion a chain of events which results in modifications to the apolipoprotein B moiety of LDL which make it recognisable to receptors on macrophages (known as scavenger or oxidised LDL receptors) also present in the sub-intimal space, which do not recognise unmodified LDL (for detailed reviews see [16-18]). Macrophage uptake of oxidatively modified LDL is unrestricted and the macrophages become foam cells, the predominant cell type of the fatty streak, the progenitor of the atheromatous plaque. LDL from individuals with angiographically proven coronary artery disease and diabetes mellitus has increased susceptability to oxidation compared to controls [19,20] lending further weight to the hypothesis.

High-density Lipoprotein and Atherosclerosis

It has been known for many years that high-density lipoprotein can protect against atherosclerosis, even when LDL levels are high [15]. It was thought that this protection was due to the pivotal role of HDL in reverse cholesterol transport [21], thereby preventing the build-up of excess cholesterol in peripheral tissues, particularly the artery wall. Recent evidence has suggested a second important role for HDL in the prevention of atherosclerosis.

The addition of HDL to LDL undergoing oxidation catalysed by redox metals or cells in culture prevents the generation of lipid peroxidation products [22,23,24], changes to the electrophoretic mobility typical of LDL [23,24], recognition by macrophage scavenger receptors [23] and the cytotoxicity of oxidised LDL [25]. Detailed studies of the effect of HDL on the redox metal catalysed oxidation of LDL carried out in our laboratory have indicated that HDL acts as a specific point in the peroxidation cascade [26]. The initial stage of fatty acid peroxidation is the rearrangement of the double bonds of polyunsaturated fatty acids and the generation of conjugated dienes which are converted via the addition of atomic oxygen to lipid peroxides which, dependent on their structure, can decompose to reactive aldehydes [18]. Chain breaking antioxidants such as the lipid-soluble vitamin-E act by preventing

Table 1. Inhibition of LDL oxidation by antioxidant vitamins and HDL

	Lipid Peroxides Generated (nmol/mg LDL)		Duration of Conjugated Diene Lag-Phase (min)
	LDL	LDL+HDL	
Pre-Vitamin Supplementation	128.7±15.3	67.1±13.1[1]	63.3±9.2
Post-Vitamin Supplementation	136.2±14.3	61.9±8.4[1]	132±39[2]

Values are mean ± standard error of the mean of 16 determinations

[1] Significantly different from LDL value P<0.001

[2] Significantly different from pre-vitamin value P<0.001

the generation of conjugated dienes [18] but appear in our studies to have no effect on the generation of lipid peroxides [26]. HDL, on the other hand, has no effect on diene generation but prevents the generation of lipid peroxides (Table 1 and Figure 1). Lipid peroxide formation is inhibited in a time-independent but HDL concentration dependent mechanism.

HDL does not appear to act by chelating Cu^{2+} (Table 2), nor are lipid peroxides formed on LDL transferred to HDL as has been suggested previously, the concentration of lipid peroxides on HDL being the same in the presence or absence of LDL [27]. The percentage reduction in LDL lipid peroxides by HDL is dependent on the rate of generation of the lipid peroxides within the LDL ($r = 0.765$,P<0.001). Thus, the faster the generation of lipid peroxides, the greater is the protection by HDL, indicating that the process is dependent in the rate of generation of substrate(s). The evidence therefore points to an enzymic mechanism of HDL action in the prevention of LDL oxidation which *in vivo* may provide an additional mechanism whereby HDL acts to prevent atherosclerosis.

Paraoxonase and LDL-Oxidation

Several enzymes are associated with HDL eg lecithin: cholesterol-acyltransferase (LCAT, E.C.2.3.1.43), phospholipase (E.C.3.1.4.50) and paraoxonase. As part of our ongoing investigations into the mechanism(s) whereby HDL is able to prevent LDL oxidation we have studied the ability of paraoxonase to inhibit lipid peroxide generation during the Cu^{2+} catalysed oxidation of LDL. Purified paraoxonase significantly reduced lipid peroxide generation when 20ug of paraoxonase was added to 1.5mg of LDL (Table 3) in comparison to HDL which was added at a concentration of 1.5mg to 1.5mg of LDL.

Table 2. Free Cu^{2+} concentration in the presence of various lipoproteins

	Free Copper uM
PBS	4.57 ± 0.93
LDL	0.73 ± 0.005 (P>0.001)
HDL	3.17 ± 0.61 (P>0.05)
LDL + HDL	1.21 ± 0.27 (P>0.001)

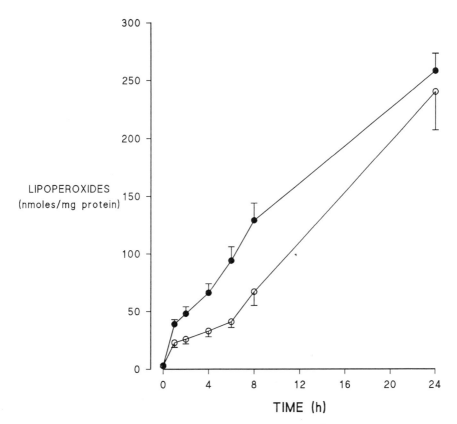

Figure 2. Inhibition of lipid peroxide generation by HDL during the Cu^{2+} catalysed oxidation of LDL.
● LDL alone, O LDL plus HDL

Table 3. Inhibition of lipid peroxide generation in LDL by HDL and paraoxonase

	LDL Lipid-Peroxides (nmol/mg LDL)	Protection Index (% per mg protein)
LDL (n=33)	315±28.6	-
LDL+HDL (n=16)	213±25.7[1]	21.3
LDL+paraoxonase (n=17)	237.7±46.3[1]	1250

[1] Significantly different from LDL P<0.01

The amount of protection against LDL oxidation afforded by HDL was related to the paraoxonase activity of the HDL up to an activity of 40U/ml [27]. In a study of the effect of HDL from 17 different subjects on LDL oxidation there was a positive trend (r=0.465, P<0.06) for the amount of protection against lipid peroxide generation to be dependent on the HDL paraoxonase activity (Figure 3).

Thus it would appear that HDL associated paraoxonase can account at least in part for the protective effect of HDL against LDL-oxidation. Whether any of the other HDL associated enzymes also contribute to this protective effect is currently under investigation in our laboratory.

CONCLUSION

Although the exact substrate specificity is unknown, serum paraoxonase activity may play an important role in preventing atherosclerosis by inhibiting LDL lipid-oxidation and thus the harmful modifications to LDL which lead to its uptake by macrophages. Both HDL and HDL-associated paraoxonase appear to act at a specific point in the lipid peroxidation cascade. They have no effect on conjugated diene formation but act to prevent lipid peroxide formation. How this is achieved is at present unknown, however, as paraoxonase is a hydrolytic enzyme, it is possible that the lipid peroxides are hydrolysed to harmless products. It has yet to be determined whether the low and high activity allozymes contribute differently to this process. It is possible that the low activity allozyme is less efficient at preventing LDL oxidation than the high activity allozyme leading to an increased susceptability to develop atherosclerosis in low activity phenotype individuals, this possibility is currently being investigated in our laboratory.

The generation of free radicals and lipid peroxidation has been implicated in the aeteology of several diseases [28]. HDL is widely distributed in tissue fluid and it is

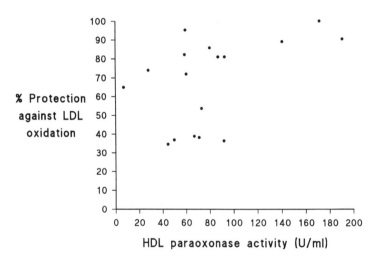

Figure 3. Relationship between HDL paraoxonase activity and protection of LDL against oxidation. Reprinted from reference 27 with permission

therefore possible that HDL has a more widespread metabolic role as an antioxidant in addition to its well documented effect on the development of atherosclerosis

ACKNOWLEDGEMENTS

Work reported in this manuscript is supported by the British Heart Foundation, the Medical Research Council and the North Western Regional Health Authority. The author wishes to thank Dr P.N. Durrington, Dr D. Bhatnagar, Sister J. Morgan, Ms C. Abbott and Ms S. Arrol for their invaluable contributions to the work described in this manuscript and to Ms C. Price for typing the manuscript.

REFERENCES

1. Chemnitius J-M., Losch H., Losch K. and Zech R. Organophosphate detoxicating hydrolases in different vertebrate species. Comp. Biochem Physiol. 76C, 1983; 85-93.

2. Brealey C.J., Walker C.H. and Baldwin B.C. A-esterase activities in relation to the differantial toxicity of pirimiphos-methyl to birds and mammals. Pestic. Sci. 11, 1980; 546-554.

3. Geldmacher-von Mallinckrodt M. and Diepgen T.L. The Human Serum Paraoxonase - Polymorphism and Specificity.Toxicol. Environm. Chem. 18, 1988; 79-196.

4. La Du B.N. and Eckerson H.W. The polymorphic paraoxonase/arylestease isoenzymes of human serum. Federation Proc. 43, 1984; 2338-2341.

5. Humbert R., Adler D.A., Disteche C.M., Hassett C., Omiecinski C.J. and Furlong C.E. The molecular basis of the human serum paraoxonase activity polymorphism. Nature genetics 3, 1993; 73-76.

6. Adkins S., Gan K.N., Mody M. and La Du B.N. Molecular basis for the polymorphic forms of human serum paraoxonase/arylesterase. Am. J. Hum Genet 52, 1993; 598-608.

7. Blatter M-C., Abbott C.A., Messmer S., Pometta D., Durrington P.N., Mackness M.I. and James R.W. In preparation.

8. Diepgen T.L. and Geldmacher-von Mallinckrodt M. Interethnic differences in the detoxification of organophosphates: The human serum paraoxonase polymorphism. Arch. Toxicol Suppl 9, 1986; 154-158.

9. Furlong C.E., Richter R.J., Chapline C. and Crabb J.W. Purification of Rabbit and Human Serum Paraoxonase. Biochem. 30, 1991; 10133-10140.

10. Blatter M-C., James R.W., Messmer S., Barja F. and Pometta D. Identification of a distinct human high-density lipoprotein subspecies defined by a lipoprotein-associated protein, K-85. Eur. J. Biochem 211, 1993; 871-879.

11. Mackness M.I., Walker C.H. and Carlson L.A. Low 'A'-esterase activity in the serum of patients with fish-eye disease. Clin. Chem. 33, 1979; 587-588.

12. Mackness M.I., Peuchant E., Dumon M-F, Walker C.H. and Clerc M. Absence of 'A'-esterase activity in the serum of a patient with Tangier diseaseClin. Biochem. 22, 1989; 475-478.

13. Mackness M.I., Harty D., Bhatnagar D., Winocour P.H., Arrol S., Ishola M. and Durrington P.N. Serum paraoxonase activity in FH and IDDM. Atherosclerosis 86, 1991; 193-199.

14. McElveen J., Mackness M.I., Colley C.M., Peard T., Warner S. and Walker C.H. Distribution of paraoxonase hydrolytic activity in the serum of patients following myocardial infarction. Clin. Chem. 32, 1986; 671-673.

15. Durrington P.N. Hyperlipidaemia: Diagnosis and Management. Wright, London. 1989.

16. Steinberg D., Parthasarathy S., Carew T.E., Khoo J.C. and Witztum J.L. Beyond cholesterol-modifications to low-density lipoprotein that increase its atherogenicity. New Eng. J. Med. 320, 1989; 915-924.

17. Witztum J.L. and Steinberg D. J. Role of oxidised low-density lipoprotein in atherogenesis Clin. Invest. 88, 1991; 1785-1792.

18. Esterbauer H., Gebicki J., Puhl J. and Jurgens G. The role of lipid peroxidation and antioxidants in oxidative modification of LDL. Free Rad. Biol. Med. 13, 1992; 341-390.

19. Regnstrom J., Nilsson J., Toruvall P., Laudou C. and Hamsten A. Susceptibility to low-density lipoprotein oxidation and coronary atherosclerosis in man. Lancet 339, 1992; 1183-1186.

20. Babiy A.V., Gebicki J.M., Sullivan D.R. and Willey K. Increased oxidisability of plasma lipoproteins in diabetic patients can be decreased by probucol therapy and is not due to glycation. Biochem. Pharmacol 43, 1992; 995-1000.

21. Assmann G., Schulbe H., Fune H. von Eckardstein A. and Seedorf U. In Miller N.E. (ed) High-density lipoproteins, reverse cholesterol transport and coronary artery disease. Amsterdam. Exerpta Medica 1989; 46-59.

22. Mackness, M.I., Arrol S. and Durrington P.N. Paraoxonase prevents accumulation of lipoperoxides in low-density lipoprotein.FEBS Letts. 286, 1991; 151-154.

23. Parthasarathy S., Barnett J. and Fong L.G. High-density lipoprotein inhibits the oxidative modification of low-density lipoprotein. Biochem. Biophys. Acta 1044, 1990; 275-283.

24. Ohta T., Takata S., Morino Y. and Matesuda I. Protective effect of lipoproteins containing apoprotein A-I on Cu^{2+} catalysed oidation of human low density lipoproteinFEBS Letts. <u>257</u>, 1989; 435-438.

25. Navab M., Imes S.S., Hama S.Y., Hough G.P., Ross L.A., Bork R.W., Valentine A.J., Berliner J.A., Drinkwater D.C., Laks H. and Fogelman A.M. Monocyte transmigration induced by modification of low-density lipoprotein in cocultures of human aortic wall cells is due to induction of monocyte chemotactic protein 1 synthesis and is abolished by high-density lipoprotein. J. Clin. Invest. <u>88</u>, 1991; 2039-2046.

26. Mackness, M.I., Abbott, C.A, Arrol S. and Durrington P.N. The role of high-density lipoprotein and lipid-soluble antioxidant vitamins in inhibiting low-density lipoprotein oxidation. Biochem. J. 1993, in press.

27. Mackness, M.I., Arrol S., Abbott, C.A., and Durrington P.N. Protection of low-density lipoprotein against oxidative modification by high-density lipoprotein associated paraoxonase. Atherosclerosis 1993, in press.

28. Lunec J. and Blake D. In: Cohen R.D., Lewis B., Alberti K.G.M.M. and Denmar A.M. (eds) The Metabolic and Molecular Basis of Acquired Disease (Vol 2). Bailliere Tindall, London. 1990; 189-212

HYDROLYSIS OF XENOBIOTIC FATTY ACID ESTERS BY CARBOXYLESTERASES OF HUMAN SKIN

Eberhard Heymann, Felicitas Noetzel, Rita Retzlaff, Gabriele Schnetgöke, and Sonja Westie

Physiological Chemistry
University of Osnabrück
D-49069 Osnabrück, Germany

ABSTRACT

Human skin contains carboxylesterases capable of degrading amphiphilic xenobiotic esters. Most of these activities are found in the subcutaneous fat tissue and are of the serine hydrolase type, but activation experiments with Ca^{2+} point to a possible involvement of epidermal phospholipase A2 in the hydrolysis of myristoyl esters and especially of sorbitan trioleate. Fatty acid ethyl and propyl esters, and 4-hydroxy-benzoic acid esters are degraded both in the epidermis and in the subcutis. With simple fatty acid esters an extract from stratum granulosum/stratum corneum exhibits optimal activity with an acyl chain length of 14, whereas the optimum is at shorter acyl residues with the other skin compartments.

INTRODUCTION

Many emulsifiers, natural and synthetic oils which are used for the preparation of cosmetics and dermatics are esters and may be degraded enzymatically after resorption into the skin. This also applies to the esters of 4-hydroxy-benzoic acid, which are frequently used as preservatives for cosmetic preparations.

We describe here the hydrolysis of three types of esters by human skin extracts: A, emulsifiers of creams and ointments, B, simple fatty acid esters that are used as oily

Esterases, Lipases and Phospholipases, Edited by M.I. Mackness
and M. Clerc, Plenum Press, New York, 1994

components in such preparations, and C, the parabenes that are used as preservatives. Using extracts from differing layers of human skin it should be possible to evaluate the localization of the carboxylesterases involved. Since human skin contains a number of differing carboxylesterases that participate in the degradation of simple xenotiotic esters [1], the use of inhibitors might allow the determination of the types of esterases involved in the hydrolysis of the long chain fatty acid esters and the 4-hydroxy-benzoic acid esters investigated here.

MATERIALS AND METHODS

Enzyme Containing Extracts

Samples of human abdominal skin were obtained from cosmetical operations in a local hospital and kept frozen until extraction of the enzymes. Cutis and subcutaneous fat were separated mechanically, minced in a blender, and homogenized in 20 mM Tris/HCl, pH 8.0, with the tissue homogenizer Ultra-Turrax (Janke and Kunkel, Stauffen, Germany) for 2 x 30 s at full speed. 3 ml buffer was used per g of cutis and 6 ml per g of subcutis. The homogenates were centrifuged for 30 min at 20.000 x g. The supernatant liquids (without fat) were used as cutis extract or subcutis extract, repectively. As we showed earlier [1] these extracts contain the bulk of esterase activity with 4-nitrophenyl acetate.

Extracts of three differing layers of the cutis were obtained after separation of dermis and two epidermal layers according to Skerrow [2], using trypsin at $4°C$. The separated cutis fractions were washed with isotonic 10% calf serum to inhibit the trypsin, followed by washing three times with isotonic NaCl/phosphate buffer, pH 7.4. The separated cellular fractions of stratum corneum/stratum granulosum, stratum basale/stratum spinosum, and dermis were homogenized and extracted as described above for the whole cutis. Protein was estimated with a modified Folin phenol method [3].

Assay of ester hydrolysis

All fatty acid ester substrates (10 mM) were dissolved in methanol. Each assay mixture contained 10 μl substrate, 10 μl 0.5 M Tris/HCl, pH 8.0, and 80 μl of one of the extracts described above. After vigorous shaking of the assay mixtures all substrates appeared to be evenly emulsified. The assay mixtures were incubated for several hours at $37°C$. The enzymatic hydrolysis was stopped by adding 400 μl methanol containing 10 nmole of pentadecanoic acid as internal standard. The samples were than freeze-dried using a Speed Vac Concentrator (Savant, Farmingdale, New York). The fatty acids were extracted with 100 μl methanol and 10 μl 0.5% $KHCO_3$ in water. After centrifugation for 5 min. at 10.000 x g the clear supernatants were freeze-dried again. The remaining fatty acid salts were derivatized with phenacyl-bromide according to Durst et al. [4], using a commercial set from Pierce (Rockford, Illinois). The resulting fatty-acid-phenacyl-esters were dissolved in 80% acetonitrile and quantitated by HPLC, (column with Hibar LiChroCART man-fix 50-4 from Merck, Darmstadt, Germany), using a gradient from 80-100% acetonitrile [5]. The amount of released fatty acid was calculated from the ratio of the peak area to that of the internal standard. It was necessary to run a blank without substrate for every skin extract and incubation time. In most cases, the amount of fatty acids found in these blanks was less than 10% of that found in the substrate assay. One milliunit (mU) of enzyme activity represents the nmoles of fatty acid released from the substrate per minute.

The degradation of 4-hydroxy-benzoic acid esters (parabenes) was followed by an estimation of the released 4-hydroxy-benzoic acid by HPLC. The assay mixtures contained varying amounts of cutis or subcutis extracts, 0.5 mM substrate, and 0.25% acetonitrile in 500 μl 50 mM Tris/HCl buffer pH 8.0. After incubation for several hours at 37°C the reaction was stopped with 100 μl 10%-sulfuric acid and 1 ml saturated NaCl solution. The released acid was then extracted with ether (3 x 1 ml). The combined ether extracts were dried with Na_2SO_4 and transferred into 500 μl acetonitrile. This solution was diluted tenfold with the HPLC elution solvent (2 mM cetyltrimethylammonium-bromide in methanol/0.0154 M phosphate buffer = 7/13 v/v [6]). After centrifugation aliquots were applied to the reverse phase HPLC-column (Hibar LiChroCART man-fix 50-4) and eluted with the solvent mentioned above. The peaks were recorded at 246 nm. Thin-layer chromatography on fluorescent silica gel with chloroform/acetic acid (88/12) allowed a simple semi-quantitative detection of the hydrolysis of parabenes.

RESULTS AND DISCUSSION

Hydrolysis of Fatty Acid Esters of Simple Aliphatic Alcohols

All extracts from differing layers of human skin show hydrolytic activity towards simple aliphatic esters of long chain fatty acids. With skin extracts obtained from several individuals the specific activities vary, but are always in the order of 0.1 - 1 mU/mg with ethyl and propyl esters of lauric and myristic acid. With the esters of palmitic, stearic and oleic acid the activities are considerably lower. Whereas the absolute values of the specific activities vary, the relative profiles of activity distribution always are very similar in the corresponding extracts from individual skin samples. Therefore, the activity profiles shown in Fig. 1 represent values obtained with extracts from the abdominal skin of an individual subject. The results with cutis extract (not shown in Fig. 1) are very similar to those with the extract from subcutis (Fig. 1A). Dermis extract (also not shown) exhibits the lowest relative specific activities with palmitoyl and oleoyl esters, activities with stearoyl esters are below the detection limit of the method. We conclude that the enzymes which hydrolyze these fatty acid esters are localized mainly in epidermis and subcutis, but not in the dermis.

With four of the five investigated cellular extracts the activity profiles are generally similar (compare Fig. 1A, B): The activity with the esters of saturated fatty acids decreases with increasing chain length of the acid, and the results with oleoyl esters are similar to those with palmitoyl esters. A comparable dependency on the acyl

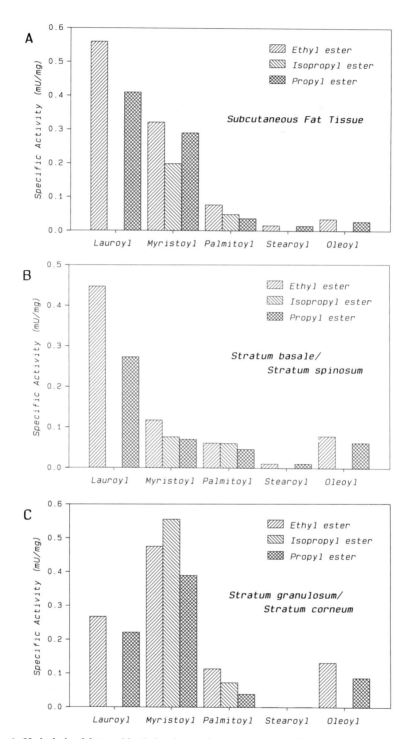

Figure 1. Hydrolysis of fatty acid ethyl and propyl esters by extracts from human skin compartments. A, extract from subcutis; B, stratum basale/stratum spinsosum; C, stratum granulosum/stratum corneum. Of the isopropyl esters only those of myristic and palmitic acid were assayed.

Table 1. Hydrolysis of palmitic acid esters by human cutis extract

	Spec. Activity (mU/mg)
PEG-sorbitan-monopalmitate	0.46
Glyceryl-1-monopalmitate	0.21
Glyceryl-2-monopalmitate	0.20
Ethyl-palmitate	0.08
Isopropyl-palmitate	0.06
Propyl-palmitate	0.05

chain length has been described for the hydrolysis of monoglycerides by liver carboxylesterases [7]. It may be interpreted as a result of the decreasing solubility of such esters with increasing chain length.

With the extract from stratum granulosum/stratum corneum, however, the acitivity profile is different (Fig. 1C): myristoyl esters are degraded faster than the corresponding lauroyl esters. This points to the existence of another carboxylesterase in this organ compartment. Acyl carnitine hydrolase of liver [8] is a relatively nonspecific carboxylesterase which exhibits the highest activity with myristoyl carnitine. We do not know, whether human stratum corneum/stratum granulosum has a prominent acyl carnitine hydrolase activity. At present, the only carboxylesterase known to have an extrordinary high activity in stratum corneum/stratum granulosum is phospholipase A2 [9].

The fatty acid ethyl esters generally seem to be better substrates for the skin carboxylesterases when compared to the propyl esters. This also points to an influence of the solubility of the esters. To verify this assumption, we assayed a series of palmitoyl esters with increasing hydrophilicity of the alcohol moiety (Tab. 1). The results indicate that our assumptions may be correct.

Hydrolysis of Amphiphilic Esters

Almost every cosmetic or dermatologic preparation contains amphiphilic esters which serve as emulsifiers or foam stabilizing agents. Many of these esters reduce the permeability of the stratum corneum [10] and may penetrate into the skin.

Table 2 shows the rates of hydrolysis of frequently used amphiphilic esters by extracts from human cutis and from subcutaneous fat. The subcutis extract has in most cases about a five to tenfold higher specific activity compared to the cutis extract. Since the cutis extract cannot be prepared totally free from subcutaneous fat tissue, this means that almost all of the hydrolytic activity for these esters is localized in the subcutis.

However, it should be noted that the ratios of subcutical and cutical rates are relatively low with sorbitan-trioleate and with the two myristoyl esters. Possibly an additional enzyme in the cutis is involved in the hydrolysis of these esters. This may be the esterase of stratum corneum/stratum granulosùm which prefers myristoyl esters (Fig. 1C).

Table 2. Hydrolysis of amphiphilic esters by extracts from cutis and subcutis at pH 8.0/37°C

	Cutis extract (mU/mg)	Subcutis extract (mU/mg)	Ratio Subcutis/ Cutis
Sorbitan-trioleate	0.017	0.040	2.4
Sorbitan-monostearate	0.021	0.20	9.5
Sorbitan-monooleate	0.051	0.25	4.9
Sorbitan-monopalmitate	0.14	0.64	4.8
Glyceryl-1-monoricinoleate	0.16	1.00	6.2
Sorbitan-monolaurate	0.14	1.15	8.2
Isopropyl-myristate	0.27	1.20	4.4
Glyceryl-2-monopalmitate	0.20	1.49	7.4
Glyceryl-1-monopalmitate	0.21	1.78	8.5
Glyceryl-monomyristate	0.55	2.02	3.7
PEG-sorbitan-monolaurate	0.50	2.32	4.6
PEG-sorbitan-monopalmitate	0.46	2.78	6.0

The pH-dependency of the activity of the subcutical carboxylesterase is shown in Fig. 2. The optimal activity is found in the range of pH 8-9 which is normal for the typical nonspecific carboxylesterases (EC 3.1.1.1) [11], for monoglyceride lipases (EC 3.1.1.23) [7], and for acyl carnitine hydrolases (EC 3.1.1.28) [8]. The relatively flat incline of the pH-curve allows the possibility that a lysosomal enzyme with an optimum around pH 5 might also be involved in the degradation of monoglycerides in human skin.

If cutis extract is preincubated with 0.1 mM paraoxon at 37°C/pH 8.0 for 5 min the esterolytic activities decrease to 8% with PEG-sobitan-monopalmitate, to 13% with glycerol-monomyristate, to 20% with glycerol-monopalmitate, to 47% with sorbitan-monooleate, and to only 70% with sorbitan-trioleate (compare Fig. 3). This indicates that at least the two fatty acid sorbitan esters are partially hydrolyzed by an enzyme which does not belong to the group of serine hydrolases. Monoglyceride lipases [7, 12], acyl carnitine hydrolases [8], and the most prominent nonspecific carboxylesterases of liver [7, 11] are serine hydrolases (also called B-esterases according to Aldridge [13]). B-esterases have been demonstrated in extracts from human cutis [1] as multiple isoelectric bands in the pH-range of 5.5 to 6.0. These esterase bands can be found in keratinocytes and in subcutaneous fat cells (C. Tschoetschel and E. Heymann, unpublished). Whereas almost no other nonspecific esterase bands can be detected in subcutaneous fat, the epidermis contains at least two types of other esterases which are insensitive to organophosphate inhibition. Phospholipase A2 which occurs in human epidermis in relatively high amounts [9] is not a B-esterase but has a Ca^{2+} dependent active site instead. Some activation

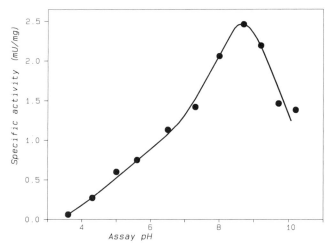

Figure 2. pH-dependency of the hydrolysis of glycerol-monomyristate by subcutis extract. Acetate buffer is used in the range of pH 3.6-5.6; imidazole buffers for pH 6.2 and 7.3; Tris buffers at pH 8.0 and 8.7; and borate buffers above pH 9.

experiments with Ca^{2+} should give hints as to whether this phospholipase might be involved in the degradation of the xenobiotic fatty acid esters investigated here. The inhibition experiments with paraoxon described above indicate that some of the sorbitan esters and the myristoyl esters might be candidates for hydrolysis by carboxylesterases other than B-esterases. As is shown in Fig. 3, the degradation of sorbitan-trioleate can be stimulated about threefold by addition of 5 mM Ca^{2+} to the assay mixture. After inhibition of the B-esterases with paraoxon the remaining 70% of sorbitan-trioleate activity can be stimulated fourfold by 5 mM Ca^{2+} (right-hand-column in Fig. 3).

Sorbitan-trioleate is not a very well-defined substrate, because it contains both furan and pyran rings at its hydrophilic site and the acyl residues may be condensed at varying positions; it also may contain dioleate. Sorbitan dioleate with a furan ring and the acyl residues in position 5 and 6 is sterically similiar to phosphatidyl ethanolamine.

Though the effect is not very strong it should be noted that the remaining isopropyl-myristate cleaving activity after paraoxon can also be stimulated twofold by the addition of 5 mM Ca^{2+} (Fig. 3). No Ca^{2+} activation can be seen with glycerol-monopalmitate as substrate. The addition of 5 mM EDTA shows no significant effect in any case. This is not surprising, because the Ca^{2+}-concentrations in our tissue extracts should be very low.

Hydrolysis of Parabenes

Both cutis extract and subcutis extract hydrolyze the methyl-, ethyl-, n-propyl-, and n-butyl-esters of 4-hydroxy-benzoic acid. The results with cutis extract are shown in Fig. 4. Subcutis extract gives about a 1.5-fold higher specific activity, but the relative distribution of the four activities is essentially the same as shown in Fig. 4. An activity profile which prefers methyl and butyl esters over the corresponding ethyl and propyl esters has never been found with purified carboxylesterases [7, 11]. Thus, we conclude that at least two different skin esterases are involved in the hydrolysis of the

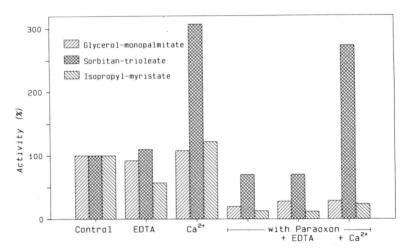

Figure 3. Activation and inhibition of carboxylesterase activity. The three right-hand colums have been measured after preincubation of cutis extract with 0.1 mM paraoxon for 5 min at 37°C, pH 8.0; the three left-hand colums are without paraoxon. EDTA or Ca^{2+}, respectively, is added to give a 5 mM assay concentration.

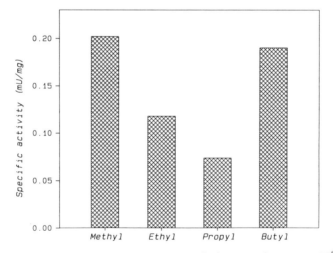

Figure 4. Hydrolysis of 4-hydroxy-benzoic acid esters by human cutis extract at 37°C, pH 8.0.

parabenes. Both are B-esterases, because a preincubation with 0.1 mM paraoxon completely inhibits the hydrolysis of all parabenes.

When assayed at pH 5.0 (acetate buffer), both the methyl and the butyl ester of 4-hydroxybenzoic acid give only traces of activity of less than 5% of those found at pH 8.0 (Fig. 4). We conclude that lysosomal enzymes are not involved in the hydrolysis of the parabenes.

REFERENCES

1. Heymann, E., Hoppe, W., Krüsselmann, A., and Tschoetschel, C., Organophosphate sensitive and insensitive carboxylesterases in human skin, Chem.-Biol. Interactions 87: 1993; 217-226.

2. Skerrow, C.J. and Skerrow, D., A survey of methods for the isolation and fractionation of epidermal tissue and cells, in: "Methods in Skin Research"; Skerrow, D. and Skerrow, C.J., eds., John Wiley, Chichester (1985).

3. Peterson, G.L., Determination of total protein, Meth. Enzymol. 91: 1983; 95-119.

4. Durst, H.D., Milano, M., Kitka, J., Conelly, S.A., and Grushka, E., Phenacyl esters of fatty esters via crown ether catalysts for enhanced UV detection in liquid chromatography, Analyt. Chem. 47: 1975; 1797-1801.

5. Borch, R.F., Separation of long chain fatty acids as phenacyl esters by HPLC, Analyt. Chem. 47: 1975; 2437-2439.

6. Terada H. and Sakabe, Y., Studies on the analysis of food additives by HPLC, J. Chromatogr. 346: 1985; 333-340.

7. Mentlein, R., Suttorp, M., and Heymann, E., Specificity of purified monoacylglycerol lipase, palmitoyl-CoA hydrolase, palmitoyl-carnitine hydrolase and nonspecific carboxylesterase from rat liver microsomes, Arch. Biochem. Biophys. 228: 1984; 230-246.

8. Mentlein, R., Reuter, G., and Heymann, E., Speficicity of two different purified acylcarnitine hydrolases from rat liver, their identity with other carboxylesterases, and their possible function, Arch. Biochem. Biophys. 240: 1985; 801-810.

9. Bergers, M., Verhagen, A.R., Jongerius, M., and Mier, P.D., A new approach to the measurement of phospholipase A2 in tissue homogenates and its application to skin, Biochim. Biophys. Acta 876: 1985; 327-332.

10. Bronaugh R.L. and Maibach, H.I., eds., "Percutaneous Absorption", Marcel Dekker, New York (1985).

11. Heymann, E., Carboxylesterases and amidases, in: "Enzymatic Basis of Detoxication", Vol. II, pp 291-323, Jakoby, W.B., ed., Academic Press, New York (1980).

12. Mentlein, R., Berge, R.K., and Heymann, E., Identity of purified monoacylglycerol lipase, palmitoyl-CoA hydrolase and aspirin-metabolizing carboxylesterase from rat liver microsomal fractions, Biochem. J. 232: 1985; 479-483.

13. Aldridge W.N. and Reiner, E., "Enzyme Inhibitors as Substrates. Interaction of Esterases with Esters of Organophosphorus and Carbamic Acids", North Holland, Amsterdam (1972).

HUMAN MONOCYTE/MACROPHAGE SERINE ESTERASE-1: EXPRESSION IN HAEMATOPOIESIS AND TRANSCRIPT DEFICIENCY IN ALVEOLAR MACRO-PHAGES DERIVED FROM PATIENTS AFFECTED BY PULMONARY FIBROSIS

F. Zschunke,[1]   J. Barth,[2]   H.Kreipe,[3]   H.J. Radzun [1]

[1] Georg-August University Göttingen, Center of Pathology, Dep. I
   Robert-Koch Str. 40, D-37075 Göttingen, Germany
[2] Christian-Albrecht University Kiel, Int. Medicine I, Michaelisstr. 11
   24105 Kiel, Germany
[3] Christian-Albrecht University Kiel, Inst. for Pathology, Michaelisstr. 11
   24105 Kiel, Germany

INTRODUCTION

Human monocytes and macrophages contain an esterolytic activity that hydrolyses alpha naphthyl acetate (ANA). The sensitivity of the monocyte esterase to the inhibitor diisopro-pylfluorophosphate classifies this enzyme as a carboxylesterase with a serine residue at the active site which is involved in hydrolysis. ANA cleaving enzymes are also present in other haematopoietic cells. Isoelectric focusing of proteins from purified haemapoietic cells reveals different patterns of ANA cleaving enzymes which are specific for each cell type. Monocytes contain at least five ANA cleaving enzymes with a predominant enzyme activity at pH 5.9 [1].

We have recently cloned and characterized the pH 5.9 monocyte esterase and we could demonstrate that this enzyme is exclusively expressed in cells of the monocytic lineage [2]. Sequencing has shown that the cDNA codes for an esterase with the consensus active site of serine esterases and we have designated the cloned enzyme **Human Monocyte/ Macrophage Serine Esterase-1 (HMSE1).**

HMSE1 cDNA was used to study the expression of the transcript in purified haemopoietic cells derived from healthy donors and in blood cell lines. Alveolar macrophages display very strong expression of the HMSE1 transcript while monocytes separated from blood contain moderate amounts of the transcript. Granulocytes, T-lymphocytes and promyelocytes do not express the HMSE1 transcript [3]. In a detailed study of a great variety of different cell lines only those of monocytic origin displayed the transcript [4].

To gain insight into the regulation of HMSE1 expression in disease, myelomonocytic cells from donors with leukaemic diseases and alveolar macrophages from patients suffering from idiopathic pulmonary fibrosis respectively were analysed. The investigation of leukaemic cells revealed that myelomonocytic leukaemias lack the transcript while cells from monocytic leukaemias contain moderate amounts of the HMSE1 transcript. In contrast to the very strong expression of HMSE1 in alveolar macrophages derived from healthy donors those derived from patients suffering from idiopathic pulmonary fibrosis fail to express the HMSE1 transcript. Despite the absence of the transcript, an ANA cleaving activity is still demonstrable enzymecytochemically in these cells. The chromosomal locus of the HMSE1 gene was assigned to the long arm of chromosome 16 [5].

Materials and Methods

Alveolar macrophages were obtained during diagnostic bronchoscopy by alveolar lavage with five 20 ml ali-quots of saline. Mononuclear leukocytes from peripheral blood were isolated by density gradient centrifugation after sedimentation of erythrocytes [6]. Monocytes were collected from the interphase by glass adherence [7]. Peritoneal macrophages were derived from lavage fluids during diagnostic laparoscopy. The cellular enzyme activity was demonstrated on cytospins using ANA as a substrate and pararosoaniline as a coupler.

Total RNA was extracted from purified cells or cell lines and sedimented by guanidine isothiocyanate/$CsCl_2$ gradient centrifugation. The RNA was separated on 1.2% formaldehyde agarose gels (10 µg total RNA per lane) and blotted on positively charged nylon membranes by capillary transfer.

Northern blot hybridization was done with cDNA probes (25 ng) which were labeled with ^{32}P-dCTP to an activity of about 5×10^7 cpm by the random primer method. Filters were prehybridized for 2 hours in 30 ml of hybridization solution (50% formamide, 5 x SSC, 5 x Denhardt's solution, 50 mM sodium phosphate buffer pH 6.5, 0.5% SDS, 5% dextransulfate, 0.1 mg/ml sheared salmon sperm DNA). Denatured ^{32}P labeled cDNA probe was added and the filters were incubated for 24 hours at 42 °C. Washing steps were carried out in 2 x SSC two times at room temperature for five minutes each, 2 x SSC containing 1% SDS two times for thirty minutes at 60 °C followed by a 0.1 x SSC wash at room temperature. Wet filters were sealed in plastic foil for autoradiography. The radioactive HMSE1 cDNA was stripped from the filters by a brief immersion in boiling water and the filters were rehybridized with ß-actin cDNA to compare the amounts and the quality of the RNA aliquots.

Results

Enzymecytochemically stained blood monocytes from healthy donors showed a diffuse ANA cleaving activity. A much higher enzyme activity was observed in samples of alveolar macrophages derived from healthy donors as well as in those from patients suffering from pulmonary fibrosis.

Monocytes from healthy donors exhibit moderate amounts of the HMSE1 transcipt. RNA purified from alveolar macrophages of healthy donors display a very strong expression of the HMSE1 transcript. RNA probes derived from alveolar macrophages of patients suffering from pulmonary fibrosis show a complete absence of the HMSE1 message. Leukeamic cells from myelomonocytic leukaemias also do not display the transcript while leukocytes from monocytic leukemias show moderate amounts of the transcript. All results were verified by investigating several RNA probes derived from different donors and by rehybridization of the blots with ß-actin cDNA.

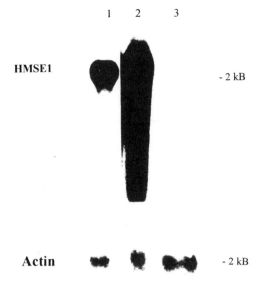

Fig. 1. Example of a Northern blot analysis for HMSE1 transcript in alveolar macrophages from a healthy donor (lane 2) and of a patient suffering from pulmonary fibrosis (lane 3). Transcript from blood monocytes derived from a healthy donor is shown in lane 1. The filters were reprobed with ß-actin cDNA.

Table 1. Expression of HMSE1 in haematopoietic cells derived from healthy donors and in permanent cell lines. All data were confirmed by the investigation of several individual probes as described above.

HMSE1 expression (healthy donors and cell lines)		
no transcript	**moderate expression**	**strong expression**
granulocytes (blood)	monocytes (blood)	alveolar macrophages
lymphocytes (blood)		
peritoneal macrophages		
T-lymphocytes (HUT-78)		
promyelocytes (HL-60)		

Table 2. Expression of HMSE1 in haematopoietic cells derived from donors with diseases. All data were confirmed by the investigation of several individual probes as described above.

HMSE1 expression (from donors with diseases)	
no transcript	**moderate expression**
alveolar macrophages (pulmonary fibrosis)	monocytic leukaemias
myelomonocytic leukaemias	

Discussion

We have analysed the expression of HMSE1 transcript in a variety of purified haematopoietic cells and cell lines. It is shown that the transcript is expressed exclusively in cells of the monocytic lineage. Differentiated macrophages display high amounts of transcript while pro-myelocytes do not express it. HMSE1 expression increases with the maturation of monocytic cells into macrophages with the exception of peritoneal macrophages. The lack of trans-cript in these cells is in contrast to the strong expression in alveolar macrophages a finding which points to the diversity of properties of tissue macrophages. Beside macrophages from the peritoneum alveolar macrophages derived from patients suffering from pulmonary fibrosis fail to express the transcript. Thus, HMSE1 expression studies provide a new and highly specific method to investigate monocyte esterase under normal and pathological conditions.

An ANA cleaving enzyme activity deficiency was described in a subpopulation of blood mo-nocytes derived from patients with lymphoproliferative neoplasia and gastrointestinal carcino-mas [8] and in a patient with chronic myelomonocytic leukemia [9]. We could demonstrate a HMSE1 transcript deficiency in alveolar macrophages of patients suffering from pulmonary fibrosis. Alveolar macrophages of healthy donors show strong expression of this enzyme. The HMSE1 transcript deficiency is in contrast to the ANA cleaving activity which is still demonstrable enzymecytochemically in these cells, a finding also observed in peritoneal macrophages. This discrepancy is explained by the fact that many different esterases and also enzymes other than esterases cleave the synthetic substrate ANA [10]. Consequently, alveolar macrophages and peritoneal macrophages must contain further ANA cleaving enzymes in addition to HMSE1.

The loss of HMSE1 transcript in patients suffering from pulmonary fibrosis may be explained either by an accelerated degradation of the transcript or by a decreased transcriptional activi-ty. The transcription of the HMSE1 gene might be down regulated or abolished by a muta-tional event such as allelic loss. Mutations concerning the long arm of chromosome 16 have been described for patients with acute myelomonocytic leukaemia [11].

It has been determined that HMSE1 is located at the cell membrane [12]. Our results from cDNA sequencing confirm the existence of a potential transmembrane region [2]. The en-zyme contains a potential signal sequence which alows the transport through the cell mem-brane. With a set of synthetic peptides we could show that HMSE1 has no proteolytic activity [13]. The enzyme does not cleave xenobiotics , amide esters, thioesters, or long chain fatty acid esters [12]. So far as is known by now, HMSE1 seems to hydrolyse only organic esters with short carbon chain length.

The biological function of HMSE1 remains unknown. With respect to the very strong expres-sion of HMSE1 in normal alveolar macrophages and the lack of transcript in alveolar macro-phages derived from patients suffering from pulmonary fibrosis it seems very likely that this esterase has an important function in the physiology of the lung. The moderate expression in blood monocytes and the lack of transcript in peritoneal macrophages also points to the fact that the biological function of HMSE1 is restricted to certain subpopulations of the mono-cytic lineage and reflects the diversity of macrophages. As an ectoenzyme, HMSE1 is optimal situated to interact with the extracellular environment including the membranes of the al-veolar endothelium. The involvement of HMSE1 in the pathobiology of pulmonary fibrosis will be studied in further detailed investigations.

References

[1] Radzun, H.J., Parwaresch, M.R., Kulenkampff, Ch., Staudinger, M., Stein, H.: Lysosomal acid esterase: Activity and isoenzymes in separated normal human blood cells. *Blood 55* (1980), 891-897

[2] Zschunke, F., Salmassi, A., Kreipe, H., Parwaresch, M.R., Radzun, H.J.: cDNA cloning and characterization of Human Monocyte/Macrophage Serine Esterase-1. *Blood, 78* (1991), 506-512.

[3] Zschunke, F., Salmassi, A., Kreipe, H., Parwaresch, M.R., Radzun, H.J.: Heterogenous expression and putative structure of human monocyte/macrophage serine esterase 1. *Res. Immunol. 143* (1992), 125-128.

[4] Uphoff, C.C., Gignac, S.M., Metge, K., Zschunke, F., Radzun, H.J.,Drexler, H.G.: Expression of the monocyte-specific Esterase gene in leukemia-lymphoma cell lines. *Leukemia 7* (1993), 58-62.

[5] Becker-Follmann, J., Zschunke, F., Parwaresch, M.R., Radzun, H.J., Scherer, G.: Assignment of the human monocyte/macrophage serine esterase 1 (HMSE1) to human chromosome 16q13-q22.1 and of its homologue to the proximal esterase cluster on mouse chromosome 8. *Cytogenet. Cell Genet. 58* (1991), 1997.

[6] Boyum, A.: Isolation of mononuclear cells and granulocytes from human blood. *Scand. J. Clin. Lab. Invest. 21 (suppl.)* (1968), 77-109

[7] Bennett, W.E., Cohn, Z.A.: The isolation and selected properties of blood monocytes. *J. Exp. Med. 123* (1966), 145-157

[8] Markey, G.M., McCormick, J.A., Morris, T.C.M., Alexander, H.D., Nolan, L., Morgan, L.M., Reynolds, M.E., Edgar, S., Bell, A.L., McCaigue, M.D., Robertson, J.H.: Monocyte esterase deficiency in malignant neoplasia. *J. Clin. Pathol. 45* (1990), 282-286

[9] Takahashi K., Mishima, K., Ichikawa, Y., Watanabe, K., Komatsuda, M., Arimori, S.,: Report of a case with chronic myelomonocytic monocyte esterase deficiency. *Tokai J. Exp. and Clin. Med. 12* (1987), 275-281

[10] Heymann, E., in *Enzymatic Basis of Detoxication*, ed. Jakoby, W.P. Academic Press, New York (1980) 291-323.

[11] Le Beau M.M., Diaz, M.O., Karin, M., Rowley, J.D.: Metallothionein gene cluster is split by chromosome 16 rearrangements in myelomonocytic leukemia. *Nature 313* (1985), 709 - 711

[12] Saboori, A.M., Newcombe, D.S.: Human Monocyte Carboxylesterase. *J. Biol. Chem. 265* (1990), 19792 - 19799

[13] Salmassi, A., Kreipe, H., Radzun, H.J., Lilischkis, R., Charchinajad-Amoey, M., Zschunke, F., Buck, F., Lottspeich, F., Parwaresch, M.R.: Isolation of monocytic serine esterase and evaluation of its proteolytic activity. *J. Leukocyte Biol. 51* (1992), 409-414

INTERACTIONS BETWEEN PESTICIDES AND ESTERASES IN HUMANS

C.H.Walker

School of Animal and Microbial Sciences
Department of Biochemistry & Physiology
University of Reading
Whiteknights, PO Box 228
Reading, Berks
RG6 2AJ

INTRODUCTION

The toxicity of many pesticides is dependent upon their interaction with various types of esterase. Indeed esterases have an important role in determining the selective toxicity of certain insecticides [1-3]. A particular case of this is the development of resistance to insecticides by insects. For example, strains of the aphid *Myzus persicae* which are resistant to organophosphorus insecticides (ops), have very high levels of a carboxylesterase, which can detoxify the active oxon forms of these compounds [4,5].

The toxicity of pesticides can be dependent upon interactions at either the toxicokinetic level or at the toxicodynamic level. In the first case, many pesticides (organophosphorus, pyrethroid, and carbamate insecticides) are esters, and are subject to detoxification by esterase hydrolysis. In the second case, acetylcholinesterases represent the site of action of organophosphorus and carbamate insecticides. (Nerve gases such as soman and tabun, also act at this site). Also, neuropathy target esterase represents the site of action of certain organophosphorus insecticides such as mipafox, leptophos, and methamidophos. In both situations, interactions between esterases and pesticides (or pesticide metabolites) are determinants of toxicity.

This paper will review the interaction of pesticides with esterases, with particular reference to humans. The question of esteratic detoxification will be dealt with first, to be followed by a consideration of acetylcholinesterases and neuropathy target esterases as sites of action for insecticides. Finally there will be a discussion of the use of esterases as the basis for biochemical biomarkers.

Esterases, Lipases and Phospholipases, Edited by M.I. Mackness
and M. Clerc, Plenum Press, New York, 1994

DETOXIFICATION OF PESTICIDES BY ESTERASES

A convenient classification of esterases for the toxicologist is one proposed by Aldridge [6]. This is based upon their interaction with organophosphates such as paraoxon. 'A' esterases metabolites (detoxify) organophosphates; 'B' esterases are inhibited by them. The discussion of the role of esterases in detoxification of pesticides will now follow this classification, taking 'A' esterases first.

'A' Esterases

The classification of this group of enzymes has become more complex with recent advances in knowledge [7]. The following types of enzymes are currently recognised under the IUB classification.

Aryldialkylphosphatases (EC 3.1.8.1) hydrolyse organophosphates.

Diisopropylfluorophosphatases (EC 3.1.8.2) hydrolyse nerve gases such as soman and tabun.

Arylesterases (EC 3.1.12) hydrolyse phenylacetate, and some forms also hydrolyse the organophosphate paraoxon. A form that hydrolyses both of these substrates has been purified from human and rabbit [7,8,9].

With regard to the human health hazards presented by organophosphorus compounds - there has been concern about the effects of both insecticides and nerve gases. Thus, there is an interest in the protective role of the 'A' esterases. Considering insecticides first, mammals have substantial levels of serum 'A' esterase, as measured by assays with the substrates paraoxon, pirimiphos-methyl oxon, diazoxon, and chlorpyriphos oxon [2,10,11,12]. Birds have little or no serum 'A' esterase activity [2,10,11,12]. In the case of serum paraoxonase activity, most mammals showed activity in the range 24-79 nmols/minute/ml serum, the value for humans being 41 nmols/minute/ml serum. Much of this activity was associated with high density lipoprotein [2]. Thus, the capacity of human serum to hydrolyse paraoxon is comparable to that of most other mammals [2]. However, there is a marked polymorphism in human populations in regard to serum paraoxonase activity. Several studies of European and North American Caucasian populations have demonstrated a bimodal distribution of serum paraoxonase activity [13,14,15]. One study has suggested a trimodal distribution. The polymorphic distribution of serum paraoxonase varies considerably between ethnic groups [13,15]. In European populations, some 53% of individuals show low paraoxonase activity [13,15]. In other ethnic groups, far smaller percentages of the population belong to this low activity category.

It seems probable that individuals with low serum paraoxonase activity will be more susceptible to the toxicity of organophosphorus insecticides than individuals with high activity [15]. In the case of agricultural or industrial workers who are exposed to these compounds, account should be taken of this factor in risk assessment.

The nerve gases soman, sarin, tabun and related compounds are detoxified by a group of hydrolyses classified as diisopropylfluorophosphatases [16,17]. These appear to be distinct from the aryldialkylphosphatases mentioned earlier. They tend to hydrolyse P-F and P-CN bonds, and usually have little or no ability to hydrolyse paraoxon. They are sometimes described as acid anhydrolases [16,17]. In the case of soman, hydrolysis by the human plasma enzyme(s) is highly stereoselective [18,19]. The relatively non-toxic P(+)

enantiomers are hydrolysed much more rapidly then the more toxic P(-) enantiomers, thus limiting the protection that these enzymes can give against soman [18,19]. Studies have been conducted to obtain preparations of aryldialkylphosphatases from various sources that can be used as decontaminants of nerve gases [17].

Carboxylesterases (EC 3.1.1.1)

Carboxylesterases are examples of 'B' type esterases, being subject to inhibition by organophosphates. A number of different forms of these enzymes have been identified in rodents [20,21,22]. Four different forms have been isolated from rat liver microsomes and their substrate specificities characterised [20]. The carboxylesterases of mice are tissue-specific, with a range of ten forms identified in the liver and kidney, only some of which are expressed in other tissues (only three of them in serum). In man, the isolation of two forms of human liver carboxylesterase has been reported [23,24].

The rat liver microsomal carboxylesterases that have been purified each show wide substrate specificities for a variety of lipophilic esters, both exogenous (ie. xenobiotic esters) and endogenous. There is some overlap in substrate specificity between the forms; nevertheless each of the major forms shows a characteristic substrate preference over a range of compounds [20].

The role of these esterases in the detoxification of pesticides is not very well understood. In the case of organophosphorus compounds (Ops), they are hardly ideal detoxifying enzymes since the enzyme is inhibited during the course of hydrolysis. The Ops act as suicide substrates, and the reactivation of the enzyme proceeds very slowly or not at all. Despite this, carboxylesterases sometimes have a significant role in detoxifying Ops, mainly because they act as 'sinks' for the active oxon forms. An interesting example is the resistance of clones of the aphid *Myzus persicae* to Ops, due to the presence of very large quantities of a carboxylesterase (E4)[4,5]. This species does not appear to possess any 'A' esterase activity, which could provide the basis for a more efficient resistance mechanism. Similarly, serum carboxylesterases of rodents act as scavengers against soman [25], where it appears that the protection is against the highly toxic (P-) enantiomers, in contrast to the situation with diisopropylfluorphosphatases referred to above.

Apart from their interactions with organophosphates, carboxylesterases can detoxify other types of pesticides such as pyrethroids, and the carboxylester bonds of the organophosphorus insecticide malathion. Of the carboxylesterases purified from rat liver microsomes, one showed significant activity towards malathion. This form, which has a pI of 6.2/6.4, is referred to as ES4 by Mentlein *et al.* [20] and RL1 by Hosokawa *et al.* [21,22] (Table 1). Activity towards pyrethroid substrates has been attributed to either or both of the enzymes referred to as ES4 or ES10 [25,26] (Table 1).

Two human forms of carboxylesterase have been purified from human liver cytosol. One of these forms (pI 5.2-5.8) showed significant activity towards malathion (Table 1); the other showed much lower activity [23]. In a preliminary study upon a carboxylesterase purified from human liver microsomes, greater activity was shown towards malathion than the pyrethroid L-cyhalothrin [27].

In summary, there is evidence for the existence of forms of human liver carboxylesterase which effectively detoxify malathion. The selectivity of malathion between mammals and insects depends upon rapid detoxification by mammalian carboxylesterases before it can be activated to malaoxon. Insects usually lack an effective carboxylesterase for malathion.

Pyrethroids (especially *cis* isomers) are hydrolysed more slowly than malathion by human and rat carboxylesterases. Monooxygenase attack may be more important than hydrophilic attack in the detoxification of pyrethroids by mammals.

ESTERASES AS SITES OF ACTION FOR PESTICIDES

Organophosphorus insecticides and nerve gases owe their toxicity, primarily, to their ability to inhibit acetylcholinesterase. They have their effects upon cholinergic synapses of both the peripheral and central nervous systems. The inhibition of acetylcholinesterases is due to phosphorylation of serine at the catalytic site of the enzyme. Reactivation of the enzyme is at best slow. The rate of reactivation depends on this structure of the phosphoryl moiety. If ageing of the phosphorylated enzyme occurs, there is no reactivation. Certain antidotes such as pyridine-2 aldoxime methiodide (P2-AM) can reactivate the phosphorylated enzyme.

Carbamate insecticides also act as anticholinesterases. However, the carbamated enzyme undergoes more rapid spontaneous reactivation then does the phosphorylated enzyme.

Apart from their action as anticholinesterases, a few organophosporus compounds can cause delayed neuropathy (organophosphate-induced delayed neuropathy, OPIDN) [28,29]. These compounds may or may not be anticholinesterases. Examples of neuropathic agents include mipafox, leptophos, and methamidophos (all are insecticides with anticholinesterases activity), diisopropylfluorphosphatase (DFP), which is also an anticholinesterase, and triorthocresol phosphate (TOCP), which has negligible activity against acetylcholinesterase [29].

Typically, the symptoms of OPIDN appear 2-3 weeks following exposure to an Op. These are sensory-motor disturbances affecting the distal extremeties, and are associated with a degeneration of myelinated nerve fibres. At the molecular level, the initial event is a phosphorylation of neuropathy-target esterase (NTE) which in most cases, is followed by ageing [28,29]. In the case of TOCP, DFP, mipafox, and leptophos, symptoms do not appear until ageing has occurred. Indeed, for these compounds, ageing appears to be necessary before damage to myelin occurs [28]. Methamidophos provides an exception to this rule - an Op whose residues do not age, but which will nevertheless cause OPIDN[30].

It should be emphasised that very few Op insecticides have been shown to cause OPIDN. The most familiar examples are mipafox, leptophos, haloxon, cyanofenphos, and methamidophos. Of the species tested for OPIDN, the domestic fowl was found to particularly sensitive, and is routinely used to test for this toxicity [28].

Current work is directed towards the purification and characterisation of NTE, and elucidation of the sequence of events leading from inhibition of the enzyme to degeneration of myelin [28].

PLASMA 'B' ESTERASES AS BIOCHEMICAL BIOMARKERS

Blood 'B' esterases can provide valuable evidence of exposure to Ops. They have not, as yet, provided good evidence of toxic effect because there is no simple relationship between inhibitors of blood 'B' esterases and inhibition of acetylcholinesterase of the nervous systems (eg. brain) across a range of compounds. The situation is complicated by the fact that some Ops are active in their original form, whereas others require activation (eg. in the liver).

In human studies, red blood cell acetylcholinesterase (AChE) and plasma butyryl cholinesterase (B ChE) have been used for this purpose [31]. Typically, cholinesterase activity is determined by the Ellman method. These procedures have been used to screen agricultural and industrial workers exposed to these compounds. Samples are taken from individuals before exposure to establish the normal range of values. Then the samples are taken from time to time after exposure has commenced. If levels of esterase activity fall below certain baselines, the individuals affected are withdrawn from work.

There has been debate about the percentage fall in esterase activity which indicates significant exposure [31]. It has been suggested that 20% and 30% reduction of red blood cell AChE or plasma BChE respectively, is good evidence of exposure [31]. Others suggest that a larger percentage fall is required to give a reliable indication. As yet, inhibition of plasma carboxylesterase has not been used as a biochemical biomarker, however, although it has been used in studies with birds [32,33].

There is a fundamental problem with this type of assay, based as it is on activity per unit volume. Blood 'B' esterases of individuals are subject to temporal variation. Studies with birds have shown that carboxylesterases can be subject to diurnal variation, and serum BChE to developmental variation [32,33]. Also, avian serum carboxylesterase levels can increase after exposure to Ops. Thus, it is difficult to define reliable control values or ranges, even for individual subjects. In this situation there is a case for purifying esterases, raising antibodies to them and developing ELISA assays which will give rapid and specific determination of quantities of esterase per ml of blood. With this information it should be possible to determine specific activities of blood esterases following immunoprecipitation with appropriate antibodies. Specific activities should be more constant than activities expressed per unit volume of plasma or red cells [33,34].

DISCUSSION

Many pesticides are lipophilic esters, and esterases frequently have a role in their metabolism. 'A' esterases have a critical role in detoxifying both Op insecticides and nerve gases. Recent work has indicated that a number of different esterases perform this function and that there are differences between the enzymes metabolising these two groups of compounds. In the case of serum aryldialkylphosphatases, there is evidence of polymorphism in human populations, which suggests that certain individuals may be at relatively high risk with regard to poisoning by Op insecticides.

Carboxylesterases can detoxify a wide range of insecticides, although relatively little is known about the significance of this in respect of human toxicity. The detoxification of malathion by a carboxylesterase is the main reason for the relatively low toxicity of this compound to mammals generally. Also, carboxylesterases can act as 'sinks' for the active forms of Ops which bind to them. It is of interest that the highly toxic (P-) enantiomers of the nerve gas soman, are deactivated in this way (the enantiomers are metabolised only slowly by diisopropylflurophosphatases of the blood of humans and other mammals). Carboxylesterases also have a role in the detoxification of pyrethroid insecticides. However, the activities of carboxylesterases isolated from rat and human liver show only low activities towards these substrates, suggesting that this is not a very effective mechanism of detoxification.

Certain esterases are targets for organophosphorus insecticides and nerve gases. The primary site of action of most of these compounds is the AChE of the nervous system. However, certain Op compounds act upon NTE, causing delayed neuropathy. The full characterisation of this enzyme, and the elucidation of its mode of action are still awaited.

Finally, blood 'B' esterases serve a valuable function as biochemical biomarkers of exposure to Ops. The development of immunochemical assays (e.g. ELISA assays) could improve the sensitivity and specificity of assays for these enzymes.

Table 1. Activities of rat and human carboxylesterases towards pesticides

Species	Purity	Source	Designation	Substrates	Acitivity mols/mg protein/min	Reference
Rat	P	Liver Microsomes	RL1 pI 6.2/6.4 (ES4)	Malathion	0.20	Hosokawa [21] et al. (1987) Hosokawa [22] et al. (1990)
Rat	P	Liver Microsomes	RH1 pI 6.0 (ES10)	Malathion	0.011	Hosokawa [22] et al. (1990)
Rat	P	Liver Microsomes	ES4 and/or ES10[c]	Malathion t-Permethrin c-Permethrin	[7.6][a] [0.070][a] [0.012][a]	) Suzuki) and) Miyamoto) (1978) [26]
Human	P	Liver Cytosol	pI 5.2-5.8	Malathion	3.96	Ketterman et al. (1989) [23]
Human	F	Liver Cytosol	pI 5.3-5.8	Malathion	0.015[b]	Ketterman
Human	F	Liver Cytosol	pI 4.2-4.8	Malathion	0.005[b]	1991 [24]
Human	P	Liver Microsomes		Malathion	0.42	Walker et al. (1993) [27]
Human	P	Liver Microsomes		L-Cyhalo-thrin	0.052	

P= Purified, F= Partially Purified
[a] Activity calculated from 60 min incubation time. May be an underestimate, since may not be linear.
[b] Mean of 8 individual values quoted in paper.
[c] Mentlein et al.(1987) [20].

REFERENCES

1. Walker, C.H. and Oesch, F.. Enzymes and selective toxicity, *in:* "Biological Basis of Detoxification", J. Caldwell and W.B. Jakoby, eds., Academic Press (1983).
2. Mackness, M.I., Thompson, H.M., Hardy, A.R. and Walker, C.H. Distinction between 'A' esterases and arylesterases. *Biochem. J.* 245: 1987; 293-296.
3. Brooks, G.T.. Pathways of enzymatic degradation of pesticides. *Environmental Quality and Safety No 1,* 1972; 106-163. George Thieme/Academic Press.
4. Devonshire, A.L. The properties of a carboxylesterase from the peach-potato applied *Myzus persicae,* and its role in conferring insecticides resistance. *Biochem. J.* 167: 675-683.
5. Devonshire, A.L. Role of esterases in resistance of insects to insecticides. *Biochem. Soc. Trans.* 19: 1991; 755-759.
6. Aldridge, W.N. Serum esterases I. *Biochem. J.* 53: 1953; 110-117.
7. Walker, C.H. The classification of esterases which hydrolyse organophosphates: recent developments. *Chem Biol. Inter.* 87: 1993; 17-24.
8. Furlong, C.E., Richter, R.J., Chapline, J. and Crabb, W.C. Purification of rabbit and human serum paraoxonase. *Biochemistry* 30: 1991; 10133-10140.
9. Gan, K.N., Smolen, A., Eckerson, H.W. and La Du, B.N. Purification of human serum paraoxonase/arylesterase. Evidence for one esterase catalysing both activities. *Drug. Metab. Dispos.* 19: 1991; 100.
10. Brealey, C.J., Walker, C.H. and Baldwin, B.C. 'A' esterase activities in relation to the differential toxicity of pirimiphos methyl to birds and mammals. *Pestic. Sci.* 11: 1980; 546-554.
11. Walker, C.H., Brealey, C.J., Mackness, M.I. and Johnston, G. Toxicity of pesticides to birds; the enzymic factor. *Biochem. Soc. Trans.* 19: 1991; 741-745.
12. Machin, A.F., Anderson, P.H., Quick, M.P., Waddell, D.R., Skibniewska, K.A. and Howells, L.C. The metabolism of diazinon in the liver and blood of species of varying susceptibility to diazinon poisoning. *Xenobiotica* 1: 1976; 104.
13. Geldmacher von Mallinckrodt, M. and Diepgen, T.L.. Human serum paraoxonase - polymorphism classification, *in:* Enzymes hydrolysing organophosphorus compounds. E. Reiner, W.N. Aldridge and F.C.G. Hoskin, eds., Ellis Horwood, Chichester, 1989; pp. 15-29.
14. Mackness, M.I. Possible medical significance of human serum 'A' esterases, *in:* Enzymes hydrolysing organophosphorus compounds. E. Reiner, W.N. Aldridge and F.C.G. Hoskin, eds., 1989; pp. 202-213.
15. Mackness, M.I. 'A' esterases: enzymes looking for a role? *Biochem. Pharmac.* 38: 1989; 385-390.
16. Hoskin, F.C.G., Chettur, G., Mainer, S., Steinmann, K.E., De Frank, J.J., Galo, B.J., Robbins, F.M. and Walker, J.E. Soman hydrolysis and detoxification by a thermophilic bacterial enzyme, *in:* Enzymes hydrolysing organphosphorus compounds. E. Reiner, W.N. Aldridge and F.C.G. Hoskin, eds., Ellis Horwood, Chichester, 1988; pp 53-64.
17. Broomfield, C., Little, J., Lenz, D. and Ray, R. Organophosphorus acid anhydride hydrolases that lack chiral specificity, *in:* Enzymes hydrolysing organophosphorus compounds. E. Reiner, W.N. Aldridge and F.C.G. Hoskin, eds., Ellis Horwood, Chichester, 1989; pp 79-89.
18. de Jong, P.A., Van Pijk, P. and Benshop, H.C. Stereoselective hydrolysis of soman and other chiral organophosphates by mammalian phosphorylphosphatases, *in:* Enzymes hydrolysing organphosphorus compounds. E. Reiner, W.N. Aldridge and F.C.G. Hoskin, eds., Ellis Horwood, Chichester, 1989; pp 65-78.

19. De Bissdorp, C.J.J.V., Quaeyhaegens, F.J.L. and Van Stockens, M.A.H. Degradation of soman in sera of some large domestic mammals including man. *in:* Enzymes hydrolysing organophosphorus compounds. E. Reiner, W.N. Aldridge and F.C.G. Hoskin, eds., Ellis Horwood, Chichester, 1989; pp 98-107.

20. Mentlein, R., Ronai, A., Robbi, M., Heymann, E. and Deimling, O.V. Genetic identification of rat liver carboxylesterases isolated in different laboratories. *Biochem. Biophys. Acta* 913: 1987; 27-38.

21. Hosokawa, M., Maki, T. and Satoh, T. Multiplicity and regulations of hepatic microsomal carboxylesterases in rats. *Mol. Pharmacol.* 31: 1987; 579-584.

22. Hosokawa, M., Maki, T. and Satoh, T. Characterisation of molecular species of liver microsomal carboxylesterases of several animal species and human. *Arch. Biochem. Biophys.* 277: 1990; 219-227.

23. Ketterman, A.J., Bowles, M.R. and Pond, S.M. Purification and characterisation of human liver carboxylesterases. *Int. J. Biochem.* 21: 1989; 1303-1312.

24. Ketterman, A.J. Polymorphism of human liver carboxylesterase. *Biochem. Soc. Trans.* 19: 1991; 306S.

25. Sterri, S.H. and Fonnum, F. Carboxylesterase - the soman scavenger on rodents: homogeneity and hormone influence, *in:* Clinical and Experimental Toxicology of Organophosphates and Carbamates, B. Ballantyne, T.C. Morris, eds., Butterworth Heinemann, 1989; 155-164.

26. Suzuki, T. and Miyamoto, J. Purification and properties of pyrethroid carboxylesterases in rat liver microsomes. *Pestic. Biochem. Physiol.* 8: 1978; 186-198.

27. Walker, C.H., Smith, G. and Wolf, C.R. Unpublished results.

28. Johnson, M.K.. The delayed neuropathy caused by some organophosphorus esters: mechanisms and challenge. *Crit. rev. Toxicol.* 3: 1975; 289-316.

29. Johnson, M.K. Molecular events in delayed neuropathy: experimental aspects of neuropathy target esterase, *in:* Clinical and Experimental Toxicology of Organophosphates and Carbamates, B. Ballantyne, T.C. Morris, eds., Butterworth Heinemann, 1992; 90-114.

30. Johnson, M.K. Anomalous biochemical responses in tests of delayed neuropathic potential of methamidophos, its resolved isomers and of some higher O-alkyl homologues. *Arch. Toxicol.* In press (1992).

31. Duncan, R.C. and Griffith, J. Screening of agricultural workers for exposure to anticholinesterases, *in:* Clinical and Experimental Toxicology of Organophosphates and Carbamates, B. Ballantyne, T.C. Morris, eds., Butterworth Heinemann, 29: 1992; 421-429.

32. Thompson, H.M., Walker, C.H. and Hardy, A.R. Avian esterases as indicators of exposure to pesticides - the factor of diurnal variation. *Bull. Environ. Contam. Toxicol.* 41: 1988; 4-11.

33. Thompson, H.M. and Walker, C.H. Blood esterases as indicators of exposure to organophosphorus and carbamate insecticides, *in:* Proceedings of International Meeting 'Non Destructive Biomarkers in Vertebrates', M.C. Fossi, C. Leonzio, Lewis Publications, 1993; 35-60.

34. Walker, C.H. Biochemical responses as indicators of toxic effects of chemicals in ecosystems. *Toxicol. Letts.* 1992; 527-533.

THE ROLE OF ESTERASES IN THE PATHOGENESIS OF DISEASE AND THEIR USE AS DIAGNOSTIC TOOLS

Coordination:
Wolfgang Junge

Zentrallaboratorium
Friedrich-Ebert-Krankenhaus
Friesenstraße 11
D-24534 Neumünster

Discussants:
E. Heymann (Germany), W. Junge (Germany), K. Kutty (Canada)
H. van Lith (Netherland), M. Mackness (UK), G.M. Markey (UK)
E. Reiner (Croatia), M. Robbi (Belgium), E. Schmidt (Germany)
CH. Walker (UK), E. Zschunke (Germany)

Esterase classification

The present classification of carboxylesterases by the latest Enzyme Commision Committee on Enzyme Nomenclature is neither unambigous nor satisfying. As pointed out by E. Reiner this becomes particularly evident within the two groups of enzymes reacting with organophosporous compounds, i.e. phosphotriester hydrolases (EC 3.1.8) and the "true" carboxylesterases like EC 3.1.1.1 or cholinesterase (3.1.1.8). Organophosphates are substrates of the former group and -formally- also of the latter one, the difference being only the incapability of the acyl residue to leave the active site in the case of carboxylesterases and therefore act as suicide substrates. The substrate specificity of the phosphoric triester hydrolases is still under discussion, particularly with regard to their reaction with carboxylesters. In this context it should be recalled that the phenyl acetate hydrolysing activity of human serum has not yet been separated from the paraoxon hydrolysing activity in a preparative manner.

Microsomal liver carboxylesterase (EC 3.1.1.1)

W. Junge reported on the use of liver microsomal carboxylesterase (MCE) as a serum marker for the diagnosis of necrotic liver diseases. Because MCE is firmly bound to the membranes of the endoplasmic reticulum of the liver parenchymal cell, the enzyme appears in serum only after desintegration of the membranes as a consequence of cell death. High serum levels of MCE where observed in various clinical conditions, predominantly in cardio - pulmonary diseases causing an acute liver congestion. Serum MCE activities developed in parallel to transaminases (ALT, AST) and to another marker enzyme of liver cell death, i.e. mitochondrial glutamate dehydrogenase. From a clinical study involving more than 200 patients it was concluded that determination of MCE provides no additional information concerning the diagnostic or prognostic aspects of the diseases.

Esterases, Lipases and Phospholipases, Edited by M.I. Mackness
and M. Clerc, Plenum Press, New York, 1994

Serum cholinesterase (EC 3.1.1.8)

E. Schmidt reported briefly on results obtained by her coworker Nichaus on the change of cholinesterase phenotypes in liver transplant recipients. In view of the increased risk of transplantation patients having to undergo repeat surgery because of complications it was recommended to determine the cholinesterase phenotype in every liver recipient. In this context the question arose as to why most laboratories use different methodologies for measuring cholinesterase activity and determining its phenotype.

Acetylcholinesterase (EC 3.1.1.7)

An intersting contribution on the potential role of acetylcholinesterase in the development of Alzheimer's disease (AD) was given by M. Robbi. Brains of patients with AD contain deposits of amyloid substance mainly formed by a peptide called ßA4. This peptide derives from several structurally related precursors (APPs), containing a large ectodomain, a transmembrane domain and a small cytosolic domain: half of ßA4 is embedded in the membrane and the remaining 28 amino acid residues are in the extracellular domain. The ectodomains contain a collagen binding site and some forms contain in addition a region that has Kunitz protease inhibitor (KPI) activity. APPs are proteolytically processed by two different pathways: one mechanism generates ßA4, the other one releases the ectodomains by cleavage in the middle of ßA4. Despite the decrease of acetylcholinesterase (AChE) in AD, the enzyme accumulates within amyloid where it shows higher proteolytic activity than in normal axons. AChE is linked to amyloid by a collagen-like tail and might be attracted by the collagen binding site of APPs; AChE could then liberate the ectodomains of APPs and reduce the production of ßA4. M. Robbi stressed,that this hypothesis is supported by transfection studies in Hela cells and by some forms of Mendelian inheritance of the disease. Affinity purified AChE liberates the ectodomains of APPs expressed in cells transfected with the cDNA of APPs; the ectodomains containing KPI activity inhibit their own release by inhibiting the protease activity of AChE, and might therefore increase the production of ßA4. Some mutations located close to the normal processing site of APPs cause massive deposition of ßA4. However, although the potential role of AChe in AD should be further studied as it has been clearly established that several genes are involved in the development of the disease.

Human Monocytic esterase (no EC number)

Human monocytic esterase has been in the focus of hematologists for many years. The enzyme activity as detected histochemically and the analyses of the isoenzyme pattern after isoelectric focussing has been used to differentiate blood cells, e.g. for the diagnosis of leukaemias. Zschunke cloned the cDNA of a human monocyte serine esterase (HMSE) and could show that only monocytes and macrophages contain this enzyme (see also the authors contribution on page ...). In this context, the observation made by G.M. Markey and coworkers who determined the overall-esterase activity in peripheral monocytes of blood donors is of interest. Among this population 0.8 % had so called monocytic esterase deficiency (MED). This condition is defined by the absence of esterase activity in more than 85 % of peripheral monocytes. The incidence of this anomaly increased significantly in patients with Hodgkin's (9%) and Non-Hodgkins lymphoma (5%), chronic lymphatic leukaemia (13%), gastro-intestinal cancer(10%) and autoimmune disease (4%). Further experimental evidence points also to a linkage between MED and neoplastic diseases: lactoferrin does not stimulate esterase negative monocytes to lyse cells of a erythroleukaemic cell line , whereas it causes a 5 -11 fold increase in esterase positive cells.

Serum esterase isoenzymes

A completely different aspect of esterases was discussed by H. van Lith, who studied electrophoretic serum esterase patterns in inbred strains of rats and rabbits. He found that differences in the responsiveness of serum cholesterol levels to dietary cholesterol was often associated with a typical expression of serum esterase isoenzymes. For example, if the cholesterolaemic response of rats was low, an esterase zone with high anodal mobility developed in 6 of 7 strains. This isoenzyme is called ES-1 in rats. In animals responding with a high degree of hypercholesterolaemia, this band did not develop in 2 out of 3 strains. Similar results were observed in inbred strains of rabbits. There is presently no idea on the underlying metabolic process which controls the variability of serum esterase expression neither is the tissue origin of esterase ES-1 known. As suggested by some discussants the purification and biochemical characterisation of this esterase should be attempted.

The round table discussion illustrated the wealth of facets in esterase research. Future investigations on this abundant enzyme group are certainly very promising with respect to the understanding of their physiological role.

CROSS-REACTION OF A HUMAN ANTI-PARAOXONASE MONOCLONAL ANTIBODY WITH A RAT HIGH DENSITY LIPOPROTEIN PEPTIDE

Marie-Claude Blatter Garin , Sylvia Messmer , Daniel Pometta
and   Richard W. James

Department of Medicine, University Hospital, Geneva, Switzerland

INTRODUCTION

During recent studies, monoclonal antibodies raised against human high density lipoproteins (HDL) identified a protein with a molecular mass of 45 kD (range 42-48 kD) and with a pI of 4.5-4.9. Aminoterminal sequence analysis showed the protein to correspond to human paraoxonase [1] (aryldialkylphosphatase E.C. 3.1.8.1), an enzyme which can hydrolyse and detoxify organophosphate compounds such as the activated form of the insecticide parathion [2]. The enzyme has been recently cloned from human and rabbit cDNA libraries [3]. Paraoxonase seems to be a highly conserved protein: 85 % of amino acids are identical and up to 94 % are similar between the 2 species. Despite the fact that it exhibits relatively high activity in the sera of all mammals [4], the natural substrate(s) as well as the physiological function(s) of paraoxonase remain obscure . The potential importance of paraoxonase in protecting against organophosphate toxicity is recognised, but it is not known whether this relates to the true function of the enzyme. The association of paraoxonase with HDL [5-7] has led to suggestions that the enzyme could have a function related to lipid metabolism. In recent studies, we demonstrated that human paraoxonase is a marker of a distinct subpopulation of HDL composed essentially of 3 proteins: paraoxonase, apolipoprotein (apo) AI, and to a lesser extent, clusterin, also called apo J [1].

In the present study, we consolidate these results by examining the serum distribution of paraoxonase in another species, the rat. Rat paraoxonase was identified, and its lipoprotein association characterised using a monoclonal antibody raised against human paraoxonase.

RESULTS AND DISCUSSION

The protein composition of ultracentrifuged rat HDL was analysed by Coomassie staining of SDS-PAGE gels (Figure 1.1) and isoelectrofocused (IEF) gels (Figure 1.3). The

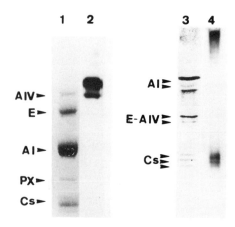

Figure 1. Ultracentrifuged rat HDL (d=1.063-1.23 g/ml; 50 ug) were subjected to SDS-PAGE (lanes 1 and 2) or isoelectrofocusing (lanes 3 and 4) and either Coomassie stained to reveal the protein pattern (lanes 1 and 3) or electrotransferred to nitrocellulose sheet for Western blotting with the mab F41F2-K (lanes 2 and 4) Bound IgG were revealed with 125I labelled rabbit anti-mouse IgG and the nitrocellulose sheet exposed for autoradiography. AIV: apo AIV, E: apo E; AI: apo AI, Cs: apos Cs.

main rat HDL apolipoproteins are apo AI, apo E, apo AIV, apos Cs and PX, a HDL-associated rat serum protein identified in this laboratory [8]. Rat HDL were also analysed by Western blot using the monoclonal antibody anti-human paraoxonase F41F2-K. The mab recognized specifically a double band of 45-50 kD (Figure 1.2), with a pI of 4.6-5.2 (Figure 1.4). The IEF pattern of the protein recognised by our mab is typical of a glycoprotein and is very similar to the pattern we observed for the human paraoxonase. The mab did not recognize any of the established rat apos, and did not cross-react with other serum proteins. Thus the 45-50 kD doublet corresponds to rat paraoxonase.

Fractionation of rat HDL by immunoaffinity chromatography on an anti-paraoxonase affinity column is shown in figure 2. Unfractionated HDL (left part of the gel), the non bound fraction (middle part of the gel) and the bound fraction (right part of the gel) were analysed by Coomassie staining of SDS-PAGE gels. The column was able to retain all the paraoxonase protein (determined by Western blot) as well as enzyme activity determined using paraoxon and phenylacetate as substrates. The bound fraction was highly enriched in paraoxonase, but other rat apos were also present, notably apos AI, AIV and E

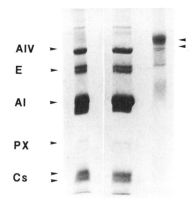

Figure 2. Unfractionated rat HDL (50ug, left part of the gel), non bound fraction (50 ug; middle part of the gel) and bound fraction (50 ug; right part of the gel) of the immunoaffinity column anti-paraoxonase were analysed by Coomassie stained SDS-PAGE. Arrows indicate the position of rat paraoxonase.

(confirmed by Western blot). These results suggest that, as for its human counterpart, rat paraoxonase is present in a specific rat HDL subpopulation and thus could be considered as a marker of a specific HDL subclass.

Preliminary studies showed a very strong correlation between enzyme activity and the 45-50 kD band recognized by the mab F41F2-K, further confirming its identity with the enzyme paraoxonase.

In conclusion, we have identified a rat plasma protein associated with HDL which cross-reacts with a monoclonal antibody against human paraoxonase. It appears to be a marker for a minor subpopulation of rat HDL. The rat could thus furnish an interesting and useful model for the study of the physiological function of paraxonase, given the existence of different rat strains with defined metabolic disorders.

REFERENCES

1. Blatter M.-C., James R.W., Messmer S., Barja F., and Pometta D., Identification of a distinct human high-density lipoprotein subspecies defined by a lipoprotein-associated protein, K45. Identity of K-45 with paraoxonase, *Eur. J. Biochem.*, 211: 1993; 871-879.

2. La Du B.N., Human serum paraoxonase/arylesterase, *Pharmacogenetics of Drug Metabolism*, 1992, pp. 51-91, Pergamon Press, Inc., New York.

3. Hasset C., Richter R.J., Humbert R., Chapline C., Crabb J.W., Omiecinski C.J., and Furlong C.E., Characterization of cDNA clones encoding rabbit and human serum paraoxonase: the mature protein retains its signal sequence, *Biochemistry,* 30: 1991; 10141-10149.

4. Aldridge W.N., and Reiner E., *Enzyme inhibitors as substrates,* Chap. 12 A-esterases, 1975, pp. 176-189, North-Holland/American Elsevier, New York.

5. Don M.M., Masters C.J., and Winzor D.J., Further evidence for the concept of bovine plasma arylesterase as a lipoprotein, *Biochem. J.,* 151: 1975; 625-630.

6. Mackness M.I., Hallam S.D., Peard T., Warner S., and Walker C.H., The separation of sheep and human serum A-esterase activity into the lipoprotein fraction by ultracentrifugation, *Comp. Biochem. Physiol.,* 82B: 1985; 675-677.

7. Mackness M.I., and Walker C.H., Multiple forms of sheep serum A-esterase activity associated with the high-density lipoprotein, *Biochem. J.,* 250: 1988; 539-545.

8. Blatter M.-C., James R.W., Borghini I., Martin B.N., Hochstrasser A.-C., and Pometta D., A novel high-density lipoprotein particle and associated protein in rat plasma, *Biochim.Biophys.Acta,* 1042:1990; 19-27.

CARBOXYL ESTER LIPASE IN ACINAR CELL CARCINOMA OF PANCREAS

M. Narahari Reddy

Pathology Department
Medical College of Wisconsin and
Children's Hospital of Wisconsin
Milwaukee, WI 53226
USA

INTRODUCTION

The association of subcutaneous fat-necrosis with pancreatic carcinoma or pancreatitis has been known for almost 50 years [1,2]. Although the importance of lipase production as well as its dissemination and lipolysis of the involved tissues has been well established in this syndrome [2,3], the exact identification of lipase(s) that are involved and their relationship to the type of pancreatic tumor is not clear.

The acini of normal pancreas produce at least two different lipases: triglyceride lipase (TGL), also called pancreatic lipase, and carboxyl ester lipase (CEL) [4]. The characteristic properties of CEL are: hydrolysis of short and long chain triglycerides at both 1, (3) and 2 positions; stimulation by bile salts; and high sensitivity to serine-enzyme inhibitors such as diisopropylphosphorofluoridate (DFP). CEL has been purified from pancreatic tissue and juice of rat, porcine, bovine and human [4]. It is a glycoprotein with a molecular weight of 64,000 to 100,000, depending on the source. Its physiological role is not clearly known, but it seems to play a role in the hydrolysis of dietary cholesteryl esters and esters of Vitamins A, E, D_3 [4]. However, the role of the CEL in the pathogenesis of pancreatic disorders or its use as a marker in the diagnosis of these, is not known.

I present here the activity of CEL in pancreatic acinar cell carcinoma, its properties in relation to the normal pancreas and its role in lipid metabolism using transplantable exocrine acinar cell carcinoma of rat pancreas. The transplantable exocrine acinar cell carcinoma of pancreas in rat was originally established by Reddy and Rao [5]. Histologically, both the primary and transplanted tumors are moderately well-differentiated adenocarcinomas of acinar cell origin. This tumor is maintained in the laboratory in F344 rats. The morphological, immunofluorescent and biochemical studies with some specific enzymes of this transplantable carcinoma have clearly established that it is a functionally active acinar type.

Esterases, Lipases and Phospholipases, Edited by M.I. Mackness
and M. Clerc, Plenum Press, New York, 1994

MATERIALS AND METHODS

Acinar cell carcinoma of the pancreas of rat (RRC-I) was obtained from Janardan K. Reddy, M.D., Pathology Department, Northwestern University Medical School, Chicago. The tumor was transplanted subcutaneously into the inguinal areas of weanling F-344 male rats using 1 to 2 mm-sized pieces. After the transplant grew to 2 to 3 cm in diameter, the tumor tissue was collected and kept frozen at -70C. Blood from normal rats and carcinoma-bearing animals was collected before removal of the tumor tissue and stored at -70C. Other pancreatic tumor tissues that have been established using experimental animals (acinar cell carcinoma-II of Rao and Reddy (RRC-II) [5,6] and of Longnecker-II (LN-II) [6]) were also obtained from Dr. J.K. Reddy. Tissues and serum samples of pancreatic ductal carcinomas of hamsters [7] were obtained from Dr. M.K. Reddy, Pathology Department, Northwestern University Medical School.

Pancreatic tissue (normal and tumor) was homogenized in a phosphate buffer (50 mM), pH 6.2, containing NaCl (0.1 M), EDTA (1 mM), hexamethylphosphoramide (2 mM), octyl-B,D-glucopyranoside (1 mM), benzamidine.HCl (1 mM), leupeptin (0.5 µg/ml), aprotenin (0.5 µg/ml) and chymostatin (0.5 µg/ml). Extraction buffer containing these compounds is found to be necessary to minimize aggregation and proteolytic degradation of CEL. To the crude homogenate, Triton X-100 (0.2%, final) was added and stirred for 20 min at 4°C to solubilize the enzyme. The homogenate was centrifuged for 20 min at 4°C and 21,000 x g. The resulting supernatant was used for purification, at pH 6.2, by affinity (hydroxylapatite followed by heparin-Sepharose) and then by gel filtration (Ultrogel-AcA34) chromatography. Inclusion of hydrophobic interaction chromatography using octyl-Sepharose before the gel filtration step provided higher purification (> 98%) and improved the recovery of CEL.

CEL activity was determined as esterolytic activity using p-nitrophenyl butyrate and also as lipolytic activity using triolein, in the presence (3 mM, final) and absence of a bile salt, taurocholate. The esterase activity was determined, at pH 7.2, as the rate of liberation of p-nitrophenol which was monitored spectrophotometrically at 20C and 400 nm [8]. Results were expressed as µ moles of P-nitrophenol liberated min^{-1}, ml^{-1} using an extinction coefficient for P-nitrophenol of 16,300 M^{-1}, cm^{-1}. Lipase activity was determined at pH 8.0 using ^{3}H-trioleolylglycerol emulsified with lecithin as substrate [9], and results were expressed as µ moles of free fatty acids liberated ml^{-1}, hr^{-1}. Specific activities were expressed as units per mg of protein.

RESULTS AND DISCUSSION

CEL activity in the circulation of normal animals (n=10) was almost absent (esterase = 2 ± 2 and lipase = 5 ± 4). However, in serum samples of the tumor bearing animals (n=10), at 5 weeks after transplantation (tumor weight = 9120 ± 1780 mg), the enzyme activity was present at high levels (esterase = 701 ± 63 and lipase = 2290 ± 745). The CEL activity was increased with tumor growth. As the tumor size was increased, from 1 day to 26 days after transplantation, both the tumor weight in the animals as well as the enzyme activity in the circulation were increased (Weight: 18 ± 3 mg at one day to 6957 ± 1197 mg at 26th day; CEL: from 2 ± 1 to 580 ± 38 as esterase, and 9 ± 10 to 1041 ± 106 as lipase). Further, surgical removal of the tumor from the animals resulted in the reversal of the enzyme activity to almost basal levels (esterase = 23 ± 7; lipase = 23 ± 8 at 14 days after surgical removal of the growing tumor), substantiating the dependency of CEL with tumor burden . Similar to the CEL, triglyceride lipase levels in the circulation also showed a similar trend to the tumor burden. However, the highest activity of the triglyceride lipase was only about 2.5X that of normal, suggesting that the CEL production by this tumor was several fold higher than the triglyceride lipase.

In addition to the RRC-I, production of CEL by other established pancreatic tumors (Ex: RRC-II and LN-II of acinar type from rat [6] and ductal tumor of hamster [7]) was evaluated, in relation to the normal pancreas of rat and hamster. Tissues were homogenized as per the procedure given under methods and the resulting crude supernatant after centrifugation was used to measure the CEL activity as esterase. Relatively high CEL activity was present in the normal tissues of both these animal species (276 ± 32 from rat pancreas and 191 ± 28 from hamster pancreas). However, the activity in rat acinar tumors was several times higher (RRC-II = 3636 ± 248 and LN-II = 1718 ± 310). Differences in the activities of these tumors probably reflect levels of differentiation since the LN-II tumor is less well differentiated compared to the RRC tumors [7]. In contrast to acinar tumors, ductal tumor had very low CEL activity (14 ± 6), even though its levels in the normal hamster pancreas was almost similar to the normal rat pancreas. Absence of the enzyme activity in ductal tumors suggests that CEL is exclusively associated with the acinar cell and is the reason why the subcutaneous fat necrosis and disseminated lipolysis is a specific characteristic feature of the acinar cell carcinomas [2,3].

CEL was purified from the normal pancreas and acinar tumors to homogeneity. In these steps, use of protease inhibitors and stabilizing agents as well as the inclusion of hydrophobic interaction chromatography resulted in > 98% purity, as judged by SDS - PAGE, with a 2500X increase in specific activity with 30% yield. The molecular weight of CEL from normal pancreas and tumor tissue was identical (68,000 ± 2,000 K Da). Further, enzyme from both sources required a bile salt (sodium taurocholate) for maximal activation (1 mM for 50% maximal activity). Substrate affinity (K_M) studies showed that CEL from tumor tissue had about 2.5X increased affinity compared to the one from normal pancreas (3.15 ± 1.62 mM from tumors and 1.22 ± 0.39 mM from normal pancreas). To understand the enzymic mechanism of CEL, especially to know about the participation of serine at its active site, purified CEL was incubated with diisopropylphosphorofluoridate (DFP) at 10^{-4}M and 20-22C. When the remaining activity was measured, a 50% loss of activity was observed in 4 to 5 minutes for CEL from the normal pancreas as well as from the tumor tissues. Inactivation of CEL by DFP, due to the formation of an active-site complex (DFP-CEL), was confirmed by autoradiography experiments using ^{3}H-DFP. In this, purified CEL was incubated with ^{3}H-DFP. The unincorporated radioactivity was removed by precipitation with 10% TCA, followed by washing 3X with TCA. Analysis of the resulting protein by SDS-PAGE and autoradiography revealed incorporation of DFP into a single protein band of about 68,000 K Da with radioactivity. Further, quantitation of incorporated radioactivity gave about 0.8-0.9 moles of radioactive label/mole of CEL. Therefore, CEL is a serine-enzyme and it is different from the Triglyceride Lipase.

The presence of CEL in different organs of the rat was studied using the Western blot technique. Aliquots of homogenates from different organs of normal rat (pancreas, heart, kidney, liver, spleen, testes), acinar tumors (RRC and LN-II carcinomas), pancreatic ductal tumor of hamster, and purified CEL (from RRC tumor and normal pancreas) were subjected to SDS-PAGE followed by transfer to an Immobilon membrane (Millipore). CEL's presence was probed using anti CEL-porcine antibody. Positive bands were present in the samples from the normal pancreas of rat and hamster as well as in samples from the acinar cell tumors. However, these positive bands were absent in the homogenates from other organs as well as from the pancreatic ductal carcinoma of hamster, indicating the exclusive presence of CEL in the acini of pancreas.

To understand the effect of acinar cell tumor on lipid metabolism, serum samples from normal and tumor bearing animals (at 28 days after transplantation; tumor weight = 7,100 ± 1,100 mg) were analyzed for the levels of different lipoproteins. Samples were collected from the animals after fasting for 16 hours. In the serum of tumor bearing rats, compared to the controls, activities of triglyceride lipase and CEL were increased 2.5 and 57X, respectively. Associated with this higher activity of lipolytic enzymes, levels of triglycerides were decreased 20% and total cholesterol 51%. However, no significant change was seen in the high density lipoproteins (HDL). These results were further confirmed by lipoprotein electrophoresis and density-gradient centrifugation studies using samples that were prestained with sudan black. The plasma from tumor bearing animals, compared to the normals, had substantially lower levels of very low density lipoproteins as well as low density lipoproteins.

The requirement of apolipoproteins for the activity of CEL in the hydrolysis of triglycerides from lipoproteins was evaluated using rat lymph chylomicrons. Chylomicrons radiolabelled in the triglyceride moiety were obtained after feeding the animals with ^{3}H-triolein. Apolipoproteins from chylomicrons were removed by treating with pronase and then purified by column chromatography using Sephedex G-50. Chylomicrons with and without apolipoproteins were incubated with CEL in the presence of taurocholate and the rate of liberation of free fatty acids (as radioactivity) was monitored. The results showed that the liberation of fatty acids was similar to the chylomicrons with and without apolipoproteins and therefore, the presence of apolipoproteins is not required for the hydrolysis of triglycerides from chylomicrons by CEL.

CEL from normal pancreas as well as from tumors can hydrolyze phospholipids. Phospholipase Activity of CEL was determined by incubating the purified enzyme with rat lymph chylomicrons for 60 minutes at 37C, and the products were analyzed by thin layer chromatography. The results indicated the presence of lysolecithin in the reaction products. Further, the lysolecithin formation was proportional to the incubation period as well as to the concentration of the enzyme. CEL from normal pancreas and from tumors gave similar results. This observation is physiologically important since circulating lipoproteins have a layer of phospholipids at their outer surface, and CEL due to its phospholipase activity can reach the inner core of triglycerides by hydrolyzing the phospholipids without the assistance of other lipolytic enzymes.

Thus, these results have clearly indicated that the transplantable acinar cell carcinoma of rat pancreas produces high levels of CEL, and consequently animals with tumor have high enzyme activity in their circulation. Because of the presence of high activity in the circulation and with its capability to hydrolyze a variety of lipids, CEL may have an important role in pathogenesis of metastatic fat necrosis, a common feature of the acinar cell carcinomas.

REFERENCES

1. Blauvelt, H. A case of acute pancreatitis with subcutaneous fat necrosis, Br. J. Surg. 34: 1946; 207-213.
2. Auger, C. Acinous cell carcinoma of the pancreas with extensive fat necrosis, Arch. Pathol. 43: 1947; 400-405.
3. Hruban, R.H., Molina, J.M., Reddy, M.N. and Boitnott, J.K. A neoplasm with pancreatic and hepatocellular differentiation presenting with subcutaneous fat necrosis, Am. J. Clin. Pathol. 88: 1987; 639-645.
4. Rudd, E.A. and Brockman, H.L. Pancreatic carboxyl esterlipase (cholesterol esterase), In: "Lipases", pp. 185-204, B. Borgstrom and H.L. Brockman, eds, Elsevier, New York (1984).
5. Reddy, J.K. and Rao, M.S. Transplantable pancreatic carcinoma of the rat, Science, 198: 1977, 78-80.
6. Longnecker, D.S. Experimental model of exocrine pancreatic tumors of pancreas, In "The Exocrine Pancreas: Biology, Pathobiology, and Diseases", pp. 443, V.L.W. Go, ed., Raven Press, New York, 1986.
7. Scarpelli, D.G. and Rao, M.S. Transplantable ductal adenocarcinoma of the syrian hamster pancreas, Cancer Res. 39: 1979; 452-458.

8. Erlanson, C. Purification, properties, and substrate specificity of a carboxylesterase in pancreatic juice, Scand. J. Gastroenterol. 10: 1975; 401-408.

9. Reddy, M.N., Maraganore, J.M., Meredith, S.C., Heinrikson, R.L. and Kezdy, F.J., Isolation of an active-site peptide of lipoprotein lipase from bovine milk and determination of its amino acid sequence, J. Biol. Chem. 261: 1986; 9678-9683.

BIOCHEMISTRY OF LIPASES AND RELATED LIPOLYTIC ENZYMES

HEPATIC TRIGLYCERIDE LIPASE AND LIPOPROTEIN LIPASE ACTION IN VITRO AND IN VIVO

Lawrence Chan, Joachim Wölle, Jing-Yi Lo, Eva Zsigmond, and Louis C. Smith

Departments of Cell Biology and Medicine
Baylor College of Medicine
One Baylor Plaza
Houston, TX 77030

INTRODUCTION

Hepatic triglyceride lipase (HTGL) and lipoprotein lipase (LPL) are two evolutionarily related enzymes that play key roles in lipoprotein metabolism. Hepatic triglyceride lipase appears to be involved in the hydrolysis of intermediate density lipoprotein (IDL) triglyceride to produce low density lipoprotein (LDL), and that of high density lipoprotein (HDL)-2 triglyceride and phospholipid to produce HDL-3 [1,2]. It may also be required for the uptake of HDL triglyceride and cholesteryl esters by the liver [3-5]. Lipoprotein lipase is essential for the metabolism of the triglyceride-rich lipoproteins, chylomicron and very low density lipoproteins (VLDL). The LPL-mediated hydrolysis of these lipoproteins produces chylomicron remnants and IDL, respectively, releasing necessary components for the production of HDL-2. Thus, both enzymes are involved in HDL metabolism, and HL activity is inversely, whereas LDL activity is directly correlated with plasma HDL levels [6].

Hepatic triglyceride lipase, LPL and the digestive enzyme, pancreatic lipase, show substantial sequence homology and similarity in gene structure. They are members of a lipase superfamily of genes.

ROLE OF N-LINKED GLYCOSYLATION IN HTGL ACTION

Both HTGL and LPL are glycoproteins. Two putative N-linked glycosylation sites are conserved in these two lipases in mammalian species (Figure 1). Carbohydrate seems to be required for the production of functional HTGL and LPL [7-9]. Tunicamycin treatment markedly reduces the production of active HTGL from rat hepatocytes [10] and LPL from rat preadipocytes [11], ob/ob adipocytes [12] or 3T3-L1 adipocytes [13]. Using in vitro expression of site-specific human LPL mutants in COS cells, Semenkovich et al. [14] found that the N-terminally situated glycosylation site (Asn-43) is essential for both enzyme activity and secretion. In contrast, using site-specific mutants involving the two N-linked glycosylation sites in rat HTGL expressed in Xenopus laevis oocytes, Stahnke et al. [15] found that a mutant lacking the homologous N-linked glycosylation site (Asn-57) was 2.7 times more active than wild-type although it was poorly secreted into the medium. Their experiments suggest a fundamental difference in the structural role of carbohydrates between LPL and HTGL, a somewhat surprising finding in view of the high degree of structural and functional similarities between the two enzymes. We have re-examined this issue [16] and will summarize our findings below.

We have expressed human HTGL in vitro using the expression vector pEE14 [17,18]. This is a stable expression system in Chinese hamster ovary cells. The vector is under the control of human cytomegalovirus 5' sequences and VS40 3' sequences and contains the glutamine synthetase selectable marker gene. We used this system because pilot experiments using transient expression in COS cells produced too low a level of HTGL for biochemical analysis. There are four putative N-linked glycosylation sites in human HTGL at Asn-20, 56, 340 and 375. Of these four sites, two are homologous to the corresponding sites in human LPL, Asn-43 and 359 (Figure 1). We expressed the wild-type HTGL as well as site-specific mutants of each of the four putative N-linked glycosylation sites. All four Asn residues were mutated individually to Ala residues. In addition, we also produced an Asn-56→Gln mutant. We found that there is consistently high HTGL activity in the media and easily detectable activity in the cellular extracts in the stable CHO cell lines expressing wild-type HTGL and substitution mutants that individually involve three of the four putative N-linked glycosylation sites. Two different substitution mutants involving Asn-56 were both devoid of HTGL activity whether the assay was performed on the media or cellular extracts. By Northern blot analysis, HTGL mRNA was detected in these Asn-56 mutants. Therefore, the absence of HTGL activity in the Asn-56 mutants could be caused by the failure of production of the mutant enzymes at the translational level or by the production of catalytically inactive enzymes by these cell lines.

We next assayed for the amount of immunoreactive HTGL produced by each cell line by Western blot analysis. Both cellular extract and media were analyzed separately. The results of our experiments indicate the following: for the immunoreactive HTGL secreted into the media, (i) the three mutant hHTGL proteins that were secreted into the culture medium (Asn-20→Gln, Asn-340→Gln and Asn-375→Gln) were all smaller than the wild-type HTGL protein, suggesting that these sites were actually glycosylated in the wild-type enzyme, and (ii) cell lines transfected with the two Asn-56 substitution mutants produced no detectable immunoreactive hHTGL in the medium. For the intracellular HTGL (i.e., cellular extracts), immunoreactive HTGL was clearly present in the wild-type cell line as well as in all four N-linked glycosylation site mutants including the Asn-56→Gln and Asn-56→Ala mutants which failed to secrete immunoreactive hHTGL protein in the media. Therefore, HTGL protein was produced in all cases. We calculated the specific activities of wild-type and mutant HTGLs produced by the CHO cells (Table 1).

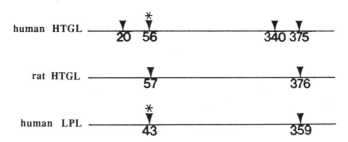

Figure 1. Alignment of human and rat hepatic lipase and human lipoprotein lipase. A schematic representation of the asparagine-linked glycosylation sites of the three lipases reveals two conserved sites. Asn residues found to be essential for the expression of human HTGL and LPL are denoted by asterisks.

Table 1. Wild-type and mutant HTGL enzyme specific activities

Expression Construct	Specific Activity			
	Experiment I		Experiment II	
	Cell	Media	Cell	Media
Asn-20/Gln	62 ± 8.2	2465 ± 150*	65 ± 8.1**	2721 ± 106*
Asn-56/Ala	0	0	0	0
Asn-56/Gln	0	0	0	0
Asn-340/Gln	69 ± 7.2**	2465 ± 82*	67 ± 4.1*	2921 ± 182*
Asn-375/Gln	37 ± 6.3**	2634 ± 155*	38 ± 4.6*	2418 ± 104**
HTGL Wild-type	53 ± 6.3	2008 ± 115	51 ± 1.4	2223 ± 70

CHO cells were stable transfected with 20 μg of DNA/dish. HTGL protein was partially purified for each construct by a heparin-sepharose column from two T175 flasks. Specific activities and distribution of HTGL mass of two independent experiments were presented. Values are presented as means ± S.D. *P<0.01, **P<0.05, compared to wild-type construct. This table is taken from Wölle et al. [16].

Although there was substantial difference in the amount of HTGL produced by different cell lines in Experiments I and II, the specific activities were very similar. The specific activities of wild-type HTGL and the three mutant HTGLs that were secreted into the media differed by less than 25%. The specific activities of the intracellular enzymes were low, being 1.5 - 2.5% that of the secreted enzymes except for the Asn-56 mutants which were totally inactive. Furthermore, the specific activity of the intracellular form of Asn-375→Gln mutant was approximately 55-75% that of the other active intracellular forms in both experiments. In every case (except for Asn-56 mutants), the secreted form of hHTGL was consistently much more active than the intracellular form. These results are consistent with the interpretation that maturation of the carbohydrate chains is required for maximal HTGL enzyme activity [8]. Furthermore, calculation of relative concentrations of immunoreactive HTGL indicates that in the wild-type enzyme and all the active Asn mutants, ~60% of the enzyme was found extracellularly and ~40% intracellularly, indicating that proper glycosylation of Asn-20, Asn-340 or Asn-375 was not required for efficient enzyme secretion. Since both the Asn-56→Ala and Asn-56→Gln mutants were detected intracellularly but were totally undetectable in the media, we conclude that glycosylation of Asn-56 is required for HTGL secretion.

STUDIES ON HUMAN LPL EXPRESSED IN VITRO AND IN VIVO

In parallel with our experiments on HTGL structure-function relationships, we performed experiments on the structure-function relationship of LPL in vitro and its metabolic role in vivo. Our laboratory has had an intense interest in the structural basis of LPL action. Based on numerous experiments on site-specific mutants of human LPL in vitro [19-21], and on observations in patients with familial LPL deficiency and Type I hyperlipoproteinemia [14,22,23], we concluded that LPL has a three dimensional structure very similar to that of pancreatic lipase, reported by Winkler et al. [24]. One interesting feature in the structural similarity of the three lipolytic enzymes is the conservation of most of the cysteine residues [25]. In human LPL, there are four pairs of cysteine residues, which are in disulfide linkage; all are conserved in all mammalian and nonmammalian species examined. We mutated these cysteine airs in turn, and found that mutant human LPLs with removal of the first three cysteine pairs seem to be inactive whereas disruption of the last pair seemed to leave the enzyme with residual activity. Finally, to study the function of LPL in vivo, we produced transgenic mice which express human LPL driven by the metallothionein promoter [26]. Our results indicate that LPL overexpression is associated with lower VLDL levels.

CONCLUSIONS

In the last five years, with the availability of cloned HTGL and LPL sequences, we have examined the structure-function relationship and functional role of these enzymes in vitro and in vivo. We believe that further exploration of HTGL and LPL action using molecular approaches will continue to help unravel the metabolic role of these important enzymes in lipoprotein metabolism.

ACKNOWLEDGMENTS

We thank Dr. Hans Jansen of Erasmus Universiteit Rotterdam, The Netherlands, for supplying us with an anti-HTGL antibody for our studies, and Ms. Sally Tobola for expert secretarial assistance. The work described in this study was supported by grants from the National Institutes of Health (HL-16512, HL-27341), March of Dimes Birth Defects Foundation, and Juvenile Diabetes Foundation International.

REFERENCES

1. Rao, S. N., Cortese, C., Miller, N.E., Levy, Y., and Lewis, B., Effects of heparin infusion on plasma lipoproteins in subjects with lipoprotein lipase deficiency. Evidence for a role of hepatic endothelial lipase in the metabolism of high-density lipoprotein subfractions in man, FEBS Lett. 150, 1982, 255-259.
2. Kinnunen, P.K.J., Hepatic endothelial lipase: isolation, some characteristics, and physiological role, in: "Lipases," B. Borgstrom and H.L. Brockman, eds., Elsevier/North-Holland, New York, 1984, 307-328.
3. Kuusi, T., Kinnunen, P.K.J., and Nikkila, E.A., Hepatic endothelial lipase antiserum influences rat plasma low and high density lipoproteins in vivo, FEBS Lett. 104, 1979, 384-388.
4. Jansen, H., van Tol, A., and Hulsman, W.C., On the metabolic function of heparin-releasable liver lipase, Biochem. Biophys. Res. Commun. 92, 1980, 53-59.
5. Bamberger, M., Lund-Katz, S., Phillips, M., and Rothblat, G.H., Mechanism of the hepatic lipase-induced accumulation of high-density lipoprotein cholesterol by cells in culture, Biochemistry 24, 1985, 3693-3701.
6. Blades, B., Vega, G.L., and Grundy, S.M., Activities of lipoprotein lipase and hepatic triglyceride lipase in postheparin plasma of patients with low concentrations of HDL cholesterol. Arteriosclerosis and Thrombosis 13:, 1993, 1227-1235.
7. Verhoeven, A.J.M., and Jansen, H., Secretion of rat hepatic lipase is blocked by inhibition of oligosaccharide processing at stage of glucosidase I, J. Lipid Res. 31, 1990, 1883-1893.
8. Verhoeven, A.J.M., and Jansen, H., Secretion-coupled increase in the catalytic activity of rat hepatic lipase, Biochim. Biophys. Acta 1086, 1991, 49-56.
9. Ben-Zeev, O., Doolittle, M.H., Davis, R.C., Elovson, J., and Schotz, M.C., Maturation of lipoprotein lipase. Expression of full catalytic activity requires glucose trimming but no translocation to the cis-Golgi compartment, J. Biol. Chem. 267, 1992, 6219-6227.
10. Leitersdorf, E., Stein, O., and Stein, Y., Synthesis and secretion of triacylglycerol lipase by cultured rat hepatocytes, Biochim. Biophys. Acta 794, 1984, 261-268.
11. Chajek-Shaul, T., Friedman, G., Knobler, H., Stein, O., Etienne, J., and Stein, Y., Importance of different steps of glycosylation for the activity and secretion of lipoprotein lipase in rat preadipocytes studied with monensin and tunicamycin, Biochim. Biophys. Acta 873, 1985, 123-134.
12. Amri, E.-Z., Vannier, C., Etienne, J., and Ailhaud, G., Maturation and secretion of lipoprotein lipase in cultured adipose cells. II. Effects of tunicamycin on activation and secretion of the enzyme. Biochim. Biophys. Acta 875, 1986, 334-343.
13. Olivecrona, T., Chernik, S.S., Bengtsson-Olivecrona, G., Garison, M., and Scow, R., Synthesis and secretion of lipoprotein lipase in 3T3-L1 adipocytes: demonstration of inactive form of lipase in cells, J. Biol. Chem. 262, 1987, 10748-10759.
14. Semenkovich, C.F., Luo, C.-C., Nakanishi, M.K., Chen, S.-H., Smith, L.C., and Chan, L., In vitro expression and site-specific mutagenesis of the cloned human lipoprotein lipase gene, J. Biol. Chem. 265, 1990, 5429-5433.
15. Stahnke, G., Davis, R.C., Doolittle, M.H., Wong, H., Schotz, M.C., and Will, H., Effect of N-linked Glycosylation on hepatic lipase activity. J. Lipid Res. 32, 1991, 477-484.

16. Wölle, J., Jansen, H., Smith, L.C., and Chan, L., Functional role of N-linked glycosylation in human hepatic lipase: asparagine-56 is important for both enzyme activity and secretion, J. Lipid Res. 34, 1993, 2169-2176.

17. Stephens, P.E., and Cockett, M.I., The construction of a highly efficient and versatile set of mammalian expression vectors, Nucleic Acids Res. 17, 1989, 7110.

18. Cockett, M.I., Bebbington, C.R., and Yarranton, G.T., The use of engineered E1A genes to transactivate the hCMV-MIE promoter in permanent CHO cell lines, Nucl. Acids Res. 19, 1990, 319-325.

19. Faustinella, F., Chang, A., Van Biervliet, J.P., Rosseneu, M., Vinaimont, N., Smith, L.C., Chen, S.-H., and Chan, L., Catalytic triad residue mutation (Asp 156→Gly) causing familial lipoprotein lipase deficiency: coinheritance with a nonsense mutation (Ser 447→Ter) in a Turkish family, J. Biol. Chem. 266, 1991, 14418-14424.

20. Ishimura-Oka, K., Faustinella, F., Kihara, S., Smith, L.C., Oka, K., and Chan, L., A missense mutation (Trp[86]→Arg) in exon 3 of the lipoprotein lipase gene: a cause of familial chylomicronemia, Am. J. Hum. Genet. 50, 1992, 1275-1280.

21. Ishimura-Oka, K., Semenkovich, C.F., Faustinella, F., Goldberg, I.J., Shachter, N., Smith, L.C., Coleman, T., Hide, W.A., Brown, W.V., Oka, K., and Chan, L., A missense (Asp[250]→ Asn) mutation in the lipoprotein lipase gene in two unrelated families with familial lipoprotein lipase deficiency, J. Lipid Res. 33, 1992, 745-754.

22. Faustinella, F., Smith, L.C., Semenkovich, C.F., and Chan, L., Structural and functional roles of highly conserved serines in human lipoprotein lipase: evidence that Serine 132 is essential for enzyme catalysis, J. Biol. Chem. 266, 1991, 9481-9485.

23. Faustinella, F., Smith, L.C., and Chan, L., Functional topology of a surface loop shielding the catalytic center in lipoprotein lipase, Biochemistry 31, 1992, 7219-7223.

24. Winkler, F.K., D'Arcy,, A., and Hunziker, W., Structure of human pancreatic lipase, Nature 343, 1990, 771-774.

25. Hide, W.A., Chan, L., and Li, W.-H., Structure and evolution of the lipase super-family, J. Lip. Res. 33, 1992, 167-178.

26. Zsigmond, E., and Chan, L., Transgenic mice expressing lipoprotein lipase, Clin. Res. 40, 1992, 291A.

LIPOPROTEIN LIPASE AND LYSOSOMAL ACID LIPASE: TWO KEY ENZYMES OF LIPID METABOLISM

Detlev Ameis and Heiner Greten

Medizinische Kern- und Poliklinik
Universitäts-Krankenhaus Eppendorf
20246 Hamburg, F.R.G.

INTRODUCTION

Lipases play a central role in the metabolism of triglyceride-rich lipoproteins in plasma and have therefore been extensively studied at the biochemical and molecular level. Primary structures from various species have been established for lipoprotein lipase (LPL), hepatic lipase (HL), pancreatic lipase (PL), and lysosomal acid lipase (LAL). Comparison of cDNA and genomic structures revealed a gene family of neutral lipases consisting of LPL, HL, and PL. Another gene family of acid lipases comprised LAL, gastric lipase, and lingual lipase.

The essential physiological function of lipolytic enzymes in lipid metabolism is illustrated by inborn errors of metabolism leading to clinically recognized syndromes. Absent activity of LPL results in a syndrome termed familial chylomicronemia (Type I hyperlipoproteinemia), while HL deficiency causes an imbalance of plasma lipoprotein patterns. LAL deficiency is associated with two distinct inherited disorders, Wolman disease and Cholesteryl Ester Storage Disease. Based on cDNA and genomic structures, lipase deficiencies are currently being assessed at the molecular level. Missense and nonsense mutations as well as deletions and insertions are being observed in lipase genes, indicating considerable molecular heterogeneity.

LIPOPROTEIN LIPASE

Dietary lipids are metabolized in the circulation by LPL, a neutral hydrolase, which is synthesized by parenchymal cells of various tissues, in particular of heart, skeletal muscle, and adipose [1,2]. Following synthesis, LPL is transported to the luminal surface of vascular endothelial cells. In this location, it is bound to the endothelial surface by ionic interaction with glucosaminoglycans, i.e. heparan sulfate and dermatan sulfate [3,4] (Fig. 1). Free fatty

Esterases, Lipases and Phospholipases, Edited by M.I. Mackness
and M. Clerc, Plenum Press, New York, 1994

acids liberated by the hydrolytic activity of LPL are being taken up by the surrounding tissue for further storage and metabolism. Besides it's enzymatic activity as a lipase, a small fraction of LPL is bound to circulating chylomicrons, and in vitro mediates the uptake of chylomicron remnants into hepatocytes [5].

cDNA clones for LPL of various species have recently been isolated (reviewed in [6]). The genomic cloning of some of the neutral lipases has revealed highly related genomic structures of these enzymes and proposed gene duplication and recombination events in the evolution of the gene family [7]. Based on sequence comparisons, LPL was established as member of a gene family of neutral lipases consisting of LPL [7-9], HL [10,11] and PL [12].

An inherited deficiency of LPL results in familial chylomicronemia (type I hyper-lipoproteinemia), a disorder of lipid metabolism characterized by a massive increase in plasma chylomicrons of fasting subjects and a consecutively pronounced increase in plasma triglycerides [13]. This alteration of lipid degradation is associated with a variety of clinical characteristics, including recurrent abdominal pain due to episodes of pancreatitis, hepatosplenomegaly, eruptive cutaneous xanthomas, and retinal lipemia. Detailed studies of plasma lipoproteins typically reveal chylomicronemia, often combined with increased levels of very low density lipoproteins (VLDL) in the fasted state. Plasma triglyceride concentrations are usually above 17 mmol/l; total plasma cholesterol tends to be mildly elevated with decreased cholesterol concentrations in the low density lipoproteins (LDL) and high density lipoprotein (HDL) subfractions of plasma lipoproteins. Restriction of dietary fat consumption to less than 20 g is usually sufficient to lower triglyceride levels and thereby control related clinical symptoms.

Familial LPL deficiency is frequently detected in childhood. The disorder is inherited in an autosomal recessive manner; the homozygous form has an estimated frequency of 1 in 1 million inhabitants [14]. For the molecular analysis of LPL deficient patients the following strategy is employed (Fig. 2). Established cell lines of fibroblasts, Epstein-Barr-virus-

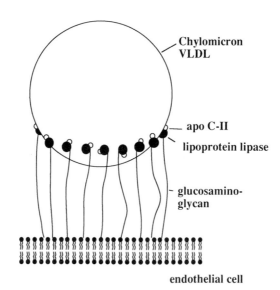

Figure 1. Lipoprotein lipase system. The lipid bilayer of an endothelial cell is shown, with glucosamino-glycans, i.e. heparan sulfate or dermatan sulfate, protruding from the cell surface. Lipoprotein lipase is bound to glucosaminoglycan molecules by ionic interaction. In the presence of apolipoprotein C-II, an obligate cofactor for enzymatic activity, lipid substrates such as chylomicrons or VLDL particles are hydrolyzed.

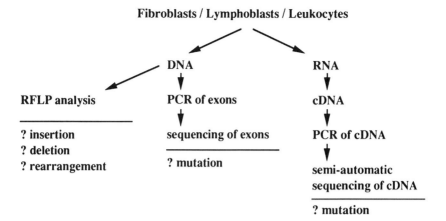

Figure 2. Strategy for the molecular analysis of lipases deficiencies.

transformed lymphoblasts or circulating leukocytes from lipase deficient subjects are obtained to extract DNA and RNA. Genomic DNA is used for analysis of restriction-fragment-length-polymorphisms (RFLPs), addressing the question of larger genomic insertions, deletions or rearrangements. DNA is also subjected to PCR amplification of individual exons with subsequent DNA sequence analysis. This is aimed at detecting point mutations resulting in amino acid exchanges or premature stop codons. As an alternative approach, total cellular RNA is reverse-transcribed into cDNA, PCR-amplified and sequenced semi-automatically using fluorochrome-labeled oligonucleotide primers.

Employing these techniques in patients with LPL deficiency, a number of point mutations and deletions distorting the regular LPL structure have been described [15,16]. Some missense or nonsense mutations have been shown by expression in eukaryotic cells to abolish lipase activity [17-21], while another mutation lead to reduced lipolytic activity with consecutive chylomicronemia in pregnancy [22].

LYSOSOMAL ACID LIPASE

Circulating low density lipoproteins are internalized by target tissues through the LDL-receptor [23]. Following endocytosis, lipoproteins are delivered to lysosomes for further degradation [24]. In this organelle, a hydrolase with a pH optimum of 4.5 to 5.0, denoted lysosomal acid lipase (LAL), is observed metabolizing exogenous triacylglycerols and cholesteryl esters (Fig. 3). Biochemical studies have described LAL activity in a wide variety of cells and species [25,26].

To further characterize the molecular properties of LAL, we have recently purified human liver LAL to apparent homogeneity [27]. The mature enzyme is a monomer of 42 kDa, and is glycosylated with high mannose sugar residues. Partial amino-terminal peptide sequences suggested post-translational processing of a precursor lipase releasing a mature LAL protein of 323 amino with a predicted non-glycosylated molecular mass of 36 kDa. Peptide sequencing allowed the synthesis of a unique 33-mer oligonucleotide and the subsequent molecular cloning of the full-length cDNA of human hepatic LAL [28]. The LAL cDNA encodes 2,626 nucleotides and a predicted LAL protein of 399 amino acids.

Sequence comparisons of human liver LAL with human fibroblast acid lipase [29] showed that the 5′ untranslated region of liver LAL extended 105 bases beyond the fibroblast sequence. Throughout the coding region, a single difference was observed in nucleotide position 46, where A is replaced by C in fibroblast lipase. This nucleotide divergence

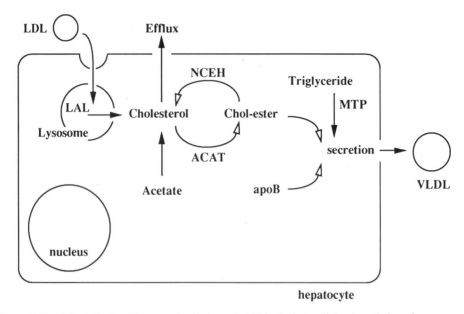

Figure 3. Physiological role of lysosomal acid lipase (LAL) in the intracellular degradation of exogeneous cholesteryl esters and neutral triglycerides. ACAT, acyl:CoA acetyl transferase; NCEH, neutral cholesteryl ester hydrolase; MTP, microsomal triglyceride transfer protein; apoB, apolipoprotein B; LDL, low density lipoprotein; VLDL, very low density lipoprotein.

resulted in threonine at position -12 in liver LAL and proline at the respective position of the fibroblast enzyme. The remaining coding region of both enzymes showed identical sequences. Four nucleotide differences were observed in the 3'non-coding region. Whether these differences represent polymorphisms or are due to tissue-specific isoforms is as yet undetermined.

cDNA clones for two acid hydrolases, gastric lipase and lingual lipase, have been described. Alignment of hepatic LAL with human gastric lipase sequences [30] showed a high degree of homology, with 59% amino acid identity throughout the length of the lipase. Human hepatic LAL and rat lingual lipase amino acid sequences [31] showed 58% identity. Fig. 4 shows the resulting extensive similarities in hydrophobicity throughout the length of the molecules, suggesting highly related tertiary structures for these enzymes. Interestingly, apart from a small region surrounding the putative active site serine in neutral lipases, no homologies were detected with neutral lipases such as hepatic lipase [32] or lipoprotein lipase [33], indicating independent gene families for acid and neutral lipases.

Using the complete human LAL cDNA as a molecular probe, Northern analyses of the LAL mRNA content of various human tissues were performed. The results demonstrated abundant expression in brain, mammary gland, kidney and adrenal gland. Placenta and HeLa cells showed intermediate levels of expression while heart, liver, and skeletal muscle expressed low levels of LAL mRNA. This pattern of expression comfirmed biochemical studies demonstrating LAL activity in different cell types and supports the assumption of a "house keeping" function of LAL in the lysosomal degradation of lipid substrates.

DEFICIENCIES OF LYSOSOMAL ACID LIPASE

Two inherited disorders are associated with a deficient LAL hydrolytic activity. In Wolman disease [34], absent LAL activity results in a pronounced accumulation of

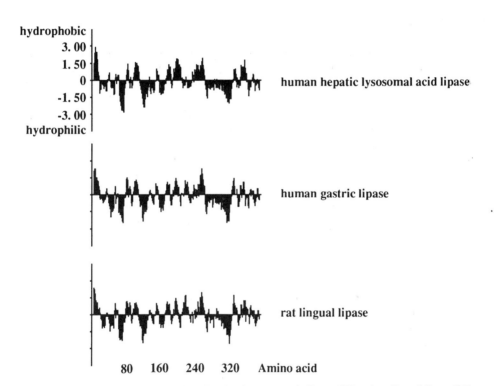

Fig. 4. Hydrophobicity plots of human hepatic LAL, human gastric lipase [30] and rat lingual lipase [31].

cholesteryl esters and triacylglycerols in lysosomes of most tissues of the body. Clinically, the disorder presents with hepatosplenomegaly, failure to thrive, steatorrhea and adrenal calcifications [26]. Wolman disease patients usually die of hepatic and adrenal failure within the first year of life. Cholesteryl ester storage disease (CESD), the other clinically recognized phenotype associated with LAL deficiency, follows a more benign clinical course [35,36]. However, hypercholesterolemia often occurs in CESD patients and premature atherosclerosis may be severe [26]. In both diseases, LAL deficiency is diagnosed in homogenates of liver biopsies or circulating leukocytes.

With the availability of cDNA clones for human LAL, the molecular investigation of LAL deficiencies is currently being undertaken. Preliminary results from DNA sequencing of LAL cDNAs of CESD patients showed two deletion mutants leading to a loss of functional LAL protein in these subjects.

CONCLUSIONS

Due to their central role in lipid metabolism lipolytic enzymes have been extensively characterized biochemically. Recently, molecular cloning has established cDNA and genomic structures of a large number of lipases. Based on information derived from primary structures of these enzymes, the molecular basis of inherited lipase deficiencies has been studied. Naturally occuring mutants observed in lipase-deficient individuals have greatly advanced our understanding of structure-function relationships of lipases and the genetics involved in these disorders. Chimeric lipases are currently being utilized to delineate functional domains at the molecular level. Animal models of lipase deficiencies are being developed, using gene targeting strategies to disrupt the respective genes. Overexpression of lipases in transgenic animals may also contribute to our understanding of the molecular biology and pathology of these central enzymes of lipid metabolism. This may eventually lead to developing enzyme replacement strategies in severely affected lipase deficient subjects.

ACKNOWLEDGEMENT

This work was supported in part by the Deutsche Forschungsgemeinschaft (Am 65/3-1).

REFERENCES

1. Augustin, J, Greten, H, The role of lipoprotein lipase - molecular properties and clinical relevance, *Atheroscl. Rev.*, 5: 1979; 91-96.
2. Garfinkel, AS, Schotz, MC. Lipoprotein lipase, in: "Plasma lipoproteins," Gotto, A.M.,Jr., ed., Elsevier, New York, 1987; 335-357.
3. Olivecrona, T, Bengtsson, G, Marklund, S-E, Lindahl, U, Höök, M, Heparin-lipoprotein lipase interactions, *Fed. Proc.*, 36: 1977; 60-64.
4. Cheng, C-F, Oosta, GM, Bensadoun, A, Rosenberg, RD, Binding of lipoprotein lipase to endothelial cells in culture, *J. Biol. Chem.*, 256: 1981; 12893-12898.
5. Beisiegel, U, Weber, W, Bengtsson-Olivecrona, G, Lipoprotein lipase enhances the binding of chylomicrons to low density lipoprotein receptor-related protein, *Proc. Natl. Acad. Sci. USA.*, 88: 1991; 8342-8346.
6. Ameis, D, Greten, H, Schotz, MC, Hepatic and plasma lipases, *Seminars Liver Dis*, 12: 1992; 397-402.
7. Kirchgessner, TG, Chuat, J-C, Heinzmann, C, Etienne, J, Guilhot, S, Svenson, K, Ameis, D, Pilon, C, d'Auriol, L, Andalibi, A, Schotz, MC, Galibert, F, Lusis, AJ, Organization of the human lipoprotein lipase gene and evolution of the lipase gene family, *Proc. Natl. Acad. Sci. USA.*, 86: 1989; 9647-9651.

8. Deeb, SS, Peng, R, Structure of the human lipoprotein lipase gene, *Biochemistry*, 28: 1989; 4131-4135.

9. Oka, K, Tkalcevic, GT, Nakano, T, Tucker, H, Ishimura-Oka, K, Brown, WV, Structure and polymorphic map of human lipoprotein lipase gene, *Biochim. Biophys. Acta*, 1049: 1990; 21-26.

10. Cai, S-J, Wong, DM, Chen, S-H, Chan, L, Structure of the human hepatic triglyceride lipase gene, *Biochemistry*, 28: 1989; 8966-8971.

11. Ameis, D, Stahnke, G, Kobayashi, J, Bücher, M, Lee, G, McLean, J, Schotz, MC, Will, H, Isolation and characterization of the human hepatic lipase gene, *J. Biol. Chem.*, 265: 1990; 6552-6555.

12. Mickel, FS, Weidenbach, F, Swarowsky, B, LaForge, KS, Scheele, GA, Structure of the canine pancreatic lipase, *J. Biol. Chem.*, 264: 1989; 12895-12901.

13. Brunzell, JD. Familial lipoprotein lipase deficiency and other causes of the chylomicronemia syndrome, in: "The metabolic basis of inherited disease," Scriver, C.R., Beaudet, A.L., Sly, W.S., and Valle, D., eds., McGraw-Hill, New York, 1989; 1165-1180.

14. Fredrickson, DS, Goldstein, JL, Brown, MS. The familial hyperlipoproteinemias, in: "The metabolic basis of inherited disease," Stanbury, J.B., wyngaarden, J.B., and Fredrickson, D.S., eds., McGraw-Hill, New York, 1978; 604-655.

15. Hayden, MR, Ma, Y, Brunzell, JD, Henderson, HE, Genetic variants affecting human lipoprotein and hepatic lipases, *Curr. Opinion Lipidol.*, 2: 1991; 104-109.

16. Hayden, MR, Ma, Y, Molecular genetics of human lipoprotein lipase deficiency, *Mol. Cell. Biochem.*, 113: 1992; 171-176.

17. Ameis, D, Kobayashi, J, Davis, RC, Ben-Zeev, O, Malloy, MJ, Kane, JP, Wong, H, Havel, RJ, Schotz, MC, Familial chylomicronemia (type I hyperlipoproteinemia) due to a single missense mutation in the lipoprotein lipase gene, *J. Clin. Invest.*, 87: 1991; 1165-1170.

18. Faustinella, F, Chang, A, van Biervliet, JP, Rosseneu, M, Vinaimont, N, Smith, LC, Chen, S-W, Chan, L, Catalytic triad residue mutation (Asp-156 -> Gly) causing familial lipoprotein lipase deficiency, *J. Biol. Chem.*, 266: 1991; 14418-14424.

19. Dichek, HL, Fojo, SS, Beg, OU, Skarlatos, SI, Brunzell, JD, Cutler, GB, Brewer, HB, Identification of two separate allelic mutations in the lipoprotein lipase gene of a patient with the familial hyperchylomicronemia syndrome, *J. Biol. Chem.*, 266: 1991; 473-477.

20. Ma, Y, Bruin, T, Tuzgol, S, Wilson, BI, Roederer, G, Liu, M-S, Davignon, J, Kastelein, JJP, Brunzell, JD, Hayden, MR, Two naturally occurring mutations at the first and second bases of codon aspartic acid 156 in the proposed catalytic triad of human lipoprotein lipase, *J. Biol. Chem.*, 267: 1992; 1918-1923.

21. Ma, Y, Liu, MS, Ginzinger, D, Frohlich, J, Brunzell, JD, Hayden, MR, Gene-environment interaction in the conversion of a mild-to-severe phenotype in a patient homozygous for a Ser172-->Cys mutation in the lipoprotein lipase gene, *Acta Med. Scand.*, 1993; -253.

22. Bruin, T, Kastelein, JJ, Van Diermen, DE, Ma, Y, Henderson, HE, Stuyt, PM, Stalenhoef, AF, Sturk, A, Brunzell, JD, Hayden, MR, A missense mutation Pro157 -->Arg in lipoprotein lipase (LPL-Nijmegen) resulting in loss of catalytic activity, *Eur. J. Biochem.*, 208: 1992; 267-272.

23. Goldstein, JL, Brown, MS. Familial hypercholesterolemia, in: "The metabolic basis of inherited disease," Scriver, C.R., Beaudet, A.L., Sly, W.S., and Valle, D., eds., McGraw-Hill, New York, 1989; 1215-1250.

24. Goldstein, JL, Dana, SE, Faust, JR, Beaudet, AL, Brown, MS, Role of lysosomal acid lipase in the metabolism of plasma low density lipoproteins, *J. Biol. Chem.*, 250: 1975; 8487-8795.

25. Fowler, SD, Brown, WJ. Lysosomal acid lipase, in: "Lipases," Borgström, B. and Brockman, H.L., eds., Elsevier, New York, 1984; 329-364.

26. Schmitz, G, Assmann, G. Acid lipase deficiency: Wolman disease and cholesteryl ester storage disease, in: "The metabolic basis of inherited disease," Scriver, C.R., Beaudet, A.L., Sly, W.S., and Valle, D., eds., McGraw Hill, New York, 1989; 1623-1644.

27. Ameis, D, Merkel, M, Eckerskorn, C, Greten, G, Purification and characterization of human hepatic lysosomal acid lipase, *Circulation*, 84: 1992; 548. (abstr)

28. Ameis, D, Merkel, M, Eckerskorn, C, Pauer, A, Greten, H, Purification, characterization and molecular cloning of human hepatic lysosomal acid lipase, *submitted.*

29. Anderson, RA, Sando, GN, Cloning and expression of cDNA encoding human lysosomal acid lipase/cholesteryl ester hydrolase, *J. Biol. Chem.*, 266: 1991; 22479-22484.

30. Bodmer, MW, Angal, S, Yarranton, GT, Harris, TJR, Lyons, A, King, DJ, Pieroni, G, Riviere, C, Verger, R, Lowe, PA, Molecular cloning of a human gastric lipase and expression of the enzyme in yeast, *Biochim. Biophys. Acta*, 909: 1987; 237-244.

31. Docherty, AJP, Bodmer, MW, Angal, S, Verger, R, Riviere, C, Lowe, PA, Lyons, A, Emtage, JS, Harris, TJR, Molecular cloning and nucleotide sequence of rat lingual lipase cDNA, *Nucl. Acids. Res.*, 13: 1985; 1891-1903.

32. Stahnke, G, Sprengel, R, Augustin, J, Will, H, Human hepatic triglyceride lipase: cDNA cloning, amino acid sequence and expression in a cultured cell line, *Differentiation*, 35: 1987; 45-52.

33. Wion, KL, Kirchgessner, TG, Lusis, AJ, Schotz, MC, Lawn, RM, Human lipoprotein lipase complementary DNA sequence, *Science*, 235: 1987; 1638-1641.

34. Patrick, AD, Lake, BD, Deficiency of an acid lipase in Wolman's disease, *Nature*, 222: 1969; 1067-1068.

35. Burke, JA, Schubert, WK, Deficient activity of hepatic acid lipase in cholesterol ester storage disease, *Science*, 176: 1972; 309-310.

36. Sloan, HR, Fredrickson, DS, Enzyme deficiency in cholesteryl ester storage disease, *J. Clin. Invest.*, 51: 1972; 1923-1926.

LIPOPROTEIN LIPASE ACTIVITY IN THE PATHOLOGICAL METABOLISM OF LIPOPROTEINS

Paul N. Durrington

University of Manchester, Department of Medicine
Manchester Royal Infirmary
Manchester M13 9WL

Many of the clinical features of defective lipoprotein lipase activity were known well before the present century, particularly the milky appearance of serum. This was as the result of the use of venesection as a means of treatment of such disorders as abdominal pain, diabetes, alcoholism, and glomerulonephritis (reference 1 cites many early reports). The syndrome of familial lipoprotein deficiency (FLLD) was not, however, first clearly described until the 1930's [2,3]. Havel and Gordon first showed that its basis was defective lipoprotein lipase activity [4]. Only recently has it also been appreciated that a similar syndrome can result from a defect in apolipoprotein CII, the circulating activator of lipoprotein lipase [5]. Lipoprotein lipase deficiency with an autoimmune basis has also been described [6].

FLLD is inherited as an autosomal recessive condition. This description may, however, be in need of revision because the carriers of the condition should be regarded as heterozygotes because they often express hyper-triglyceridaemia albeit in a less marked form and with different clinical characteristics (see later) [7]. In many countries such as Britain FLLD is rare probably affecting only around 1 in 10^6 of the population. In some populations it is more frequent such as the French Canadians and some immigrant populations from the Indian subcontinent [8].

In childhood a type I lipoprotein phenotype (increased serum chylomicron levels with no increase in VLDL) occurs in FLLD, but most adults will have a type V hyperlipoproteinaemia (increase in both serum chylomicrons and VLDL). Kinetic studies show that the principal clearance defect is in chylomicrons and large VLDL particles, whereas smaller VLDL particles tend to be cleared at relatively normal rates [9] (Fig. 1). Normally the smaller VLDL particles, secreted principally by the liver, are metabolised by lipoprotein lipase, but when this is inoperative as in FLLD hepatic lipase may be able to assist in their clearance, but not that of the larger VLDL particles and chylomicrons secreted by the gut. This

Esterases, Lipases and Phospholipases, Edited by M.I. Mackness
and M. Clerc, Plenum Press, New York, 1994

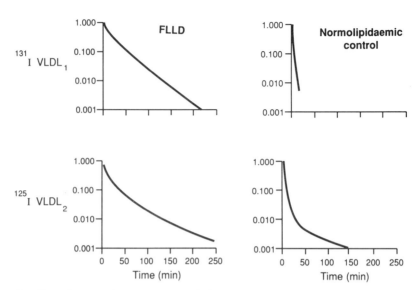

Figure 1. The plasma decay curve of large (Sf 60-400) and small (Sf 20-60) VLDL in patient with familial lipoprotein lipase deficiency and in a intravenously injected radiolabelled autologous normal control [**data from reference 9**]

explains the Type I phenotype. Adults may, however, secrete VLDL at rates too high for hepatic lipase to clear it adequately. Hepatic overproduction of VLDL may be exacerbated by obesity or by a low fat high carbohydrate diet, which, although limiting gut triglyceride-rich lipoprotein secretion, may cause induction of hepatic VLDL secretion. This may explain the usual finding of type V or type IV hyperlipoproteinaemia in most adults even after treatment.

The plasma of untreated or partially treated FLLD looks milky. It may thus be discovered by chance when blood is taken for some unrelated investigation. Examination of the optic fundus may reveal a generally pale retina with the veins as well as the arteries appearing white (lipaemia retinalis). Patients frequently develop eruptive xanthomata. These are yellow papules usually < 0.5cm in diameter with a pink base on the extensor surfaces of the arms, shoulders, buttocks and occasionally elsewhere.

The most serious manifestation is acute pancreatitis. This presents with severe abdominal pain and other features typical of acute pancreatitis from any cause. However, it may pose particular diagnostic difficulties because the lipaemia can interfere with the determination of serum amylase activity so that a normal or only moderately elevated value is reported. Laparotomy for suspected perforated peptic ulcer may therefore be undertaken. Such a procedure greatly increases the mortality of acute pancreatitis over conservative treatment. Acute pancreatitis should therefore be diagnosed in patients with severe abdominal pain and milky serum regardless of the amylase result.

Further iatrogenic complications in the management of acute pancreatis due to FLLD may ensure if clinicians are unaware of the phenomenon of pseudohyponatraemia [**10**] which may lead to the infusion of saline inappropriately, or if fat emulsions are infused intravenously in the belief that these are providing nutritional support.

Two theories have been advanced to explain the pancreatitis associated with hypertriglyceridaemia. They are not mutually exclusive. One is that when blood triglyceride levels are high the small amounts of pancreatic lipase which normally leak into the pancreatic microcirculation may generate NEFA in quantities sufficient to damage the capillaries. This will increase capillary permeability and thus the release of lipase so that a vicious cycle leading to extensive tissue damage and inflammation is created [11]. The other is that oxidation of the huge quantities triglyceride-rich lipoproteins as they circulate through the pancrease generates alo dehydes and lysolipids which damage the pancreas [12].

Abdominal pain other than that due to acute pancreatitis may also occur in hyperchylomicronaemia. Hepatosplenomegaly is common and this may lead to splenic infarction or pain due to stretching of the hepatic capsule. Hepatosplenomegaly occurs due to the removal of triglyceride-rich lipoproteins from the circulation by the reticuloendothelial system. Macrophages stuffed with lipid droplets to form foam cells are abundant in the liver, spleen and bone marrow. High concentrations of triglyceride-rich lipoproteins can interfere with the determination of serum transaminases leading to falsely elevated levels, which in association with hepatomegaly may lead to the mistaken diagnosis of chronic liver disease.

Abdominal cramping pains which can be confused with irritable bowel syndrome are also common in FLLD and may be due to bowel ischaemia because when they pass through small blood vessels the huge quantities of large triglyceride-rich lipoproteins lead to sluggish blood flow. A similar explanation has been advanced for the occurrence of transient hemiparesis, lack of concentration and other neurogical syndromes in severe hypertriglyceridaemia [12].

In adults type V hyperlipoproteinaemia is more likely to be due to conditions other than FLLD. Usually there is some less severe defect in lipoprotein lipase activity which is either associated with an elevated hepatic VLDL secretion due for example to excessive alcohol intake, diabetes mellitus, or oestrogen therapy or some additional inhibitory influence on triglyceride clearance such as diabetes, hypothyroidism, renal disease or beta-adrenoreceptor blocker treatment.

The diagnosis of FLLD may be confirmed by measuring lipoprotein lipase in an adipose tissue biopsy [13] (Fig. 2) or by demonstrating that post-heparin plasma lipase activity is unaffected by inhibitors of lipoprotein lipase such as protamine [14].

The treatment of FLLD involves the institution of a low fat diet, often as low as 20g or less each day. This reduces the production of gut triglyceride-rich lipoproteins. Drug therapy usually has a much lesser part to play than diet. Fibric acid derivatives may be tried, but because their predominant effect is to increase lipoprotein lipase activity, which is not possible in FLLD, they may be ineffective. Nicotinic acid and its derivates may be helpful in limiting hepatic VLDL production. The primary purpose of therapy is to prevent acute pancreatitis. It is seldom possible to maintain normal serum triglyceride levels, but

attacks of acute pancreatitis are usually prevented if the
concentration is held below 20mmol/l. Many patients seem
resistant to pancreatitis even when triglycerides exceed
100mmol/l. For those still subject to attacks despite
attempts at lipid-lowering, antioxidant therapy with
selenium, methionine and vitamins A, C and E may be tried,
but clinical trial evidence for the benefits of this is
currently lacking.

Why do patients with FLLD not get premature atherosclerosis?

Patients with FLLD have particularly low levels of
serum HDL cholesterol despite which premature
atherosclerosis is not a feature of the syndrome. One reason
for this may be that they also have low levels of LDL
cholesterol. Indeed the coinheritance of familial
hypercholesterolaemia with FLLD seems to nullify the effect
of the LDL receptor defect [15]. Furthermore although the
foam-filled macrophages in the reticulo-endotheial system of
patients with FLLD resemble those which occur in the
arterial subintima in fatty streaks, the triglyceride-rich
lipoproteins which can readily cross the fenestrated
vascular endothelium of the reticulo-endothelial system are
probably too large to cross the arterial endothelium. Also
in FLLD plasma fibrinogen and factor VII activity are not
increased, unlike other primary hypertriglyceridaemias in
which lipoprotein lipase activity is maintained albeit at a
decreased level. We have recently shown that lipolysis of
large triglyceride-rich lipoproteins is necessary for the
activation of coagulant factor XII, which in turn activates
factor VII [16]. Both fibrinogen and factor VII are
independent risk factors for coronary heart disease [17].

Lipoprotein lipase and hepatic chylomicron remnant removal

Chylomicron remnants and some IDL are cleared from the
circulation by the liver. However, the receptor involved has

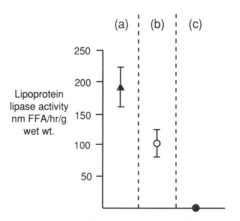

Figure 2. Adipose tissue lipoprotein lipase activity in a)
normolipidaemic healthy people b) primary hypertriglycerideaemic c)a
patient with familial lipoproein lipase deficiency.

proved elusive. Apolipoprotein E (apo E) appears to be an important ligand in mediating hepatic remnant clearance and in type III hyperlipoproteinaemia apo E polymorphisms or mutations which bind less avidly to the LDL receptor are usually expressed, most commonly the apo E_2 isoform. However, the hepatic LDL receptor is unlikely to be solely responsible for remnant clearance because in homozygous familial hypercholesterolaemia remnant clearance is relatively normal. Recently the LDL receptor-like protein (LRP) has been proposed as the remnant receptor [18]. This receptor binds a variety of proteins including alpha macroglobulin, apolipoprotein E and plasminogen activator/plasminogen activator inhibitor complexes and does so at different binding sites. It also binds lipoprotein lipase. Uptake of chylomicron remnants containing the quantity of apo E present when isolated from the circulation is disappointingly low. However, it has recently been shown that when they also contain lipoprotein lipase there is more substantial binding to liver cells [19]. Physiologically a certain amount of lipoprotein lipase detaches itself from the vascular endothelium and attaches to chylomicrons [20]. In FLLD there is no lipoprotein lipase activity and thus chylomicron remnants are probably not formed, but when there are lesser reductions in its activity remnants may be formed, but their capacity to be cleared by the LRP may be impaired if the quantity present on the remnants is decreased. This could be important in diabetes (where lipoprotein lipase activity is decreased due to insulin deficiency or resistance), hypothyroidism or when there is decreased affinity of apo E for receptors, for example in type III hyperlipoproteinaemia, which might be exacerbated by any coexistent deficiency in lipoprotein lipase.

Do partial defects in lipoprotein lipase activity contribute to the risk of atherosclerosis?

The vast majority of patients with hypertriglyceridaemia have accelerated rates of hepatic VLDL production [10]. This does not mean, however, the rate of lipolysis of VLDL is unimportant in determining serum triglyceride levels. Postprandially the increase in serum triglycerides is limited physiologically by an increase in lipoprotein lipase activity. Furthermore in obese patients although increased VLDL production is almost universal, it is matched by an increase in the rate of clearance in many patients who do not develop hypertriglyceridaemia as the result [21]. In patients with both primary and secondary hypertriglyceridaemia it is common to find that triglyceride clearance rates and lipoprotein lipase activity is subnormal (although not, of course, as low as FLLD) [7, 10, 22-25] perhaps because of genetic polymorphisms or mutations of the enzyme itself or defects in its activation due to an increaed ratio of apo CIII to CII in circulating lipoproteins [26]. Although it is disputed that hypertriglyceridaemia when it occurs in isolation is a risk factor for atherosclerosis [27], there seems little doubt that when it is combined with hypercholesterolaemia, the risk is greater than for similar levels of cholesterol in the absence of hypertriglyceridaemia and that this effect is not simply explained by the decreased HDL levels which frequently accompany hypertriglyceridaemia [28]. Several mechanisms which might involve lipoprotein lipase may account for this:

i) Preß HDL is the small, protein-rich, lipid-deficient particle which may be the initial acceptor of tissue cholesterol in the reverse cholesterol transport pathway. We have shown that it probably circulates complexed to other lipoproteins and that it is released locally in tissues in which these lipoproteins undergo lipolysis [29]. This may account for the observation that HDL_3 is produced postprandially on exercising the forearm in normal people, but not in patients with lipoprotein lipase deficiency [30].

ii) Cholesterol entering preß HDL is esterified by the enzyme LCAT. Cholesteryl ester so formed moves to the core of preß HDL converting it to HDL_3. Kinetic studies do not indicate that the rate of catabolism of the whole HDL particle is sufficiently fast for its hepatic uptake to explain reverse cholesterol transport. Two routes by which cholesteryl ester can egress from HDL appear more important: firstly cholesteryl ester may be taken up by the liver during the circulation of HDL through it without the whole particle being catabolised and secondly cholesteryl ester may be transferred from HDL to VLDL, a process mediated by the transport protein, cholesteryl ester transfer protein (CETP) [28]. This latter pathway is potentially atherogenic because cholesteryl ester may thence contribute to the LDL pool. We have shown that this transfer in increased in patients with coronary heart disease [32] and that it may be decreased by gemfibrozil [33] which is a fibric acid derivative, a class of drug the main action of which is to increase lipoprotein lipase activity [34](Fig. 4).

iii) Chylomicron remnant particles are atherogenic. Because lipoprotein lipase may be important for their hepatic catabolism, defects in lipoprotein lipase might contribute to the accumulation of remnants in the circulation in some people (see previous section).

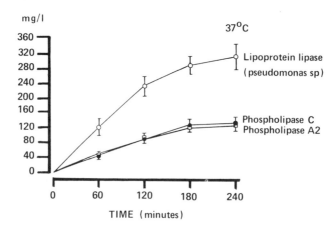

Figure 3   The increment in preß HDL concentration when normal serum is incubated with lipolytic enzymes [data from reference 29]

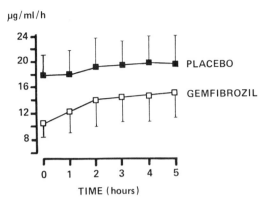

Figure 4 - Cholestryl ester transfer from HDL to VLDL/LDL measured using an in vitro method in the serum of patients with primary hypertriglyceridaemia before and after treatment with gemfibrozil. A mixed meal was consumed at time zero. [data from reference 33].

SUMMARY

FLLD represents a well defined clinical syndrome, the features of which are due to the gross accumulation of chylomicrons and large VLDL in the circulation as the result of defective catabolism. Less clearly defined are the effect of more minor decreases in lipoprotein lipase activity which are common in hyperlipidaemias associated with elevated serum triglyceride levels. Because coronary heart disease risk does not appear to be increased in FLLD, but is in many more common hypertriglyceridaemic states the conundrum exists as to whether a partial defect in lipoprotein lipase activity can increase atherosclerosis risk when a complete deficiency does not. Perhaps the low serum LDL cholesterol level and the absence of a hypercoagulable state in FLLD is important. Diminished lipoprotein lipase activity might contribute to the risk of atherosclerosis in other conditions, however, because of its role in reverse cholesterol transport (generation of preß-HDL, cholesteryl ester transfer from HDL to VLDL and hepatic chylomicron remnant clearance).

REFERENCES

1. Fisher B. Uber lipamie und cholesteremie, sowie uber des pancreas und der leber bei diabetes mellitus. Virchow's Archiv. 172:1903;30-71.

2. Burger M., Grutz O. Uber hepatosplenomegale lipoidose mit xanthomatosen veranderunger in haut un schleimhaut. Archive. Dermatol Syph 166:1932;542-475.

3. Holt L.E., Aylward F.X. Timbres H.G. Idiopathic familial lipaemia. John Hopkins Hosp. Bull 64:1939;279-314.

4. Havel R.J., Gordon R.J. Idiopathic hyperlipidaemia. Metabolic studies in an affected family. J. Clin. Invest. 39:1960; 1777-1790.

5. Breckenridge W.C., Little A., Steiner G., Chow A. Poapst M. Hypertriglyceridaemia associated with deficiency of apolipoprotein C-II. N. Engl. J. Med. 298:1978;1265-1273.

6. Glueck C.J., Kaplan A.P., Levy R.I., Greten H., Gralnick H.,Fredrickson D.S. A new mechanism of exogenous hypertriglyceridaemia. Ann. Int. Med. 7:1969;1051-1057.

7. Babirak S.P., Iverius P-H., Fjimoto W.Y., Brunzell J.D. Detection and characterisation of the heterozygote state for lipoprotein lipase deficiency. Arteriosclerosis 9:1989;326-334.

8. Hayden M., De Braekeleer M., Henderson H.E., Kastelein J. Molecular geography of inherited disorders of lipoprotein metabolism: lipoprotein lipase deficiency and familial hypercholesterolaemia in Molecular Genetics of Coronary Artery Disease. Candidate Genes and Processes in Atherosclerosis. Lewis A.J., Rotter J.I., Sparkes R.S. eds., Monogr. Hum. Genet. Karger, Basel 14:1992;350-362.

9. Demant T., Gaw A., Watts G.F. et al Metabolism of apo B-100-containing lipoproteins in familial hyperchylomicronaemia. J. Lipid. Res. 34:1993;147-156.

10. Durrington P.N. Hypertriglyceridaemia. Chapter 7 in Hyperlipidaemia. Diagnosis and Management. Wright, London 1989;135-156.

11. Havel R.J. Pathogenesis differentiation and management of hypertriglyceridaemia. Adv. Intern. Med. 15:1969;117-154.

12. Braganza J.M. Pancreatic disease: a casualty of hepatic 'detoxification'? Lancet ii:1983;1000-1002.

13. Durrington P.N., McIver J.E., Holdsworth G., Galton D.J. Severe hypertriglyceridaemia associated with pancytopenia and lipoprotein lipase deficiency. Ann. Int. Med. 94:1981;211-212.

14. Bengtsson-Olivecrona G., Olivecrona T. Assay of lipoprotein lipase and hepatic lipase Chapter 7a in Lipoprotein Analysis: A Practical Approach. Converse C.A., Skinner E.R. eds. Oxford University Press, Oxford, 1993;169-185.

15. Zambon A., Torres A., Bijouet S., Gange C., Moorjani S., Lupien P.J., Hayden M.R., Brunzell J.D. Prevention of raised 'low-density' lipoprotein cholesterol in a patient with familial hypercholesterolaemia and lipoprotein lipase deficiency. Lancet 341:1993;119-1121

16. Mitropoulos K.A., Miller G.J., Watts G.F., Durrington P.N. Lipolysis of triglyceride-rich lipoproteins activates coagulant factor XII: a study in familial lipoprotein-lipase deficiency. Atherosclerosis 95:1992;119-125.

17. Meade T.W., Mellows S.M., Brozovic M. et al Haemostatic function and ischaemic heart disease: principal results of the Northwick Park Heart Study. Lancet ii:1986;533-537.

18. Brown M.S., Herz J., Kowal R.C., Goldstein J.L. The low-density lipoprotein receptor-related protein: double agent or decoy? Curr. Opin. Lipidol. 2:1991;65-72.

19. Beisiegel U. Apolipoproteins as ligands for lipoprotein receptors. Chapter 10 in Structure and Function of Apolipoproteins ed Rosseneu M. CRC Press Inc. Boca Raton 1992;269-294.

20. Karpe F., Olivecrona T., Walldius G., Hamsten A. Lipoprotein lipase in plasma after an oral fat load: relation to free fatty acids. J. Lipid Res. 33:1992;975-984.

21. Pykolisto O.J., Smith P.H., Brunzell J.D. Determinants of human adipose tissue lipoprotein lipase: effects of diabetes and obesity on basal and diet induced activity J. Clin. Invest. 56:1975;1108-1117.

22. Auwerx J.H., Babirak S.P., Hopkanson J.E., Stahnke G., Will H., Deeb S.S., Brunzell J.D. Coexistance of abnormalities of hepatic lipase and lipoprotein lipase in a large family. Am. J. Hum. Genet. 46:1990;470-477.

23. Durrington P.N., Twentyman O.P., Braganza J.M., Miller J.P. Hypertriglyceridaemia and abnormalities of triglyceride catabolism persisting after pancreatitis. Int. J. Pancreatol. 1:1986;195-203.

24. Huttunen J.K., Ehnholm C., Kekki M., Nikkila E.A. Post-heparin lipoprotein lipase and hepatic lipase in normal subjects and in patients with hypertriglyceridaemia: correlations to sex, age and various parameters of triglyceride metabolism. Clin. Sci. Mol. Med. 50:1976;249-260.

25. Krauss R.M., Levy R.I., Fredrickson D.S. Selective measurement of two lipase activities in postheparin plasma from normal subjects and patients with hyperlipoproteinaemia. J. Clin. Invest. 54:1974;1107-1124.

26. Wojciechowski A.P., Farrall M., Cullen P. et al Familial combined hyperlipidaemia linked to the apolipoprotein AI-CIII-AIV gene cluster on chromosome 11q23-q24. Nature 349:1991;161-164.

27. Hulley S.B., Rosenman R.H., Bawol R.D., Braid R.J. Epidemiology as a guide to clinical decision. The association between triglyceride and coronary heart disease. N. Engl. J. Med. 302:1980;1383-1389.

28. Huttunen J.K., Manninen V., Tenkanen L., Heinonen O.P., Koskinen P., Frick M.H. Drug induced changes in HDL-cholesterol and coronary heart disease. Experinnces from the Helsinki Heart Study in High Density Lipoproteins and Atherosclerosis II Miller N.E. Ed, Excerpta Medica, Amsterdam 1989,191-198.

29. Neary R., Bhatnagar D., Durrington P.N., Ishola M., Arrol S., Mackness M.I. An investigation of the role of lecithin: cholesterolacyl transferase and triglyceride-rich lipoproteins in the metabolism of pre-beta high density lipoproteins. Atherosclerosis 89:1991;35-48.

30. Ruys T., Sturgess I., Shaikh M., Watts G.F., Nordestgaard B.G., Lewis B. Effects of exercise and fat ingestion on high density lipoprotein production by peripheral tissues. Lancet 334:1989;1119-1121.

31. Editorial. Cholesteryl ester transfer protein. Lancet; 338:1991 666-667

32. Bhatnagar D., Durrington P.N., Channon K.M., Prais H., Mackness M.I. Increased transfer of cholesteryl esters from high density lipoproteins to low density and very low density lipoproteins in patients with angiographic evidence of coronary artery disease. Atherosclerosis 98:1992;25-32.

33. Bhatnagar D., Durrington P.N., Mackness M.I., Arrol S., Winocour P.H., Prais H. Effects of treatment of hypertriglyceidaemia with gemfibrozil on serum lipoproteins and the transfer of cholesteryl ester from high density lipoproteins to low density lipoproteins. Athersclerosis 92:1992;49-57

34. Taylor T.G., Holdsworth G., Galton D.J. Clofibrate increases lipoprotein-lipase activity in adipose tissue of hypertriglyceridaemic patients. Lancet ii:1977;1106-1107.

GASTRIC LIPASE: CHARACTERISTICS AND BIOLOGICAL FUNCTION OF PREDUODENAL DIGESTIVE LIPASES

Margit Hamosh[1]   and Paul Hamosh[2]

[1] Department of Pediatrics

[2] Department of Physiology and Biophysics
Georgetown University Medical Center
Washington, DC 20007

INTRODUCTION

In this chapter we will discuss gastric lipase and, briefly, other preduodenal lipases with similar function. The emphasis will be on the function of these lipases in the process of fat digestion in health and disease.

STRUCTURE, CHARACTERISTICS, TISSUE ORIGIN AND ONTOGENY

The digestion of fat starts in the stomach. It is catalyzed by three enzymes of similar structure and characteristics, but of different origin. The enzymes originate either in the lingual serous glands (lingual lipase) [1], the glossoepiglotic area (pregastric esterase) [2], or the gastric mucosa (gastric lipase) [3]. These enzymes have been purified: rat lingual lipase [4-6], calf pregastric esterase [7,8], and gastric lipase from human [9,10], rabbit [11] and dog [12] gastric mucosa. Rat lingual lipase [13] and human gastric lipase [14] have been cloned and expressed in E. coli or yeast. They are glycoproteins of an approximate molecular weight of 52 kDa and consist of 377 and 379 amino acid residues with an unglycosilated molecular weight of 42.56 kD and 43.16 kD for lingual and gastric lipase, respectively [13,14]. The amino acid sequence of the two enzymes has an overall homology of 78% (Table 1). Deglycosylation does not reduce catalytic activity [11,15], however, the terminal tetrapeptide, in particular lysine-4, is essential for enzyme binding to lipid-water interfaces [16]. Rabbit and human gastric lipases have been crystallized recently [15].

The species differences in site of origin of the three preduodenal lipases are listed in Table 2. For each species there is a major site of enzyme synthesis: stomach (chief cells [20,26] and possibly mucous cells [12]), or oral (lingual serous glands [1,28,32] and oral-pharyngeal glands [7,8,30,31]).

The ontogeny of lingual and gastric lipases has been studied in the rat and human, respectively. Lingual lipase can be detected on day 20 of the 22 day gestation in the rat, whereas considerable lipase activity is present in gastric aspirates obtained at birth from premature infants of 25 wks gestation [34]. Data reported on two aborted fetuses indicate that traces of activity can be found in gastric tissue at 11 wks gestation [35]. In gastric aspirates of premature infants, lipase activity level is similar between 26-34 wks gestation, is 40% higher at 35-37 wks gestation and decreases slightly in full term infants [34]. Lipase activity levels in biopsy specimens of gastric mucosa are similar in infants, children and adults [36]. Similar findings to those previously reported [36] have been published recently [35,37].

Esterases, Lipases and Phospholipases, Edited by M.I. Mackness
and M. Clerc, Plenum Press, New York, 1994

Table 1. Amino acid composition of preduodenal lipases

Amino Acid Residue per molecule	Gastric Lipases[a]				Lingual Lipase[b] Rat	Pharyngeal Lipase[c] Bovine	
	Human		Dog	Rabbit			
Asx	49	48[d]	46	45.40	46[d]	47	51
Thr	20	19	23	19.0	20	18	22
Ser	31	26	24	28.0	23	22-23	29
Glx	38	29	26	29.9	29	30	34
Pro	23	22	24	29.2	21	25	25
Gly	47	23	24-25	24.5	2S	28	36
Ala	24	24	25	28.3	26	35	38
Val	22	24	23	26.9	24	22	25
Met	7	9	10-11	7.9	13	10	12
Ile	18	22	20	18.3	20	21	21
Leu	31	33	39	25.8	34	41	43
Tyr	19	21	20	16.1	19	14	20
Phe	18	25	21	17.6	25	23	24
Lys	12	22	18	17.3	22	17	24
His	8	10	13	6.8	10	14	16
Arg	9	10	11	10.1	7	9-10	13
Trp	n.d.[d]	9	7	5.9	9	n.d.	5
Cys	n.d.	3	3	9.1	4	4	8
Total number	379	77-379	359-360	377	380	380-382	
Total weight of residues		43,162	42,576	40,216	42,564	41,700	41,884
Ref:	[10]	[14]	[12]	[11]	[13]	[8]	[7]

a Origin: human-gastric juice, dog and rabbit- gastric mucosa
b Origin: rat lingual serous glands
c Origin:calf oro-pharyngeal tissue
d Amino acid composition derived from the cDNA sequence of the cloned genes.
n.d. = not determined

Table 2. Species difference in origin of preduodenal lipases

Species		Gastric Mucosa		Oro-pharyngeal Glands		References
Primates	Human	++++	chief cells	+	lingual serous glands	[17-21]
	Baboon	++++	?	±	?	[17, 18]
	Macaque	++++	?	±	?	[18]
Carnivores	Dog	++++	mucous cells	±		[12,18, 22-24]
Lagomorphs	Rabbit	++++	?	±		[3, 17, 18, 25, 26]
Caviidae	GuineaPig	++++	?	±		[17, 18]
Omnivores	Pig	++++	?	±		[18, 27]
Herbivores	Horse	++++	?	±		[18]
Rodents	Rat	±		++++	lingual serous glands	[1, 17, 18, 28]
	Mouse	±		++++	?	[17, 18]
Ruminants	Calf	±		++++	?	[7, 8, 29-31]
	Sheep	±		++++	?	[18, 31]
	Goat	±		++++	?	[31]

Comparison of activity levels among species is difficult, because of the use of different assay systems, different substrates (from tributyrin to triolein) and different methods for detection of the products of lipolysis.

Table 3. Characteristics of Preduodenal Lipases

	Lingual Lipase	Pregastric Esterase	Gastric Lipase
Origin	Lingual serous glands Serous demilunes of mucous glands	Glosso-epiglotic area Cell type unknown	Chief cells and mucous cells of gastric mucosa[a]
Ontogeny	end of fetal development	fetus? newborn	fetus 26wks gestation
pH Optimum	2.0- 6.5	4.5-6.5	3.0 -7.0
Stability			
pH	above pH 2.2	above pH 4.5	above pH 1.5
pepsin	stable	stable	stable
pancreatic proteolytic enzymes	labile	labile	labile
bile salts[b]	stimulation or inhibition	inhibition	stimulation or inhibition
Substrate Specificity			
Triglyceride, fatty acid chain length	MCFA≥LCFA	MCFA≥LCFA	MCFA=LCFA
fatty acid saturation	Sat < Unsat	Sat < Unsat	Sat < Unsat
fatty acid position	Sn 3 > Sn 1	Sn 3 > Sn 1	Sn 3 > Sn 1
cholesteryl ester	not hydrolyzed	not hydrolyzed	not hydrolyzed
phospholipid	not hydrolyzed	not hydrolyzed	not hydrolyzed
Reaction products	FFA,DG>MG	FFA,DG>MG	FFA,DG>MG
Product inhibition	LCFA-in vitro	LCFA-in vitro	LCFA-in vitro

a The cellular origin of gastric lipase differs among species
b Stimulation or inhibition of lipase activity is concentration and structure dependent.

Although gastric lipase is the major preduodenal digestive enzyme in the human [38,39], several studies suggest that lingual lipase might be of greater importance in the early postnatal period than at later times. Lipase activity is higher in secretions of the oesophageal pouch (2.7-130 mU/mL) than in the stomach (2.9-40.4 mU/mL) of infants with congenital oesophageal atresia [40]. Since there was no connection between oesophagus and trachea or oesophagus and stomach at the time of the study, contamination of either site by secretions from the other is excluded. Similar characteristics (pH optima and nature of reaction products) of the lipase in the oesophageal pouch of these infants [40] to those of the lipase in human serous glands [21] and differences from gastric lipase, support an oral origin of the enzyme. In addition, while non-nutritive sucking (i.e. providing a pacifier during gastric gavage feeding) did not lead to an increase in lipase activity in the stomach [41], nutritive suckling (i.e. bottle feeding as compared with gavage feeding) led to a significant increase of lipolytic activity in the stomach [42]. Since lingual lipase can be detected only toward the end of gestation [32,33], absence of lipase from the lingual-easophageal area of two fetuses aborted at 27 wks gestation [35] does not preclude presence of this enzyme and function in the newborn. The characteristics of preduodenal lipases are listed in Table 3. The reader is referred to recent reviews for comprehensive information on this topic [43-45].

GASTRIC LIPOLYSIS IN THE INFANT: EXTENT AND EFFECT OF DIETARY FAT COMPOSITION

A relatively minor role for gastric lipolysis in the overall digestion of fat has been suggested on the basis of limited in vitro lipolysis (10-20%) even under optimal conditions, as well as the erroneous belief that preduodenal lipases are specific for medium chain fatty acids (MCFA) [43,44]. This concept has to be revised in view of earlier and increasing recent evidence that extensive fat digestion occurs also when the diet contains little MCFA [42,46,47] or exclusively long chain triglycerides [48-50]. Table 4 examines this aspect in the newborn, including the premature human infant. Several conclusions can be drawn from the data presented in the table:

Table 4. Extent of fat digestion in the stomach of the newborn

Species	Age	Diet	FA < C14 (%)	TG hydrolysis (%)	References
Rat	14d	Mother's milk	48	25	Aw & Grigor (51)
Rat	9-10d	"	38	50	Fernando-W (52)
Rat	1-20d	"	6-26	50-60	Bitman et al (46)
Calf	7d	"	12	43	Siewert & Otterby (53)
Dog	14d	"	0.17	60	Iverson et al (48)
Seal	7-14d	"	0	25-56	Iverson et al (49)
Pig	1d	Menhaden Oil	0	45	Chiang et al (50)
Human*	Preterm	SMA formula	18.9+	30-45	Smith et al (42)
Human**	Preterm	SMA formula	18.9	29	Armand et al (47)
Human**	Preterm	Mother's milk	7.9++	40	Armand et al (47)

*30-33 wks gestation, 2-4 wks postnatal; **29.5 wks gestation, 4 wks postnatal
+ref 54; ++ ref 55

1. Milk MCFA are low immediately after birth and rise with duration of lactation, however, even when they amount to only 6% of total milk FA, gastric lipolysis is of the same magnitude as during later times of lactation [46]. 2. Except for the study of Aw and Grigor [51], the extent of gastric lipolysis exceeds the milk content of MCFA. 3. Milk fat containing exclusively LCFA [48] and especially rich in LC-PUFA [49] is extensively hydrolyzed in the stomach. The same applies to digestion of menhaden oil, composed exclusively of LC-PUFA. Thus, gastric lipase is able to release readily LC-PUFA [49], contrary to pancreatic lipase, which is unable to hydrolyze the ester bond of these fatty acids because of the proximity of the double bound to the carboxyl group [44,49]. 4. Gastric lipolysis is extensive even in species with a very high fat intake (seal milk contains 30-60% fat [49]). 5. The human studies show that the gastric phase of fat digestion is well developed even in premature infants in contrast to the immature intestinal lipolysis due to low pancreatic lipase and bile salt levels [43,44]. The high rate of lipolysis of milk rich in LC-PUFA is probably due to the location of theses fatty acids at the Sn-3 position of milk triglyceride [48,49]. Similar location of MCFA leads also to their preferential release in the stomach [46,52,56], an observation that started the erroneous belief that gastric lipase and pregastric esterase are specific exclusively for MCFA [57]. The preferential release of MCFA and of LC-PUFA (of both the n-3 and n-6 series) has special significance for the newborn: MCFA can be absorbed directly from the stomach of the newborn [25,51] including the preterm infant [56], and thus provide an energy source more readily accessible than carbohydrates. LC-PUFA (n-3) are essential for brain development and retinal function [58] while the n-6 LC-PUFA are the precursors of prostaglandins and lipid modulators [58]. This is especially important for the preterm infant born with very low LC-PUFA reserves (both n-3 and n-6) because fetal deposition and storage occur only during the last trimester of gestation.

The discrepancy between the relatively limited *in vitro* lipolysis by gastric lipase and the extensive *in vivo* gastric fat digestion is probably due to efficient fatty acid removal, and thus absence of potent product inhibition. Indeed, dietary proteins seem to be better fatty acid acceptors [59] than albumin, which is widely used in *in vitro* studies.

ROLE OF GASTRIC LIPOLYSIS IN FACILITATION OF SUBSEQUENT INTESTINAL FAT DIGESTION

The only food of the newborn is mother's milk, which is adapted to the specific needs of each species. Exclusive breastfeeding is also recommended for the newborn infant for the first 4-6 months of life [60]. The fat in milk is contained within milk fat globules, the triglycerides (> 98% of total milk fat) constituting the core and the more polar lipids (phospholipids and cholesterol) and proteins, the milk fat globule membrane. The latter is a strong barrier to pancreatic lipase [43,44] and to milk digestive lipase [43,44], an enzyme present in the milk of primates and carnivores [61,63]. In vitro studies have shown that initial lipolysis by lingual/ gastric lipase is essential for the subsequent action of pancreatic lipase [64,66]. The ability of these preduodenal lipases to access the triglyceride within the core of the fat globule without disrupting the membrane is probably associated with the hydrophobic nature of these enzymes and their inability to hydrolyze membrane phospholipids. Studies on the digestion of milk fat by its own digestive lipase show that this lipase is unable to penetrate the milk fat globule, thus partial digestion by gastric lipase is essential for its action. These in vitro studies [57,66,67] have been confirmed by in vivo observations [48]. It is interesting that hydrolysis of the fat in infant formulas by milk lipase is also dependent upon predigestion by gastric lipase [57,67,68]. The milk digestive lipase has similar characteristics to those of pancreatic lipase, pH optimum 8.0-8.5 and absolute dependence upon primary bile salts, but contrary to the former has no positional specificity, leading to complete hydrolysis of triglyceride. The bile salt requirement is abolished after predigestion by gastric lipase, probably because of generation of free fatty acids, which substitute for bile salts, and of partial glycerides, whose hydrolysis is bile salt independent.

FUNCTION OF GASTRIC LIPASE IN PANCREATIC INSUFFICIENCY (CYSTIC FIBROSIS, ALCOHOLISM) AND OTHER GI DISEASES

The well developed gastric lipase in the newborn, a period of physiologic pancreatic insufficiency led to the question whether this enzyme might remain fully active also in pathologic pancreatic insufficiency. Several studies have quantified gastric lipase and/ or gastric lipolysis in patients with cystic fibrosis or chronic alcoholism. To date these studies are inconclusive because of the conflicting data reported, higher [69,71], lower [72], or similar [73-75] lipase activity and/ or lipolysis to that of normal subjects. This variability could be due to the qualitative nature of some of the studies (absence of markers that permit quantification of secretion rates), differences in patient populations and differences in lipase assay techniques. Where such markers were used, it was evident that basal lipase activity and gastric pH were similar in patients and healthy subjects, the duodenal pH, however, was significantly lower during basal and postprandial conditions in the patient group [73,74]. Furthermore, this study shows that throughout a two hours postprandial period, gastric lipase amounted to over 90% of total lipase activity delivered to the ligament of Treiz in the patients as compared to 15% during basal and 8.5 and 4.7% during the first and second postprandial hours. This study also shows that in patients with pancreatic insufficiency, the total amount of unstimulated gastric lipase delivered to the intestine during the first and second postprandial hours [73] was 25-40% of the maximal amount of pancreatic lipase released in response to CCK-PZ stimulation in healthy adults [43,44]. A positive relationship was found between gastric lipase activity and fat absorption in the patient group [73]. Thus, while gastric lipase activity levels might be similar in pancreatic insufficiency to those of normal subjects , the enzyme probably has greater digestive potential because of its ability to continue to act in the intestine because of the lower pH and bile salt concentrations associated with pancreatic insufficiency [43,44]. Gastric lipase activity may or may not change with GI diseases known to affect the stomach. Thus, Helicobacter pylori infection does not affect lipase or pepsin activity, whereas studies on the lipase activity in cases of gastritis or duodenal ulcer have resulted in conflicting reports. The extent of loss of lipase activity probably depends on the degree of damage to the gastric mucosa. In acute gastritis, lipase activity is much more sensitive than pepsin to mucosal damage [77].

Table 5. Gastric Lipase activity or Gastric Lipolysis in Disease

Condition	Ratio patients: control
Cystic fibrosis	same [73-74], higher [69-71], lower [72]
Alcoholism with PI*	same [72,74], higher or same [70]
Helicobacter pylori	same [76]
Gastritis*	absent [77]*, absent or same [35]
Congenital microgastria	absent [78]
Duodenal ulcer	same [72], higher [70]

*Lipase absent during repeated attacks of gastritis in the same patient, whereas pepsin activity unchanged.

We have previously suggested [43,44] that gastric lipase and milk digestive lipase (both recently cloned), because of their limited requirements for cofactors and remarkable stability, might be much better suited for enzyme replacement therapy than the currently used pancreatic lipase preparations.

Table 6. Function of Preduodenal Lipases

Normal subjects
    Long-chain FFA released in the stomach stimulate CCK secretion, which in turn stimulates secretion of pancreatic and gastric lipases
    The FFA also facilitate the action of pancreatic lipase (lipase-colipase binding to bile salts and substrate triglyceride)
Specific role in infants
    Gastric lipolysis of milk is extensive (30-60%)
    Penetration into milk fat globules and initiation of lipolysis is a prerequisite for hydrolysis by milk digestive lipase and pancreatic lipase
    Gastric lipolysis is a significant compensatory mechanism for low neonatal pancreatic and hepatic function
    Medium-chain fatty acids released in the stomach are absorbed through the gastric mucosa
Pathologic pancreatic insufficiency
    Gastric lipolysis is maintained (at normal, lower or higher levels, depending upon the publication)
    Lipolysis continues in the duodenum because of favorable conditions for lingual and gastric lipase activity (low pH, lower pancreatic proteolytic activity and bile salt levels)
    Compensatory lipolytic activity by preduodenal lipases may account for digestion and absorption of 50% to 70% of dietary fat

Adapted from Hamosh M, Nutrition, 6 (1990) 421-430.

ROLE OF GASTRIC LIPASE IN OVERALL FAT DIGESTION IN THE HEALTHY ADULT

Initial digestion and release of free fatty acids in the stomach are essential for secretion and function of pancreatic lipase and adequate fat digestion [43,44]. Indeed, diversion of lingual lipase in rats led to almost complete inhibition of gastric lipolysis and to increased fat and bile acid secretion. Furthermore, it is clear from studies in several species including man that 50-70% of dietary fat is absorbed in the absence of pancreatic lipase, indicating that this is probably accomplished in large measure by gastric lipase. It is important to quantify the extent of gastric lipolysis and its role in the overall digestion of fat in health and in disease by carefully planned studies, rather than to merely infer to the in vivo situation from in vitro observations.

REFERENCES

1. M. Hamosh, and R.O. Scow, Lingual lipase and its role in the digestion of dietary lipid, *J Clin Invest.* 52 (1973) 88-95
2. G.H. Wise, P.G. Miller, and G.W. Anderson, Changes observed in milk "sham-fed" to dairy calves, *J.Dairy Sci.* 23 (1940) 997-1001
3. C.S. Fink, M. Hamosh, P. Hamosh, S.J. DeNigris, and D.K. Kasbekar, Lipase secretion from dispersed rabbit gastric glands, *Am J Physiol* . 248 (1985) G68-G72
4. M. Hamosh, D. Ganot, and P. Hamosh, Rat lingual lipase. *J Biol Chem.* 254 (1979) 12121-12125
5. R.B. Field, and R.O. Scow. Purification and characterization of rat lingual lipase, *J Biol Chem.* 258 (1983) 14563-14569
6. I.M. Roberts, R.J. Montgomery, and M.C. Carey, Lingual lipase: partial purification, hydrolytic properties and comparison with pancreatic lipase, *Am J Physiol.* 247 (1984) G385-G393
7. B.J. Sweet, L.C. Matthews, and T. Richardson., Purificiation and characterization of pregastric esterase of calf., *Arch Biochem Biophys.* 234 (1984) 144-150
8. S. Bernback, O. Hernell, and L. Blackberg, Purification and molecular characterization of bovine pregastric esterase, *Eur J Biochem* . 148 (1985) 233-238
9. C. Tiruppathi, and K.A. Balasubramanian, Purification and properties of an acid lipase from human gastric juice, *Biochim Biophys Acta.* 712 (1982) 692-697
10. C. Tiruppathi, and K.A. Balasubramanian, Single-step purification and amino acid and lipid composition of purified acid lipase from human gastric juice, *Indian J Biochem Biophys* .22 (1985) 111-114
11. H. Moreau, Y. Gargouri, D. Lecat, J.L. Junien, and R. Verger, Purification, characterization and kinetic properties of the rabbit gastric lipase, *Biochim Biophys Acta* 960 (1988) 286-293
12. F. Carriere, H. Moreau, and V. Raphel. Purification and biochemical characterization of dog gastric lipase, *Eur J Biochem* . 202 (1991) 75-83
13. A.J.P. Docherty , M.W. Bodmer, S. Angal et al., Molecular cloning and nucleotide sequence of rat lingual lipase cDNA, *Nucleic Acid Res.* 13 (1985) 1891-1903
14. M.W. Bodmer, S. Angal, G.T. Yarraton, et al.., Molecular cloning of a human gastric lipase and expression of the enzyme in yeast, *Biochim Biophys Acta* 909 (1987) 237-244

15. H. Moreau, C.Abergel, F. Carriere et al., Isoform purification of gastric lipases, *J Mol Biol* . 225 (1992) 147-153
16. S. Bernback, and L. Blackberg, Human gastric lipase. The N-terminal tetrapeptide is essential for lipid binding and lipase activity, *Eur J Biochem*. 182 (1989) 495-499
17. S.J. DeNigris M. Hamosh, D.K. Kasbekar, T.C. Lee, and P. Hamosh, Lingual and gastric lipases: species differences in the origin of prepancreatic digestive lipases and in localization of gastric lipase, *Biochim Biophys Acta* 959 (1988) 38-45
18. H. Moreau, Y. Gargouri, D. Lecat, J.L Junien, and R. Verger, Screening of preduodenal lipases in several mammals, *Biochim Biophys Acta* 959 (1988) 247-252
19. S.J. DeNigris, M. Hamosh, D.K. Kasbekar, C.S. Fink, T.C. Lee, and P. Hamosh., Human gastric lipase: Secretion from dispersed gastric glands, *Biochim Biophys Acta* 836 (1985) 67-72
21. M. Hamosh, and W.A.Burns, Lipolytic activity of human lingual glands (Ebner), *Lab Invest*. 37 (1977) 603-608
22. M. Hull, and R.W. Keaton, The existence of a gastric lipase, *J Biol Chem*. 32 (1917) 127-140
23. G.R. Douglas, A.J. Reinauer, W.C. Brooks, and J.H. Pratt, The effect on digestion and absorption of excluding the pancreatic juice from the intestine, *Gastroenterology* 23 (1953) 452-459
24. C.L. Kirk, S.J. Iverson, and M. Hamosh, Lipase and pepsin activities in the stomach mucosa of the suckling dog, *Biol Neonate* 59 (1991) 78-85
25. J.P. Perret, Lipolyse gastrique chez le lapereau. Origine et importance physiologique de la lipase, J Physiol. 78 (1982) 221-230
26. H. Moreau, A. Bernadac, N. Tretout, Y. Gargouri, F. Ferrato, and R. Verger, Immunocytochemical localization of rabbit gastric lipase and pepsinogen, *Eur J Cell Biol*. 51 (1990) 165-169
27. F. Volhard, Uber das Fettspaltende Ferment des Magens, *Z Klin Med*. 43 (1901) 397-401
28. I.M. Roberts, and R. Jaffe, Lingual lipase: immunocytochemical localization in the rat von Ebner gland, *Gasteroenterology* 90 (1986) 1170-1175
29. J. Toothill, S.W. Thompson, and J.D. Edwards-Webbs, Studies on lipid digestion in the preruminant calf. The source of lipolytic activity in the abomasum, *Br J Nutr*. 36 (1976) 439-447
30. H.A. Ramsey, G.H. Wise, and S.B. Tove, Esterolytic activity of certain alimentary and related tissues from cattle in different age group, *J Dairy Sci* . 39 (1956) 1312-1321
31. J.W.F. Grosskopf, Studies on salivary lipase in young ruminants, *Onderstepoort J Vet Res*. 32 (1965) 153-180
32. M. Hamosh, and A.R. Hand, Development of secretory activity in serous cells of the rat tongue, *Dev Biol*. 65 (1978) 100-113
33. I.M. Roberts, L.E. Nochomovitz, R. Jaffe, S.I. Hanel, M. Rojas, and R.A. Agostini Jr, Immunochemical localization of lingual lipase in serous cells of the developing rat tongue, *Lipids* 22 (1987) 764-766
34. M. Hamosh, J.W. Scanlon, D. Ganot, M. Likel, K. B. Scanlon, and P. Hamosh, Fat digestion in the newborn: characterization of lipase in gastric aspirates of premature and term infants, *J Clin Invest*. 67 (1981) 838-846
35. J. Sarles, H. Moreau, and R. Verger, Human gastric lipase: ontogeny and variations in children, Acta Paediatr. 81 (1992) 511-513
36. J.S. DiPalma, C. Kirk, M. Hamosh, A.R. Colon, S.B. Benjamin, and P. Hamosh, Lipase and pepsin activity in the gastric mucosa of infants, children and adults, *Gastoenterology* 101 (1991) 116-121
37. P.C. Lee, R. Borysewicz, M. Struve, K. Raab, and S.L. Werlin, Development of lipolytic activity in gastric aspirates from premature infants, *J Pediatr Gastroenterol Nutr*. In Press (1993).
38. C.K. Abrams, M. Hamosh, and T.C. Lee, et al, Gastric lipase: location in the human stomach, *Gastroenterology* 95 (1988) 1460-1464
39. H. Moreau, R. Laugier, Y. Gargouri, F. Ferrato, and R. Verger, Human preduoudenal lipase is entirely of gastric fundic origin, *Gasteroenterology*. 95 (1988) 1221-1226
40. C. Salzman-Mann, M. Hamosh, K.N. Sivasubramanian, A. Bar-Maor, O. Zinder, G.B. Avery, J.B. Watkins, and P. Hamosh, Congenital esophageal atresia: lipase activity is present in the esophageal pouch and in the stomach, *Dig Dis Sci* . 27 (1982) 124-128
41. M. Hamosh, Y.F. Smith, J. Bernbaum, N.R. Mehta, C.S. Fink, C.M. York, et al.., Fat digestion in premature infants: effect of mode of feeding on lipase activity in the stomach, . "In: Physiologic Foundations of Perinatal Care", L. Stern, W. Oh, and B Friis-Hansen, eds., New York, Elsevier, 75 (1989) 75-84
42. L.J. Smith, S. Kaminsky, and S.W. D'Souza, Neonatal fat digestion and lingual lipase, *Acta Paediatr Scand*. 75 (1986) 913-918
43. M. Hamosh., "Lingual and Gastric Lipases: Their Role in Fat Digestion", CRC Press, Boca Raton, (1990).
44. M. Hamosh, Gastric and lingual lipases in "Physiology of the gastrointestinal tract" 3rd edition, L.R. Johnson, D.H. Alpers, J. Christensen, E.D. Jacobson, and J.H. Walsh, eds. Raven Press, New York. In Press (1993).
45. Y. Gargouri, H. Moreau, and R. Verger, Gastric lipases: biochemical and physiological studies. *Biochim Biophys Acta* 1006 (1989) 255-271.

46. J. Bitman, D.L. Wood, T.H. Liao, C.S. Fink, P. Hamosh, and M. Hamosh, Gastric lipolysis of milk lipids in suckling rats, *Biochim Biophys Acta* 834 (1985) 58-64
47. M. Armand, M. Hamosh, N.R. Mehta, P.A. Angelus, and J.R. Philpott, Gastric lipolysis in premature infants, effects of diet: human milk formula. FASEB J 7 (1993) A201
48. S.J. Iverson, C.L. Kirk, M. Hamosh, and J. Newsome, Milk lipid digestion in the neonatal dog: The combined actions of gastric and bile salt stimulated lipases, *Biochim Biophys Acta* 1083 (1991) 109-119
49. S.J. Iverson, J. Sampugna, and O.T. Oftedal, Positional specificity of gastric hydrolysis of long-chain n-3 polyunsaturated fatty acids of seal milk triglycerides, *Lipids* 27 (1992) 870-878
50. S-H. Chiang, J.E. Pettigrew, S.D. Clark and S.G. Cornelius. Digestion and absorption of fish oil by neonatal piglets. *J Nutr.* 119 (1989) 1741-1743
51. T.Y. Aw, and M.R. Grigor, Digestion and absorption of triacylglycerol in 14 day old sucking rats, *J Nutr* . 110 (1980) 2133-2140
52. A.J.P. Fernando-Warnakulasurija, J.E. Staggers, S.C. Frost, and M.A. Wells, Studies on fat digestion absorption and transport in the suckling rat. I. Fatty acid composition and concentration of major lipid components, *J Lipid Res.* 22 (1981) 668-674
53. K.L. Siewert, and D.E. Otterby, Effects of *in vivo* and *in vitro* acid environments on the activity of pregastric esterase, *J Dairy Sci* . 53 (1970) 571-575
54. M. Hamosh , N.R. Mehta , C.S. Fink, J. Coleman, and P. Hamosh, Fat absorption in premature infants: Medium chain triglycerides and long chained triglycerides are absorbed from formula at similar rates. *J Pediatr Gastroenterol Nutr.* 13 (1991) 143-149
55. J. Bitman, D.L. Wood, M. Hamosh, P. Hamosh, and N.R. Mehta, Comparison of the lipid composition of breast milk from mothers of term and preterm infants, *Am J Clin Nutr.* 38 (1983) 300-312
56. M. Hamosh, J. Bitman, T.H. Liao, N.R. Mehta, R.J. Buczek, D.L. Wood et al., Gastric lipolysis and fat absorption in preterm infants: effect of MCT or LCT containing formulas, *Pediatrics* 83 (1989) 86-92
57. M. Hamosh, S.J. Iverson, C.L. Kirk, and P. Hamosh, Milk lipids and neonatal fat digestion: relationship between fatty acid composition, endogenous and exogenous digestive enzymes and digestion of milk fat. In "Fatty acids and lipids from cell biology to human disease", C. Galli, ed. Karger, Basel, In Press (1993).
58. S.M. Innis, ed. Lipids in infant nutrition. Symposium. *J Pediatr.* 120 No 4 part 2 (1992)
59. M.D. Perez, L. Sanchez, P. Aranda, J.M. Ena, R. Oria, and M. Calvo, Effect of β-lactoglobulin on the activity of pregastric lipase. A possible role for this protein in ruminant milk, *Biochim. Biophys. Acta* 1123 (1992) 151-155
60. Institute of Medicine "Nutrition during lactation", National Academy Press, Washington (1991).
61. E. Freudenberg, "Die Frauenmilch-Lipase," S. Karger, Basel, (1953).
62. L.M. Freed, C.M. York, M. Hamosh, J.A. Sturman, and P. Hamosh, Bile salt-stimulated lipase in non-primate milk: longitudinal variation and lipase characteristics in cat and dog milk, *Biochim Biophys Acta* 878 (1986) 209-215
63. L.A. Ellis, and M. Hamosh, Bile salt stimulated lipase: comparative studies in ferret milk and lactating mammary gland, *Lipids* 27 (1992) 917-922
64. M. Cohen, G.R.H. Morgan, and A.F. Hofmann, Lipolytic activity of human gastric and duodenal juice against medium and long-chain triglycerides, *Gasteroenterology* 60 (1971) 1-15
65. T.M. Plucinski, M. Hamosh, and P., Hamosh. Fat digestion in the rat: role of lingual lipase, *Am J Physiol* 237 (1979) E541-E547
66. S. Bernback, L. Blackberg, and O. Hernell, The complete digestion of human milk triacylglycerol *in vitro* requires gastric lipase, pancreatic colipase dependent lipase and bile salt stimulated lipase, *J Clin Invest.* 85 (1990) 1221-1226
67. P. Hamosh, and M. Hamosh, Differences in composition of preterm, term and weaning milk. In "New Aspects of Nutrition in Infancy and Prematurity" Xanthou M (ed.) Elsevier, Amsterdam, (1987) 129-141
68. Kirk CL, and M. Hamosh. Initial lipolysis by gastric lipase is essential for the hydrolysis of milk or formula fat by milk lipase. *FASEB J* . 5 (1991) A1288
69. M. Roulet, A.M. Weber, Y. Paradis, C.C. Roy, L. Chartrand, R. Lasalle, et al., Gastric emptying and lingual lipase activity in cystic fibrosis., *Pediatr Res.* 14 (1980) 1360-1362
70. J. Moreau, M. Bouisson, D. Balas, A. Kavand, S. Stupnik, L. Buscail, N. Vaysse, and A. Ribert, Gastric lipase in alcoholic pancreatitis, *Gastroenterology* 99 (1990) 175-180
71. K. Balasubramanian, P.L. Zentler-Munro, J.C. Batten, and T.C. Northfield, Increased intragastric acid-resistant lipase activity and lipolysis in pancreatic steatorrhea due to cystic fibrosis, *Pancreas* 7 (1992) 305-310
72. H. Moreau, J.F. Sauniere, Y. Gargouri, G. Pieroni, and R. Verger, Human gastric lipase: variations induced by gastrointestinal hormones and by pathology, *Scand J Gastroenterol..* 23 (1988) 1044-1048
73. C.K. Abrams, M. Hamosh, V.S. Hubbard, S.K. Dutta, and P. Hamosh, Lingual lipase in cystic fibrosis. Quantitation of enzyme activity in the upper small intestine of patients with exocrine pancreatic insufficiency, *J Clin Invest.* 73 (1984) 374-382

74. C.K. Abrams, M. Hamosh, S.K. Dutta, V.S. Hubbard, and P. Hamosh, The role of non-pancreatic lipolytic activity in exocrine pancreatic insufficiency, *Gastroenterology* 92 (1987) 125-129

75. B. Fredrikzon, and L. Blackberg, Lingual lipase: an important enzyme in the digestion of dietary lipids in cystic fibrosis? *Pediatr Res.* 14 (1980) 1387-1390

76. T. Lam, S.B. Benjamin, J. Gallagher, M. Bertagnolli, C. Kirk, L. Ellis, M. Hamosh, and P. Hamosh, Gastric digestive enzyme activity is not affected by Helicobacter pylori infection. *Gastroenterology* 98:A74 (1990).

77. J.S. DiPalma, C. Kirk, M. Hamosh, A.R. Colon, S.B. Benjamin, and P. Hamosh, Lipase and pepsin activity in the gastric mucosa of children and the role of lipase as a marker for gastrointestinal inflamation. *Clin Res.* 37 (1989) 367A

78. N. Scribanu, N.R. Mehta, J.S. DiPalma, and M. Hamosh, Congenital microgastria and hypoplastic upper limb association: histologic confirmation of gastric tissue, *J Pediatr. Gastroenterol Nutr.* Submitted.

PANCREATIC CHOLESTERYL ESTERASE IN HEALTH AND DISEASE

Eric Mas and Dominique Lombardo

INSERM U260
Faculté de Médecine
27, Bd Jean Moulin
13385 Marseille Cedex 5, France

INTRODUCTION

Pancreatic cholesterol esterase, also called carboxyl ester lipase or bile salt-dependent lipase (BSDL, E. C. 3. 1. 1.13), catalyzes the hydrolysis of ester substrates as well as some phospholipids and lysophospholipids[1]. In addition, the bile salt-dependent lipase is the unique enzyme in the pancreatic juice capable of hydrolyzing fat-soluble-vitamins and cholesteryl esters[2]. This latter function strongly suggests that BSDL is important for catalyzing the lymphatic absorption of dietary cholesterol and fat-soluble vitamins. BSDL from the rat, dog, pig, bovine and human pancreas has been purified to homogeneity. The molecular mass differs significantly among species[3]: human BSDL 100 kDa ; dog 87-93 kDa ; pig 80-90 kDa ; bovine BSDL occurs predominantly as a 72 kDa protein ; and that of the rat is 67 kDa. Recently, the full-length cDNA of pancreatic BSDL from different species has been cloned and sequenced[4-6]. Jacobson et al.[7] have reported that rat BSDL is a glycosylated protein which contains a potential N-linked glycosylation site at Asn at position 187, [4-6], this site is conserved in the bovine and human protein. The sugar composition of the human pancreatic BSDL agrees with the presence one N-linked oligosaccharide of complex or hybrid type and of O-linked oligosaccharides[8]. The transfer *en bloc* of the precursor dolichol-pyrophosphate oligomannoside is essential for the setting up of the active conformation and secretion of the enzyme[9]. The trimming and the processing of the high-mannose structure does not affect either the enzyme activity or its secretion[9].

O-Glycosylation sites could be clustered on tandem repeat "mucin-like" sequences[10] located at the C-terminal part of the protein. The great variability of the enzyme size is due in part to the glycosylation, most probably the O-linked one[3] and to the variable number of repeated "mucin-like" sequences [4-6].

Comparisons between the human pancreatic BSDL and the feto-acinar pancreatic protein (FAP) lead to the conclusion that the last protein is an oncofetal-type subpopulation

Esterases, Lipases and Phospholipases, Edited by M.I. Mackness
and M. Clerc, Plenum Press, New York, 1994

of the normally secreted BSDL[11, 12]. Both proteins have the same molecular weight, and close amino-acid composition, their N-terminal sequences are identical at least up to the 23rd residue. Nevertheless, their respective sugar composition differs ; that of BSDL being compatible with a complex-type or hybrid-type N-linked oligosaccharide while that of FAP is more probably related to high-mannose type. Difference in the amount of galactose and galactosamine residues can also be noted[11] and carbohydrate composition indicates a decrease in O-glycosylation. Whether the J28 epitope, which is carbohydrate dependent and characterizes the FAP, belongs to the N-or to the O-linked glycosylation of the protein is under investigation. Since the glycovariant of BSDL, FAP (i.e. the J28 epitope) is expressed during the ontogenic process of the human pancreas and in the case of oncogenic pathology of the pancreas the glycosylation of BSDL may vary with pancreatic pathologies. Therefore BSDL could be a valuable tool for the diagnosis of these pathologies.

RESULTS AND DISCUSSION

Basic structure of N-linked oligosaccharides of BSDL

The first part of the work was devoted to the determination of the structure of the N-linked oligosaccharide chains of BSDL isolated from a normal donor (i.e. without pancreatic pathology)[13]. After hydrazinolysis and reduction with NaB[^{3}H]$_4$ neutral (N) sugar chains were separated from acidic (AN) ones by paper electrophoresis. N-linked oligosaccharide fraction N and AN were fractionated using serial column chromatography on Aleuria aurantia lectin (AAL)-Sepharose and concanavalin A (Con A)-Sepharose.

On AAL-Sepharose column chromatography the N fraction (72% of N-linked sugar) was separated into two fractions : an unbound [N(AAL$^-$), 54%] and a bound one eluted with 1 mM fucose [N(AAL$^+$), 18%]. The AN fraction (28% of N-linked sugar) also was divided into an AN(AAL$^-$) fraction (19%) and an AN(AAL$^+$) fraction (9%). All but N(AAL$^+$) were separated into three fractions on Con A-Sepharose ; an unbound fraction (Con A$^-$), a weakly bound fraction eluted with 5 mM methyl α-glycoside (Con A$^+$) and a strongly bound fraction eluted with 100 mM methyl α-mannoside (Con A^{++}). For the N(AAL$^-$) fraction they represented 10%, 31% and 13% respectively. With the AN(AAL$^-$) and AN(AAL$^+$) fraction, only Con A$^-$ and Con A$^+$ fractions were obtained. The Con A$^-$ fraction represented 7% and 4% respectively for AN(AAL$^-$) and AN (AAL$^+$) fraction and 12% and 5% for AN(AAL$^-$, Con A$^+$) and AN(AAL$^+$, Con A$^+$) fractions.

The structures of oligosaccharides in each fraction were determined by sequential exoglycosidase digestions. The reaction mixtures at each step were analyzed by Bio-Gel P-4 column chromatography. All the determined structures are presented in table 1. Most of them are of the N-acetyllactosamine-type, the basic structure of which is that of oligosaccharide OL.I. Oligosaccharides OL.II, OL.VII and OL.VIII are derived from the basic OL.I structure by fucosylation or sialylation on the terminal Gal residue. Poly-N-acetyllactosamine structures are poorly represented (OL.III), and oligosaccharides OL.IV to OL.VI are intermediate from oligomannoside unprocessed structures and fully trimmed N-acetyllactosamine structures. Among all the structures only one (O.VI) is of a hybrid-type. Structures depicted in table 1 are compatible with the general trimming of glycoproteins from the endoplasmic reticulum to the trans-golgi vesicles. The carbohydrate-dependent J28 epitope which characterizes FAP is present on a subpopulation of BSDL [11,12]. These proteins differ in their sugar composition both for sugars involved in N- and in O-linked oligosaccharides. In particular the N-linked structure of FAP appears to be unprocessed[11].

Table 1. Proposed structure of N-linked oligosaccharides of the human pancreatic bile salt-dependent lipase

Fraction (N) AAL- Con A+

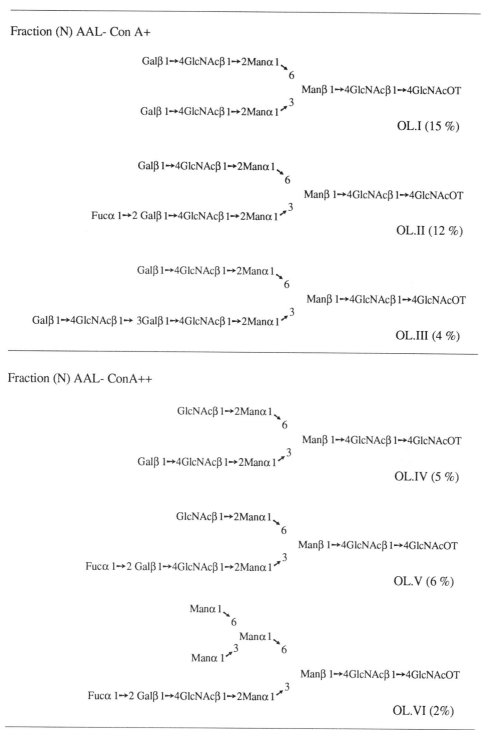

Galβ 1→4GlcNAcβ 1→2Manα 1
6
Manβ 1→4GlcNAcβ 1→4GlcNAcOT
3
Galβ 1→4GlcNAcβ 1→2Manα 1

OL.I (15 %)

Galβ 1→4GlcNAcβ 1→2Manα 1
6
Manβ 1→4GlcNAcβ 1→4GlcNAcOT
3
Fucα 1→2 Galβ 1→4GlcNAcβ 1→2Manα 1

OL.II (12 %)

Galβ 1→4GlcNAcβ 1→2Manα 1
6
Manβ 1→4GlcNAcβ 1→4GlcNAcOT
3
Galβ 1→4GlcNAcβ 1→ 3Galβ 1→4GlcNAcβ 1→2Manα 1

OL.III (4 %)

Fraction (N) AAL- ConA++

GlcNAcβ 1→2Manα 1
6
Manβ 1→4GlcNAcβ 1→4GlcNAcOT
3
Galβ 1→4GlcNAcβ 1→2Manα 1

OL.IV (5 %)

GlcNAcβ 1→2Manα 1
6
Manβ 1→4GlcNAcβ 1→4GlcNAcOT
3
Fucα 1→2 Galβ 1→4GlcNAcβ 1→2Manα 1

OL.V (6 %)

Manα 1
6
Manα 1
3   6
Manα 1
Manβ 1→4GlcNAcβ 1→4GlcNAcOT
3
Fucα 1→2 Galβ 1→4GlcNAcβ 1→2Manα 1

OL.VI (2%)

Fraction (AN) AAL+ ConA+

$$Gal\beta 1\rightarrow 4GlcNAc\beta 1\rightarrow 2Man\alpha 1\searrow_6 \qquad Fuc\alpha 1\searrow_6$$

$$\pm NeuAc \rightarrow \qquad\qquad\qquad\qquad Man\beta 1\rightarrow 4GlcNAc\beta 1\rightarrow 4GlcNAcOT$$

$$Gal\beta 1\rightarrow 4GlcNAc\beta 1\rightarrow 2Man\alpha 1\nearrow^3$$

OL.VII (5 %)

Fraction (AN) AAL- ConA+

$$Gal\beta 1\rightarrow 4GlcNAc\beta 1\rightarrow 2Man\alpha 1\searrow_6$$

$$\pm NeuAc \rightarrow \qquad\qquad\qquad Man\beta 1\rightarrow 4GlcNAc\beta 1\rightarrow 4GlcNAcOT$$

$$Gal\beta 1\rightarrow 4GlcNAc\beta 1\rightarrow 2Man\alpha 1\nearrow^3$$

OL.VIII (12 %)

Variation of the glycosylation of BSDL

In the light of the alteration of protein glycosylation in inflammatory processes and because no secreting human cell model is available, we performed a comparative analysis of the BSDL secreted in human pancreatic juice of normal donor with those of patients affected with chronic pancreatitis, a nonmalignant disease of the pancreas[14]. This work was done by means of lectins of variable specificity associated to carbohydrate analysis of pure BSDL obtained either from normal donor or from patients.

The well-defined electrophoretic migration of BSDL on SDS-PAGE ($Mr = 100$ KDa) allows us to study lectin affinity by direct blotting of pancreatic juice protein after electrophoretic migration and transfer to nitrocellulose membranes. When pancreatic glycoproteins (Figure 1) were characterized by affinity blotting with Concanavalin-A (Canavalia ensiformis lectin, Con-A), two main areas were positive in both juices; one of them most probably corresponds to protease E and kallikrein with Mr of 30-35 KDa, a second intense staining at Mr approx. 48-50 KDa corresponds to isoforms of colipase-dependent lipase. These proteins were already characterized as glycoproteins and two of them as Con-A positive. Only a faint reactivity with this last lectin can be detected where BSDL migrates. But in pathological juice (juice S) a staining can be seen at Mr compatible with that of BSDL. The lentil lectin (Lens culinaris, LcH) reacted with colipase-dependent lipase in both pancreatic juices. No staining can be seen at a migration distance corresponding to BSDL but in juice S although at lower Mr (approx. 90-95 KDa). Ulex europaeus (UEA-I) and Phaseolus vulgaris (E-PHA) lectins as expected reacted with colipase-dependent lipase, E-PHA also strongly bound to proteins associated with Mr around 30 KDa. These lectins recognized BSDL and revealed little heterogeneity between juices. The wheat germ agglutinin (Triticum vulgaris lectin, WGA) bound quite well to colipase-dependent lipase in either juice without a marked difference, nevertheless the reactivity of BSDL differed from juices ; the staining was similar in normal juice D and X, but it was intense for juice G and very faint (or absent) for juice S. After neuraminidase treatment, the staining of colipase-dependent lipase was still the same but that of BSDL decreased (juices X and G) or disappeared (juice D) (Figure 2). Reactivity of WGA to asialoproteins has already been in some cases described as weaker. The lost reactivity of

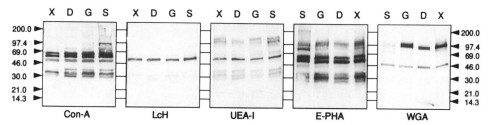

Figure 1. Lectin affinity blotting of human pancreatic juices. Affinity blotting of normal pancreatic juices (lane X and D, 5 μg) and of pathological juices (lane G and S, 5 μg) were done after SDS-PAGE and electrotransfer of proteins on nitrocellulose membranes. Membranes were then incubated with Biotin-conjugated lectins (10 μg/ml): Con-A, Concanavalin-A lectin; LcH, Lentil lectin; UEA-I, Ulex europaeus lectin; E-PHA, Phaseolus vulgaris lectin and WGA, wheat germ agglutinin.

juice D indicated that terminal sialyl residues may interfere with WGA reactivity and specificity for N-glycan core structure [15]. Data suggested that a systemic abnormality in pancreatic protein glycosylation was unlikely since only BSDL displays glycosylation heterogeneity ; the affinity patterns to lectins (at least those used here) of other pancreatic glycoproteins was not modified by the pancreas status (Figure 1).

Also shown in Figure 2, is the reactivity of peanut lectin (Arachis hypogaea lectin, PNA) which recognized specifically the O-linked glycan Gal β1-3→GalNAc. After neuraminidase treatment of nitrocellulose replicas (10 mU/ml of neuraminidase in 50 mM sodium acetate pH 5.5 buffer containing 2% BSA and 10 mM $CaCl_2$, 6h at 37°C) only BSDL was labelled with PNA, while other proteins were not significantly stained. Therefore it seems that BSDL is the only O-linked glycoprotein of the human pancreatic secretion. It is worth noting that the intensity of staining differs in each juice. Pure BSDL did not react with PNA without a prior treatment with neuraminidase and an extra incubation with fucosidase (either from bovine kidney or from bovine epididymis) helped the removal of O-glycans by O-glycosidase (not shown). Interestingly treatment of BSDL with fucosidase led to a stronger reactivity of PNA.

Lectin affinity blotting showed that the carbohydrate structure of BSDL might vary in the four samples of pancreatic juice examined. Further studies were attempted to determine the structural differences. Nevertheless, the amount of pancreatic juice, particularly that

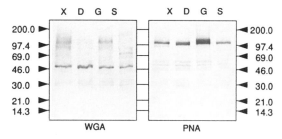

Figure 2. Lectin affinity blotting of human pancreatic juices after prior treatment of membranes with neuraminidase. Affinity blotting was done as described in Fig. 2 using WGA lectin and peanut agglutinin (PNA). Prior to lectin binding, proteins bound to membranes were desialylated by an overnight treatment with 10 mU/ml neuraminidase.

coming from patients with pancreatic inflammation was not sufficient to purify a sufficient quantity of BSDL to permit structural determinations. Therefore carbohydrate compositions of pure BSDL were determined as indirect evidence of the structural modifications of glycans linked to the protein. The BSDL was purified from pancreatic juice following the two-step procedure. Based on total BSDL activity recovered and protein amount obtained by the end of enzyme preparation, it can be determined that in normal juice (juice X) BSDL represents approx. 4 to 5 % of total protein, thus confirming previous data, but in the pathological juices the enzyme amount varied a little and represented some 3-7 % of protein in juice S and 3-5% in juice G.

As shown in table 2, the content of total carbohydrate of variant S and G of BSDL (i.e. purified from pathological juices S and G respectively) was not different from that of the variant X of the enzyme purified from the pancreatic juice of a normal donor. The amount of neutral sugar as determined by phenol-sulfuric acid was not significantly lowered in the variant S and G of BSDL [(5.5 ± 0.8) and (5.3 ± 0.5) % respectively] as compared to the variant X [(6.1 ± 0.6)%]. The sugar composition presented in this table is consistent with the presence of O- and N-linked oligosaccharides on BSDL [8]. From the carbohydrate composition (deduced from molar ratios and % of neutral sugar) shown in table 2, two main facts appear when comparing the BSDL variant from juice X to the variant from pathological juices S or G. First of all, the ratio GalNAc/Man is between 2 and 3 in the BSDL variant from the normal donor while it decreases to < 2 in both variants S and G from pathological juices. Secondly, the amount of mannose increases in the latter variants while that of N-acetylglucosamine decreases (variant S) or is identical (variant G). The amount of sialyl residues is decreased in variant G but remained unchanged in the variant S, that of fucose being identical whatever the pathological state of the pancreatic source. The identical amount of Gal, GalNAc and fucose residues found in all variants suggest that these sugars predominantly belong to the O-linked glycans. Thus the carbohydrate composition of the variant obtained from juice G appears close to that of the normal variant X, while that of variant from the juice S differs when considering one sugar involved in the N-glycosylation.

Table 2. Carbohydrate composition of variants of bile salt-dependent lipase

	Bile salt-dependent lipase		
	variant X	variant G	variant S
	Molar ratio	Molar ratio	Molar ratio
Fucose	6.2 ± 1.1	5.8 ± 0.1	4.8 ± 1.6
Galactose	7.6 ± 1.4	5.0 ± 0.1	5.5 ± 1.0
Mannose	3	3	3
N-Acetyl-Galactosamine	7.0 ± 1.2	5.3 ± 0.1	4.9 ± 1.3
N-acetyl-Glucosamine	2.1 ± 0.3	2.0 ± 0.1	0.9 ± 0.2
Sialic acid	6.3 ± 0.2	2.8 ± 0.1	4.3 ± 1.3
Total sugar (% by weight)	12.1	11.4	11.9

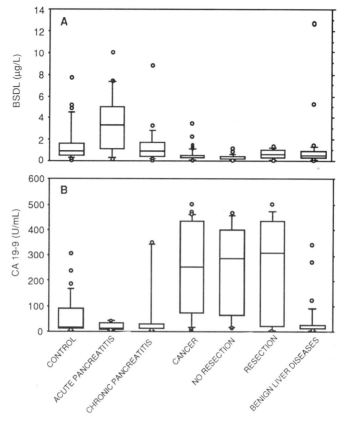

Figure 3. Box-plots of serum values of bile salt-dependent lipase and of CA 19-9 antigen.

A : serum values of bile salt-dependent lipase. B : serum values of CA 19-9 antigen. Determinations were done in sera of healthy subjects (controls) and in sera of patients suffering with an established adenocarcinoma of the pancreas (cancer) chronic or acute pancreatitis and benign liver diseases. Assays were done in sera of patients before any surgical exeresis of the tumor (no resection) or in sera of patients after the tumor resection. Each serum was measured at least in triplicate.

CONCLUSION

The results presented suggest that the glycosylation of the BSDL can be modified in pathological processes associated with the pancreas. In a normal donor, the N-linked oligo-saccharide structures of BSDL are of a complex-type. The complex-type structure correlated with a complete processing of sugar [16]. The BSDL glycovariants found in the pancreatic secretion of patient suffering from chronic pancreatitis, could be predominantly hybrid-type or high-mannose type or may be variants with decreased sialylation. The amount of sugar involved in the O-linked glycans was not modified, nevertheless, the FAP is subjected to unprocessed N-glycosylation and decreased O-glycosylation. Therefore, the difference of glycosylation in variants of BSDL may be the basis for discrimination and the quantification of BSDL in serum may be a specific tool for the diagnosis of pancreatic pathologies.

The diagnostic value of bile salt-dependent lipase for pancreatic diseases, was tested in sera of 187 patients (Figure 3). Of these patients, 76 suffered from pancreatic carcinoma, 43 from non-malignant liver diseases (cirrhosis and chronic hepatitis), 18 from acute pancreatitis and 20 from chronic pancreatitis. The remaining subjects were controls without pancreatic pathology.

Bile salt-dependent lipase was determined by a sandwich enzyme-linked immunosorbent assay using polyclonal antibodies [17]. Amylase and CA 19-9 antigen were also determined. In sera from control patients, the mean level of bile salt-dependent lipase was 1.5 µg/L. This level is similar to that of patients with benign liver diseases (1.1 µg/L) or with chronic pancreatitis (1.4 µg/L) but raised to 3.5 µg/L in patients with acute pancreatitis and decreased to 0.5 µg/L in the sera of subjects with pancreatic adenocarcinoma. 30 % of control subjects and 73 % of cancer patients had a bile salt-dependent lipase serum level below 0.5 µg/L. In acute pancreatitis, 11/16 had a level above 1.5 µg/L.

Amylase levels largely increased in acute pancreatitis but were normal in all other groups. In the case of the CA 19-9 antigen, 65 % of control patients and more than 80 % of patients with non-malignant pancreatic or liver diseases had normal levels. In sera from cancer patients, 80 % presented with high levels. Accordingly, in pancreatic cancer 36 out of 38 subjects had either low serum levels of bile salt-dependent lipase (< 0.5 µg/L) or high values (> 37 U/ml) for CA 19-9 antigen (sensitivity, 95 %). In control patients 2/30 have a low level of enzyme or an elevated level of CA 19-9 (specificity, 94 %). Therefore, alone or better, in conjunction with CA 19-9 antigen determination, the assay of serum levels of bile salt-dependent lipase help to diagnose pancreatic carcinoma and to discriminate malignant diseases from either chronic or acute pancreatitis.

REFERENCES

1. D. Lombardo, J. Fauvel and O. Guy, Studies on the Substrate Specificity of a Carboxyl Ester Hydrolase from Human Pancreatic Juice, *Biochim. Biophys. Acta.,* 611, (1980), 136-146.

2. D. Lombardo and O. Guy, Studies on the Substrate Specificity of a Carboxyl Ester Hydrolase from Human Pancreatic Juice. II. Action on Cholesterol Esters and Lipid-Soluble Vitamin Esters, *Biochim. Biophys. Acta.,* 611, (1980), 147-155.

3. N. Abouakil, E. Rogalska, J. Bonicel and D. Lombardo, Purification of Pancreatic Carboxylic-Ester Hydrolase by Immunoaffinity and its Application to the Human Bile-Salt-Stimulated Lipase, *Biochim. Biophys. Acta.* 961, (1988), 299-308.

4. J.H. Han, C. Stratowa and W.J. Rutter, Isolation of Full-Length Putative Rat Lysophospholipase cDNA Using Improved Methods for mRNA Isolation and cDNA Cloning, *Biochemistry.* 26, (1987), 1617-1625.

5. E.M. Kyger, R.C. Wiegand and L.G. Lange, Cloning of the Bovine Pancreatic Cholesterol Esterase/Lysophospholipase, *Biochem. Biophys. Res. Comm.* 164, (1989), 1302-1309.

6. K. Reue, J. Zambaux, H. Wong, G. Lee, T.H. Leete, M. Ronk, J.E. Shively, B. Sternby, B. Borgstrom, D. Ameis, and M.C. Schotz, cDNA Cloning of Carboxyl Ester Lipase from Human Pancreas Reveals a Unique Proline-Rich Repeat Unit, *J. Lipid Res.* 32, (1991), 267-276.

7. P.W. Jacobson, P.W.Wiesenfeld, and L.L. Gallo, Sodium Cholate-Induced Changes in the Conformation and Activity of Rat Pancreatic Cholesterol Esterase, *J. Biol. Chem.* 265, (1990), 515-521.

8. O. Guy, D. Lombardo, and J.G. Brahms, Structure and Conformation of Human Pancreatic Carboxyl-Ester Hydrolase, *Eur. J. Biochem.* 117, (1981), 457-460.

9. N. Abouakil, E. Mas, N. Bruneau, A. Benajiba, and D. Lombardo, Bile-Salt Dependent Lipase Biosynthesis in Rat Pancreatic AR 4-2 J Cells : Essential requirement of N-linked oligosaccharide for secretion and expression of a fully active enzyme. *J. Biol. Chem.* in press

10. T. Baba, D. Downs, K.W. Jackson, J.Tang, and C. Wang, Structure of Human Milk Bile Salt Activated Lipase, *Biochemistry.* 30, (1990), 500-510.

11. M.J. Escribano and S. Imperial, Purification and Molecular Characterization of FAP a Feto-acinar Protein Associated with the Differentiation of Human Pancreas, *J. Biol. Chem.* 264, (1989), 21865-21871.

12. E. Mas, N. Abouakil, S. Roudani, F. Miralles, O. Guy-Crotte, C. Figarella, M.J. Escribano, and D. Lombardo, Human Fetoacinar Pancreatic Protein : an Oncofetal Glycoform of the Normally Secreted Pancreatic Bile-salt Dependent Lipase. *Biochem J.* 289, (1993), 609-615.

13. T. Sugo, E. Mas, N.Abouakil, T. Endo, M.J. Escribano, A. Kobata, and D. Lombardo, The Structure of N-linked Oligosaccharides of Human Pancreatic Bile-salt Dependent Lipase. *Eur. J. Biochem.* in press

14. E. Mas, N.Abouakil, S. Roudani, J-L. Franc, J. Montreuil and D. Lombardo, Variation of the Glycosylation of Human Pancreatic Bile-salt Dependent Lipase. *Eur. J. Biochem.* in press

15. T.W.Rademacher, R.B. Parekh, and R.A. Dwek, Glycobiology. *Ann. Rev. Biochem.* 57, (1988), 785-836.

16. R. Kornfeld and S. Kornfeld, Assembly of Asparagine-linked Oligosaccharides.*Ann. Rev. Biochem.* 54, (1985), 631-664.

17. D. Lombardo, G. Montalto, S. Roudani, E. Mas, R. Laugier, V. Sbarra, and N. Abouakil, Is Bile Salt-dependent Lipase Concentration in Serum of any Help in Pancreatic Cancer Diagnosis. *Pancreas.* 8, (1993), 581-588.

PANCREATIC LIPASE, COLIPASE AND ENTEROSTATIN - A LIPOLYTIC TRIAD

Charlotte Erlanson-Albertsson

Dpt Medical and Physiological Chemistry 4
P.O. Box 94
S-221 00 Lund, Sweden

INTRODUCTION

Fat digestion occurs in the duodenum by the concerted action of pancreatic lipase and colipase [1,2]. The purpose of the present work is to discuss some properties of these two proteins and the more recently discovered peptide enterostatin [3], released from pancreatic procolipase, acting as a feed-back signal for regulation of fat intake.

PANCREATIC LIPASE

Pancreatic Lipase Belongs to a Lipase-Family

Pancreatic lipase is an old enzyme, present not only in mammals but also in birds and fish. It has up to 45% homology with lipoprotein lipase and hepatic lipase [4]. The identities extend through the entire molecules indicating that these lipases are derived from a single ancestor.

Pancreatic lipase has a serine residue at its active site, ser 152 in human lipase, forming part of the catalytic triad, involving a histidine (his 263) and an aspartic acid residue (asp 176) [5]. The catalytic triad, which is also present in hepatic lipase and lipoprotein lipase, could be exactly positioned with relevant residues in serine proteases [5], the similarities, however, being confined to the catalytic site triad only.

The exon-intron organization of canine pancreatic lipase has been determined [6] as well as the promotor region. The lipase gene is organized in 13 exon sequences. The promotor region contains CAAT and TATA boxes at positions -112 and -35, respectively,

Esterases, Lipases and Phospholipases, Edited by M.I. Mackness
and M. Clerc, Plenum Press, New York, 1994

but also has a class 2 glucocorticoid receptor binding sequence at position -97. This receptor is probably responsible for the decreased synthesis of lipase following glucocorticoid injection [7], and the increased lipase synthesis observed following adrenalectomy [7].

Pancreatic Lipase is Inhibited by Bile Salt

The natural substrate for pancreatic lipase is the long-chain triglycerides dispersed in micellar bile salt solution. However, lipase is strongly inhibited by bile salt [8], being unable to reach the bile-salt covered interface with its higher surface pressure [9]. To overcome the inhibition exerted by bile salt, lipase is activated by pancreatic colipase [10].

PANCREATIC COLIPASE

Colipase Forms a 1:1 Molar Complex with Lipase

Colipase is a pancreatic protein secreted together with lipase from the pancreas. The binding between lipase and colipase has a low affinity, Kd being 10^{-6} M in buffer, indicating that the two proteins are essentially separate, while in pancreatic juice [11]. In the intestine, a complex is formed, catalysed by long-chain fatty acids, which increase the binding 100-fold [12].

Lipase-Colipase Interaction Occurs through Ionic Bonds with a Small Contact Area

By x-ray analysis following crystallization of lipase and colipase the structure of the molar complex has been elucidated [13]. It turns out that pancreatic lipase binds to colipase with two ionic bonds in the C-terminal part of lipase involving, Lys 399-Glu 45, and Asp 389-Arg 44 in lipase-procolipase, respectively. The contact area between lipase and colipase thus is small in agreement with partition studies, which indicated that the surface of the total lipase-colipase complex was equal to the sum of the surface of colipase and lipase [11]. Also modification studies of colipase and lipase have indicated the involvement of ionic charges in the lipase-colipase interaction [14].

A Hypothetical Lipid Binding Site of Colipase is Proposed

From the crystal structure of the lipase-procolipase complex a hypothetical lipid binding site of colipase was proposed in the region opposite to the lipase binding site [13]. This region, organized in three hydrophobic fingers [13], contains the hydrophobic N-terminal region [15], as well as the three tyrosine residues (55, 58 and 59) [16]. How colipase activation of lipase might occur is now known through the crystallization of the lipase-procolipase complex in the presence of mixed phospholipid-bile salt micelles [17]. The active site catalytic triad in lipase is normally covered by an amphipathic lid [17], similar to what has been found for other lipases, *i.e.* the *Rhizomucor miehei* lipase. Through the interaction of colipase and the substrate interface the opening of the lid is effected [17], provoking a drastic conformational change in the active site loop of lipase. The importance of a combined action of colipase and a lipid interface in activating lipase is supported by spectral studies [18], showing a conformational change of lipase with a spectral shift of a tryptophan residue, induced by colipase in the presence of mixed bile salt-oleic acid micelles [18], not by colipase itself.

Colipase is Secreted as a Procolipase

Like most other pancreatic enzymes pancreatic colipase, somewhat surprisingly, was found to be secreted as a procolipase [19]. In the intestine the procolipase is activated by trypsin during formation of colipase and a pentapeptide. The pentapeptide, named *enterostatin*, was found not to be involved in the procolipase-lipase interaction supported by crystallization studies [13]. Instead it has been found to act as a satiety signal for the intake of high-fat food [20,21].

Colipase Gene is Unique

Comparison of the colipase gene with other proteins revealed no similarity above that due to random coincidence, not even for other lipase cofactors like the apolipoprotein C-II [4].The colipase gene was found to be localised on chromosome 6 [22], being different from that of the lipase gene, situated on chromosome 10.

The genomic organization of the pancreatic colipase gene was recently characterized [23]. The transcriptional unit is organized in three exon sequences, exon 1, from amino acid residue -17 to residue 11, encoding a 17 residue signal sequence and the enterostatin sequence, exon 2 from residue 12 to 52 encoding the lipase binding site, and exon 3 from residue 53 to 95 encoding part of the lipid binding region. The gene contained ~ 3.5 kb [23]. Two TATA sequences were observed in the 5'-flanking region of the colipase gene. In the promotor region there was also a glucocorticoid receptor binding sequence (TGTTGT), as was found in the lipase gene. This region is likely to be involved in the regulation of colipase synthesis by the corticoid hormone [7], which inhibits the transcription of colipase mRNA [24], thus lowering colipase synthesis.

Colipase Expression Determines Fat Digestion

The role of colipase is to specifically bind to lipase and to the triglyceride substrate, providing an anchor for the binding of lipase to its substrate [10,17]. Measurement of colipase and lipase in various animals have indicated variable ratios between lipase and colipase (Table I). In rat, for instance, the colipase and lipase ratio is below one, while in man the ratio is around one. Larger animals thus appear to have a higher ratio of colipase/lipase than smaller animals. Assuming that lipase-colipase interacts in a 1:1 molar ratio, the variable ratio may be related to dietary differences, larger animals requiring a higher percentage fat in the diet for optimal growth. Another reason for the variable colipase/lipase ratio may reflect the double role of colipase; in addition to assisting lipase during fat digestion it has a potential role in body weight regulation through its activation peptide, enterostatin, as described below.

Table 1. Pancreatic lipase and colipase activities of exocrine pancreas from mouse, rat, pig and rabbit per g tissue and from man in intestinal content

Species	Lipase units	Colipase units	Ratio colipase/lipase
Mouse	1984	726	0.37
Rat	8640	4110	0.48
Rabbit	2440	1613	0.66
Pig	33600	34800	1.03
Man	2200	2300	1.05

In experiments, where regulation of lipase and colipase was studied, it was found that a change of lipase synthesis was almost always followed by a change in colipase synthesis. Furthermore, colipase syntheses always changed more slowly than lipase [25,26]. In this view, colipase synthesis is the rate-limiting step in fat digestion. Patients specifically lacking colipase had a severe steatorrhoea [27].

Synthesis of Lipase and Colipase is Inhibited by Insulin

During the search for factors particularly affecting lipase and colipase synthesis it was found that lipase and colipase production was inversely related to insulin production [28]. Thus four hours after administration of insulin (0.5U/100 g) both lipase and colipase synthesis were significantly reduced by 80 and 72%, respectively [28]. In the opposite situation, where insulin production was decreased, as during fasting, there was an increased production of lipase and colipase. This was even more evident when making the animals diabetic. In this condition the levels of lipase and colipase increased two-fold [29], mainly by a transcriptional regulation [26,30].

The inverse relationship between colipase synthesis and insulin is probably important for the body weight regulation exerted by colipase/enterostatin.

Colipase is also Produced in the Stomach

In addition to pancreas, the stomach and duodenum were found to produce procolipase, assayed with Northern blot analysis [24]. The size of the procolipase mRNA was similar to the mRNA of procolipase (0.6 kb). Immunoreactive enterostatin was also found in the stomach and in the duodenum, where it was found to be co-localized with serotonin [31]. The tissue localization of mRNA for colipase and enterostatin was the same, that is in the antrum of the stomach, indicating that enterostatin was either released from pancreatic procolipase or produced in the same cells as procolipase. The importance of these findings are not known.

ENTEROSTATIN

Enterostatin - A Gut Hormone

Enterostatin, released in the intestine by the action of trypsin on pancreatic procolipase, is a peptide that has been found to decrease appetite, with a specificity for the fat contained in the food. The anorectic effect was observed both after central and peripheral administration and has been observed in rat [32], sheep [33] and monkey [34].

The mechanism of action is under investigation at present, but suggests a centrally localized receptor, that specifically binds to enterostatin [35]. Enterostatin after its production in the intestine is probably taken up, based on *in vitro* experiments [36], where a passage of intact peptide was observed from the mucosal side to the serosal side of rabbit ileum amounting to 1-2%. Enterostatin, produced in enterochromaffin cells of the gastric antrum and duodenum, will also be released in the circulating blood.

Enterostatin Produces an Early Satiety

An anorectic substance may either delay the onset of eating, like the amphetamines or terminate eating at an earlier time, like fenfluramine. In studies by Lin *et al.* [37], enterostatin was found to use the second mechanism, *i.e.* to terminate eating at an earlier time, instead prolonging the time of resting or sleeping. Enterostatin thus does not influence feeding behaviour by stimulation of an alternative behaviour, but reduces food intake by inducing an earlier onset of satiety.

Enterostatin also has Metabolic Effects

In the earlier experiments on enterostatin feeding behaviour was mainly studied. In the following experiments various metabolic effects of enterostatin were investigated. One of these is the inhibition of insulin secretion by enterostatin, both observed after incubation with Langerhans islets *in vitro* [38] and *in vivo* after acute as well as chronic infusion of enterostatin [39]. The inhibition of insulin secretion was observed at the second phase of insulin secretion during the islet perfusion after stimulation with glucose (Fig. 1). In the *in vivo* experiments a significant reduction of insulin secretion was observed both after acute and chronic injection of enterostatin [39].

The significance of inhibition of insulin secretion is speculative. Nevertheless, such an effect would be important for the control of body weight, promoting weight loss [39]. Since lipase and colipase synthesis is inhibited by insulin one important consequence of the enterostatin effect on insulin secretion is the stimulation of lipase/colipase synthesis. Further experiments are needed to substantiate such a relationship.

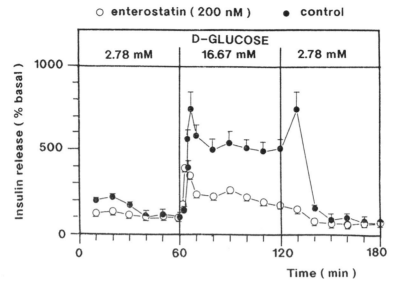

Figure 1. Effect of 200 nM concentration of enterostatin in comparison to control on glucose-stimulated insulin secretion from isolated perfused rat islets *in vitro* (•) enterostatin; (o) control, n=14. During stimulation of insulin secretion with glucose there was a significant reduction in insulin secretion by enterostatin. Values are plotted as mean ± SEM.

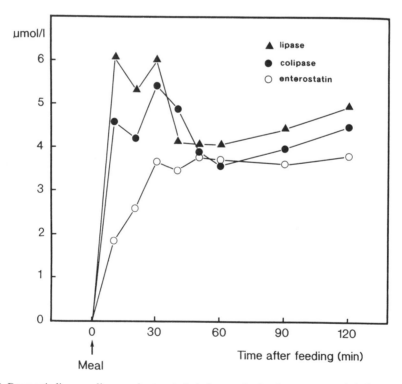

Figure 2. Pancreatic lipase, colipase and enterostatin in human duodenal contents sampled after a test meal. Values were obtained from nine healthy volunteers at various time points during two hours after a test meal. No correction for dilution of the test meal has been done. Lipase and colipase were measured enzymatically [29], while enterostatin was measured immunologically [44].

In experiments with chronic infusion of enterostatin [39], it was furthermore found that there was a decreased body weight with a reduction of epididymal and retroperitoneal fat pads. In the enterostatin-treated animals there was also a down-regulation of the hepatic glucocorticoid receptor, in response to markedly elevated serum corticosterone levels, promoting weight loss [39]. Thus enterostatin in addition to decreasing appetite interacts with systems - insulin production, steroid hormone action - that inhibits lipid synthesis, promoting body weight loss.

Lipase, Colipase and Enterostatin in Man

The production of lipase and colipase during a standard meal has recently been described [40], measuring lipase and colipase activities in intestinal samples in man after duodenal intubation.

We have performed duodenal intubation in man and measured lipase, colipase and enterostatin [42]. After a standard meal lipase and colipase activites were produced in the micromolar range and in a 1:1 ratio in agreement with previous work [40]. Enterostatin in these experiments was found to gradually increase during the collection period and reached optimal values 30 min after the beginning of the meal (Fig. 2). At time point 30 min the enterostatin concentration was similar to colipase indicating a complete activation of pancreatic procolipase. The non-parallell production of enterostatin and colipase during the initial thirty minutes may be due to an incomplete activation of pancreatic procolipase in the presence of dietary products, since procolipase in pure pancreatic juice was rapidly activated, actually the first zymogen to be activated [41] by enterokinase. Further studies are needed to clarify this non-parallellism.

CONCLUSIONS AND PERSPECTIVES

Pancreatic lipase, colipase and enterostatin form a fascinating lipolytic triad, pancreatic lipase catalysing the hydrolysis of dietary fat by the assistance of colipase, providing a binding site for lipase at the triacylglycerol substrate and enterostatin, forming a feed-back signal for regulation of fat intake. Enterostatin, by being part of pancreatic procolipase, is an example of *"briccolage of Nature"* as proposed by Nobel Prize Winner F. Jacob [43], showing how Nature works during evolution, by *producing novelties not from scratch, but on what already exists, transforming a system to give it new functions or combining several systems to produce a more elaborate one.*

Acknowledgements

Ms Ruth Lovén is thanked for typing the manuscript. The work in my laboratory has been possible mainly through grants from Swedish Medical Research Council (B93-03X-07904-07B).

REFERENCES

1. Borgström, B. and Erlanson-Albertsson, C. Pancreatic colipase, *in*: Lipases, Borgström, B. and Brockman, H.L., Elsevier North-Holland, Amsterdam 1984; 151-184.

2. Verger, R. Pancreatic lipase, *in*: Lipases, Borgström, B. and Brockman, H.L., Elsevier North Holland, Amsterdam 1984; 83-150.

3. Erlanson-Albertsson, C. Enterostatin - the pancreatic procolipase activation peptide - a signal for regulation of fat intake. *Nutr. Rev.* 50: 1992; 307-310.

4. Persson, B., Bengtsson-Olivecrona, G., Enerbäck, S., Olivecrona, T., and Jörnvall, H. Structural features of lipoprotein lipase, lipase family relationships, binding interactions, non-equivalence of lipase cofactors, vitellogenin similarities and functional subdivision of lipoprotein lipase. *Eur. J. Biochem.* 179: 1989; 39-45.

5. Winkler, F.K., D'Arcy, A., and Hunziker, W. Structure of human pancreatic lipase. *Nature* 343: 1990; 771-774.

6. Mickel, F.S., Weidenbach, F., Swarovsky, B., La Forge, K.S., and Scheele, G.A. Structure of the canine pancreatic lipase gene. *J. Biol. Chem.* 264: 1989; 12895-12901.

7. Duan, R. and Erlanson-Albertsson, C. The anticoordinate changes of pancreatic lipase and colipase activity to amylase activity by adrenalectomy in normal and diabetic rats. *Int. J. Pancreatology* 6: 1990; 271-279.

8. Borgström, B. and Erlanson, C. Pancreatic lipase and colipase. Interactions and effects of bile salts and other detergents. *Eur. J. Biochem.* 37: 1973; 60-68.

9. Verger, R. Enzyme kinetics of lipolysis. *Methods in Enzymology* 64; 1980; 340-392.

10. Erlanson-Albertsson, C. Pancreatic colipase. Structural and physiological aspects. *Biochim. Biophys. Acta* 1125: 1992; 1-7.

11. Patton, J., Albertsson, P.Å., Erlanson, C., and Borgström, B. Binding of porcine pancreatic lipase and colipase in the absence of substrate studied by two-phase partition and affinity chromatography. *J. Biol. Chem.* 253: 1978; 4195-4202.

12. Larsson, A. and Erlanson-Albertsson, C. The identity and properties of two forms of activated colipase from porcine pancreas. *Biochim. Biophys. Acta* 664: 1981; 538-548.

13. van Tilbeurgh, H., Sarda, L., Verger, R., and Cambillau, C. Structure of the pancreatic lipase-procolipase complex. *Nature* 359: 1992; 159-162.

14. Erlanson, C. Chemical modification of pancreatic lipase. Effect on the colipase-reactivated and true lipase activity. *FEBS Letters* 84: 1977; 79-82.

15. Erlanson-Albertsson, C. and Larsson, A. Importance of the N-terminal sequence in porcine pancreatic colipase. *Biochim. Biophys. Acta* 665: 1981; 250-255.

16. McIntyre, J.C., Hundley, P., and Behnke, W.D. The role of aromatic side chain residues in micelle binding by pancreatic colipase. *Biochem. J.* 245: 1987; 821-829.

17. van Tilbeurgh, H., Egloff, M.P., Martinez, C., Rugani, N., Verger, R., and Cambillau, C. Interfacial activation of the lipase-procolipase complex by mixed micelles revealed by x-ray crystallography. *Nature* 362: 1993; 814-820.

18. Erlanson-Albertsson, C. and Åkerlund, H.E. Conformational change in pancreatic lipase induced by colipase. *FEBS Letters* 144: 1982; 38-42.

19. Borgström, B., Wieloch, T., and Erlanson-Albertsson, C. Evidence for a pancreatic procolipase and its activation by trypsin. *FEBS Letters* 108: 1979; 407-410.

20. Erlanson-Albertsson, C., Jie, M., Okada, S., York, D., and Bray, G.A.. Pancreatic procolipase propeptide, enterostatin, specifically inhibits fat intake. *Phys. Behav.* 49: 1991; 1191-1194.

21. Okada, S., York, D.A., Bray, G.A., and Erlanson-Albertsson, C. Enterostatin (Val-Pro-Asp-Pro-Arg) the activation peptide of procolipase selectively reduces fat intake. *Phys. Behav.* 49: 1991; 1185-1189.

22. Davies, R.C., Xia, Y., Mohandas, T., Schotz, M.C., and Lusis, A.J.. Assignment of the human pancreatic colipase gene to chromosome 6p21.1 pter. *Genomics* 10: 1991; 262-265.

23. Fukuoka, S.-I., Zhang, D.-E., Taniguchi, Y., and Scheele, G.A. Structure of the pancreatic colipase gene includes two protein binding sites in the promotor region. *J. Biol. Chem.* 268: 1993; 11312-11320.

24. Okada, S., York, D.A., and Bray, G.A. Procolipase mRNA: Tissue localization and effects of diet and adrenalectomy. *Biochem. J.* 292: 1993; 787-789.

25. Wicker, C. and Puigserver, A. Effects of inverse changes in dietary lipid and carbohydrate on the synthesis of some pancreatic secretory proteins. *Eur. J. Biochem.* 162: 1987; 25-30.

26. Duan, R.-D. and Erlanson-Albertsson, C. The effect of pretranslational regulation on synthesis of pancreatic colipase in streptozotocin-induced diabetes in rats. *Pancreas* 7: 1992; 465-471.

27. Hildebrand, H., Borgström, B., Békássy, A., Erlanson-Albertsson, C., and Helin, I. Isolated colipase deficiency in two brothers. *Gut* 23: 1982; 243-246.

28. Duan, R.-D., Wicker, C., and Erlanson-Albertsson, C. Effect of insulin administration on contents, secretion and synthesis of pancreatic lipase and colipase in rats. *Pancreas* 6: 1991; 595-602.

29. Duan, R. and Erlanson-Albertsson, C. Pancreatic lipase and colipase activity increase in pancreatic acinar tissue of diabetic rats. *Pancreas* 4: 1989; 329-334.

30. Duan, R., Poensgen, J., Wicker, C., Weström, B., and Erlanson-Albertsson, C. Increase in pancreatic lipase and trypsinogen activities and their mRNA levels in streptozotocin - induced diabetic rats. *Dig. Dis. Sci.* 34: 1989; 1243-1248.

31. Erlanson-Albertsson, C., Mei, J., Sörhede, M., and Ohlsson, A. Enterostatin - a new gut hormone that regulates appetite. *The Faseb Journal* 7: 1993; A 89 (Abstract).

32. Shargill, N.S., Tsujii, S., Bray, G.A., and Erlanson-Albertsson, C. Enterostatin suppresses food intake following injection into the third ventricle of rats. *Brain Research* 544: 1991; 137-140.

33. Miner, J.L., Erlanson-Albertsson, C., Paterson, J.A., and Baile, C.A. Enterostatin and feed intake in sheep. (1993). Manuscript in preparation.

34. Weatherfood, S.C., Lattermann, D.F., Sipols, A.J., Chavez, M., Kermani, Z.R., York, F.S., Bray, G.A., Porte, D., Jr, and Woods, S.C. Intraventricular administration of enterostatin decreases food intake in baboons. *Appetite* 19: 1992; 225.

35. Sörhede, M., Mei, J., and Erlanson-Albertsson, C. Enterostatin - a brain-gut peptide-regulating fat intake. *J. Physiol.* 87: 1993; 273-275.

36. Huneau, J.-F., Erlanson-Albertsson, C., and Tomé, D. Absorption of enterostatin across the rabbit ileum *in vitro. The Faseb Journal* 7: 1993; A 89 (Abstract).

37. Lin, L., Mc Clanahan, S., York, D.A., and Bray, G.A. The peptide enterostatin may produce early satiety. *Physiol. Behav.* 53: 1993; 789-794.

38. Mei, J., Cheng, Y., and Erlanson-Albertsson, C. Enterostatin - its ability to inhibit insulin secretion and to decrease a lipid-enriched meal. *Int. J. Obesity*, 1993. In press.

39. Okada, S., Lin, L., York, D.A., and Bray, G.A.. Chronic effects of intracerebral ventricular enterostatin in Osborne-Mendel rats fed a high-fat diet. *Physiol. Behav.* 54: 1993; 325-330.

40. Sternby, B., Nilsson, Å., Melin, T., and Borgström, B. Pancreatic lipolytic enzymes in human duodenal contents. *Scand. J. Gastroenterol.* 26: 1991; 859-866.

41. Borgström, A., Erlanson-Albertsson, C., and Borgström, B. Pancreatic proenzymes are activated at different rates. *Scand. J. Gastroenterol.* 28: 1993; 455-459.

42. Erlanson-Albertsson, C., Mei, J., and Thesleff, P. Enterostatin in man. *Int. J. Obesity* 16: 1992; 12 (Abstract).

43. Jacobs, F. Evolution and tinkering. *Science* 196: 1977; 1161-1166.

44. Bowyer, R.C., Jehanli, A.M.T., Patel, G., and Hermon-Taylor, J.. Development of enzyme-linked immunosorbent assay for free human pro-colipase activation peptide (APGPR). *Clin. Chim. Acta* 200: 1991; 137-152.

LCAT : FROM STRUCTURE TO CLINICAL SIGNIFICANCE

M.F. Dumon, A. Berard and M. Clerc

Laboratoire Central de Biochimie, Hôpital Saint-Andre
1, Rue Jean-Burguet - 33075 Bordeaux Cedex, France

INTRODUCTION

Lecithin : cholesterol acyltransferase (LCAT ; EC 2.3.1.43) is a plasma glycoprotein enzyme of the serine esterase type which is synthesized and secreted by hepatocytes and has a key role in the metabolism of cholesterol especially in reverse cholesterol transport (Efflux) from the peripheral tissues to the liver [1].

Although it is composed of a single polypeptide unit of about 66 Kd [2] LCAT catalyses sequential reactions. It mediates the transfer of the sn-2 acyl group from phosphatidylcholine (PC) to the 3-hydroxyl group of cholesterol with the production of lyso-PC and cholesteryl ester (CF), which requires both phospholipase (PL) and acyl transferase activities. However, the phospholipase activity of LCAT has some particularities : it is $Ca2+$ independent, is not absolutely specific for the sn-2 position as are other PLA2, it is also responsible for the PC-lyso PC acyltransferase exchange reaction, it is inhibited by PC containing eicosapentaenoic acid (EPA) or docosahexaenoic acid (DHA) in the sn-2 position [3-5]. Furthermore LCAT catalyses the transacylation of intact C.E. [6]. LCAT activity is dependent on the composition of its physiological substrates, high density lipoprotein (HDL) and the major apolipoprotein (apo) component, apo A-I, which is required for optimal activation of the enzyme.

LCAT is responsible for the synthesis of virtually all of the C.E. in plasma lipoproteins in man. Immunologically reactive LCAT protein has been found on low-density lipoprotein (LDL) [7], however, it is debatable that this LCAT is active. LCAT derived C.E. on HDL are subsequently transfered by cholesterol ester transfer protein (CETP) to acceptors among the major plasma lipoprotein classes (LDL and very low density lipoprotein (VLDL)) in exchange for triglyceride [8]. Thus LCAT appears both as a "pump" of cell-cholesterol and the origin of a transfer cascade of C.E. in plasma lipoproteins. However, plasma cholesterol metabolism is much more complex and the detailed analysis of LCAT, CETP as other factors within HDL subfractions are required to understand what regulates the flow of cholesterol through plasma lipoproteins [9]. Major compositional abnormalities of plasma lipoproteins would therefore be expected in cases of LCAT deficiency.

To date the familial LCAT deficiency which was firstly described by Norum and Gjone [10], has now been identified in more than 50 patients from approximately 30 families. Massive corneal opacities presenting as arcus lipoides which appear in the third decade of life are observed in all patients and are associated with one or more of the following clinical symptoms : normochronic hemolytic anaemia (target cells, stomatocytes, schyzocytes, anulocytes), focal and segmental glomerulosclerosis (renal insufficiency, proteinuria, hematuria, uremia), and in a small number of cases premature atherosclerosis. Partial LCAT deficiency or Fish-Eye disease (FED) which was first described by Carlson [11] has presently been identified in about 10 patients who show early arcus lipoides and corneal clouding without any other clinical symptom and a total lack of premature atherosclerosis. In intermediary forms of LCAT deficiency which resemble FED, abnormal osmotic fragility of erythrocytes has been reported [12]. Familial LCAT deficiency, FED and intermediary forms are all biochemically characterized by the impairment of cholesterol esterification with drastic decreases in HDL cholesterol (HDLc) concentration and anomalies in lipoprotein composition (decrease in normal HDL, increase in fast migrating HDL, increase of triglycerides in LDL) and decreases of apo A-I and apo A-II caused by the lack and/or dysfunction of plasma LCAT [13]. LCAT activity is also reduced in some HDL deficiencies which are not genetically related to anomalies of the LCAT gene eg Tangier disease, apo A-I/A-IV/C-III deficiency, apo A-I/C-III deficiency, and in the case of single base deletion in codon 202 of apo A-I (truncated apo A-I) [14]. Corneal opacifications have been observed in these secondary forms of LCAT activity deficiency. Paradoxically, it has been found, in a peculiar Japanese case, that a marked hyper-HDL2-cholesterolaemia was associated with premature corneal opacity [15]. The variability of the clinical phenotypes corresponding to these patients with primary or secondary LCAT deficiency suggest that the disease is caused by different mutations in the LCAT gene and/or defects in the genes of LCAT activating proteins.

GENE STRUCTURE AND FUNCTIONAL REGIONS OF LCAT

Following the report of the protein sequence from amino acid (AA) analysis [17] the AA and m RNA sequence of human LCAT has been determined from cDNA and genomic cloning experiments [16]. A probable linkage of the LCAT locus in man to the alpha haptoglobin locus on chromosome 16 was first described by Teisberg and Gjone [18]. Azoulay et al [19] refined the assignement of the LCAT gene to 16 q 22. The 1550 base LCAT m RNA can be detected in liver and Hep G2 cells [16]. Preliminary data on the human LCAT gene promoter has been reported [20]. The mature protein was found to contain 416 AA residues (AAR) with a calculated polypeptide molecular weight of 47090. A hydrophobic leader sequence of 24 AA represents the signal peptide [16]. The gene is divided into six exons spanning about 4200 bp. Exon five codes for AA homologous to the interfacial active site of several lipases (a hydrophobic hexapeptide identical with the interfacial binding segment of the active site of pancreatic lipase and lingual lipase), and also codes for an amphipatic alpha helix resembling the carboxyl terminus of apo E.

A number of functional regions have been studied : (i) the active site and the location of the two essential sulphydryl groups among the six CYS residues of the protein, (ii) the sites of disulfide linkage, (iii) the location of the carbohydrate side chains, and (iiii) the site of protein-protein interaction with apolipoprotein A-I.

(i) - SER, HIS and CYS residues have been known for a long time as active residues as their chemical modification inhibits the activity of the enzyme [3]. The putative active site of LCAT including SER 181 as the acyl acceptor was identified through its homology with other serine-type esterases [21]. A covalent catalytic mechanism of action for LCAT in which SER 181 and HIS 180 residues mediate PC cleavage and both CYS residues 31 and 184 mediate cholesterol esterification was proposed [22]. Site directed mutagenesis experiments demonstrated that the serine residue at position 181 when replaced by either ALA, GLU or THR gave rise to an enzyme product that was normally secreted by transfected CHO cells but had no detectable catalytic activity [23]. Paradoxically, the double mutant in which CYS-31 and CYS-184 had both been replaced with GLY residues was fully active in the synthesis of cholesteryl-esters [24]. The catalytic mechanism of LCAT on LDL has been described as the ability of LCAT to catalyse acyl transfer to lyso-PC on the surface of LDL [25, 26]. The physiological role of this last reaction suggests a new function of LCAT in completing the repair of oxidatively damaged PC by re-esterifying the lyso-PC. The discovery of an intrinsic PLA2 activity of apo B100 increases interest in this phenomenon [27].

(ii) - Two disulphide bridges between CYS 50 - CYS 74 and CYS 313 - CYS 356 have been identified. These bridges produce loops of 25 and 44 AAR respectively [21].

(iii) - Post-translationally, LCAT undergoes N-glycosylation at ASN 272 and three other ASN residues at positions 20, 84 and 384 [21]. Analysis of the plasma enzyme indicated that almost all of the large carbohydrate moiety of LCAT (approximately 25% by weight) was N-linked. Inhibitors of the subsequent processing of the N-linked high mannose chains were without effect on either the secretion rate or the catalytic activity of LCAT [28]. Site directed mutagenesis used to generate LCAT species in which individual attachment sites for N-linked oligosaccharide residues were replaced with residues that prevent the attachment of carbohydrate, produced mutants at three of four sites that retained significant acyltransferase activity and phospholipase activity in the absence of cholesterol. However mutation at site ASN 272 converted LCAT to a phospholipase generating free fatty acids not cholesteryl esters [29].

(iv) - It is now well established that the chemical and physical properties of the particles and the nature of their lipid/water interface are critical in determining the rate at which the molecular substrates react with LCAT [25]. The major protein of HDL, apo A-I, is the most potent activator of LCAT [30]. Activation by recombinant normal and mutated apo A-I suggests that the importance of the carboxyl terminus of apo A-I for LCAT activation is related to its ability to bind to lipid and/or to form a discoidal substrate for LCAT and the interaction of several domains of apo A-I are required for the activation of LCAT [31]. In vitro, some naturally occuring mutants of apo A-I, which have common altered secondary structures were found to have altered LCAT cofactor activity [13]. The use of synthetic peptide analogues to localize the LCAT activating domain in apo A-I emphasised the importance of the 22 mer repeats of apo A-I and suggested that the tandemly repeated alpha-helices located between residues 66 and 121 in the native apo A-I are essential [30]. The epitopes that span AA residues 95-121 of mature apo A-I are the most likely to be involved in the activation of LCAT [32]. The altered conformation of apo A-I in interstitial fluid may render it a poor activator of LCAT [33]. Glycated apo A-I activates LCAT to a level significantly lower than that of normal apo A-I. It is tempting to postulate that this abnormal activation may be associated with a reduction in reverse cholesterol transport and subsequently with the accelerated development of atherosclerosis in diabetic patients [34]. Neither the molecular mechanism nor the position of the apo A-I activation sites are known. This is also true for other apolipoprotein activators (apo A-IV, apo C's) of LCAT.

A model of the secondary structure of LCAT has been proposed [21] based upon the primary structure, chemical modification, homology with other proteins, predictive algorithms and the enzymatic assays of Jauhiainen and Dolphin [22]. The predictive algorithms especially suggest that the amphiphilic regions in LCAT, like those in the apolipoproteins, are bounded by PRO and also suggest that the lipid-associating determinants of LCAT are probably different from those that regulate the association of apolipoproteins with phospholipid surfaces. The tertiary structure of LCAT is unknown.

GENETIC AND PHENOTYPIC HETEROGENEITY IN FAMILIAL DISORDERS WITH LCAT ACTIVITY DEFICIENCY

LCAT activity deficiency caused by hereditary disorders may be of primary or secondary origin. They are caused by a deficiency in the enzyme mass and/or activity as well as by a deficiency in the substrate (mainly HDL) or the lack of activators or the presence of inhibitors.

The primary LCAT deficiencies are currently divided into two types : (i) the chemical or familial LCAT deficiency with early corneal clouding and one or more of the complementary clinical features initially described by Norum and Gjone, (ii) the partial LCAT deficiency with isolated corneal clouding (FED) initially described by Carlson.

To date the classical LCAT deficiency has been reported in about 30 homozygotes or compound heterozygotes from unrelated families and 12 countries in Europe, north America and Asia. 14 cases among these have been analysed at the gene level [38]. The data of which are summarized below.

Table 1. Genetic and phenotypic features from classic homozygote or compound heterozygote patients currently reported

Ref.	Anomaly of LCAT	LCAT mass	LCAT activity	Corneal clouding	Anaemia	Renal involvment	Premature atherosclerosis
[35]	ARG 147 ➔ TRP	# 0	# 0	+	?	+	0
[36]a	GLY INSERT (EXON 4)	# 0	# 0	+	?	+	0
b	ASP 128 ➔ LYS	↘	0	+	?	+	0
c	MET 292 ➔ ILE	↘	↘	+	?	0	0
[37]	MET 293 ➔ ILE	?	?	+	+	?	0
[38]	MET 252 ➔ LYS	↘	# 0	+	+	+	?
[39]a	LEU 209 ➔ PRO	# 0	0	+	0	+	0
b	ALA 93 ➔ THR + ARG 158 ➔CYS	?	↘↘	+	+	+	0
c	ARG 135 ➔ TRP GLN 376 ➔THR➔END	0	0	+	+	+	0
d	THR 321 ➔ MET	# 0	# 0	+	+	?	0
e	TYR 83 ➔ END	0	0	+	+	+	0
[40]	TYR 83 ➔ STOP TYR 156 ➔ ASN	# 0	# 0	+	±	+	0
[41]	?	# 0	# 0	+	±	±	0

0 = trace

Due to the presence of a structurally normal gene in a patient with confirmed familial LCAT deficiency and in-vitro transfection studies with mutated LCAT genes which reproduce in the medium (as in the plasma of patients) the same defects of LCAT mass and/or activity [35-40], the structural anomalies identified in the LCAT gene may be strongly considered as causative of the LCAT deficiency. Though the nature of the defects in enzyme action caused by the mutations is unclear, interesting speculations may be proposed :
- disruption of the interaction of LCAT with its lipid substrate as a consequence of the TYR 156 ➔ ASN substitution in the AA 154 - 167 amphipathic helix of LCAT [40].
- non functional, if secreted, truncated LCAT caused by the TYR 83 ➔ STOP mutation [40].
- altered conformation which could destabilise and inactivate the enzyme as a consequence of a 3 bp insertion which introduces a GLY residue at position 141 [36].

- addition of a positive charge resulting from ASN 228 ➔ LYS mutations which occurs in a stretch of AA that are conserved among species [36].
- substantial change in the secondary and tertiary structure of the enzyme resulting from the ARG 147 ➔ TRP substitution which generates an unstable protein either inside or outside the cell, or one when is retained by the rough endoplasmic reticulum [35].

Therefore several different mechanisms appear to be involved in the reduced mass and/or activity of the plasma enzyme. Current evidence supports the idea that several domains are important for LCAT function [39]. However, the absence of LCAT protein from the plasma is observed in most cases. It implicates a loss of LCAT activity and therefore represents the origin of the different clinical features observed in familial LCAT deficiency. These features result from the defect of cholesterol efflux and/or metabolic abnormalities of the modified plasma lipoproteins following from the drastically reduced LCAT activity.

Since Carlson designated as FED, the prominent corneal opacifications with alpha-LCAT deficiency originally observed in three Swedish patients [11] and Frölich reported a case of hypoalphalipoproteinemia resembling FED [42], 5 new homozygous FED patients have been described [43 - 49]. The LCAT genetic defect has still to be identified in most of them (table II).

Table 2. Genetic and phenotypic features from FED homozygotes and compound heterozygote patients currently reported

Ref.	Anomaly of LCAT	LCAT mass	Alpha-LCAT activity	Corneal clouding	Anaemia	Renal involvment	Premature atherosclerosis
[43]a	THR 123 ➔ ILE	↘	0	+	0	0	0
b	THR 123 ➔ ILE	↘	0	+	0	0	0
[44]	THR 123 ➔ ILE THR 347 ➔ MET		0	+	0	0	0
[45]a	PRO 10 ➔ LEU	↘	0	+	0	0	0
b	PRO 10 ➔ LEU	↘	0	+	0	0	0
[46-48]	LEU 300-DELET	↘	N	+	0	0	0
[49]	THR 123 ➔ ILE	↘	0	+	0	0	0

N = Normal specific activity

The THR 123 ➔ ILE mutation is located outside the catalytic centre of LCAT but in a highly hydrophobic region. It effects the ability of LCAT to bind to or to be active in small substrate particles. The net ability of plasma containing this mutated LCAT to esterify radiolabelled cholesterol is unaffected [43,49]. In a compound heterozygote with the mutations THR 123 ➔ ILE and THR 347 ➔ MET a normal cholesterol esterification rate was observed, which indicated an intact catalytic site but a defective interaction of LCAT with HDL particles [44]. The PRO10 ➔ LEU mutation that effects the N terminal hydrophobic end far from the putative catalytic site of LCAT also suggests that the N terminal region is involved in binding HDL particles and may effect cofactor or substrate binding rather than the enzymatic transfer mechanism [45]. In contrast to significant residual alpha-LCAT activity and an approximately normal specific LCAT activity. The data suggest that the defect in this case results from synthesis of an enzyme which is not normally secreted [47]. Finally, the presence of significant amounts of cholesteryl esters in most FED plasma and the ability of LCAT to be active in larger particles suggest that the alpha-LCAT defects currently described in FED patients effect its ability to bind or to be active in small substrate particles such as pre-beta-HDL [43].

Corneal opacification may be induced by several different lipoprotein disorders associated with HDL deficiency as well as by primary LCAT deficiency. This indicates a cooperative process involving LCAT and apoproteins contained in lipoprotein substrates, which is frequently invoked to explain the functional defect of mutated LCAT. These HDL deficiencies are due to increased catabolism of HDL eg in Tangier disease [50, 51], or to a DNA inversion including parts of the structural genes for apo A-I and C-III [52] and a deletion of the complete A-I, A-IV, C-III gene cluster on chromosome 11 [53] or to specific apo A-I variants [14, 54]. The degree of corneal opacity ranges from late and mild cloudiness in Tangier disease to severe opacification and does not correlate with the degree of HDL reduction or the risk of premature atherosclerosis. Mild fasting hypertriglyceridaemia is present in HDL deficiencies especially those involving the absence of apo C-III. The homozygous base deletion in the fourth exon of apo A-I that results in a shift of the reading frame [54] and leads to a mutant apo A-I with 229 instead of 243 AA demonstrates the importance of the activation properties of the carboxyl terminal region of Apo A-I. Whatever may be the consequences of these anomalies (the inabililty to form a subclass of small sized HDL, the absence of a normal amount of circulating HDL for the maintainance of regular LCAT plasma concentrations, the enhanced catabolism of HDL) a decreased mass and activity of LCAT is observed [54].

In addition to the deficiencies considered above, numerous other factors effect the function or properties of LCAT (apo A-IV, C-I, C-II, C-III, triglyceride rich lipoproteins, serum amyloid A and paraoxonase) [55-58]. They strengthen, the highly genetic and phenotypic heterogeneity of familial disorders with abnormal LCAT activity.

CONCLUDING REMARKS

LCAT deficiencies may be classified into three types [59] :
(i) Those with the apparent absence of both LCAT activity and protein mass.
(ii) Those with functionally defective LCAT.
(iii) Those with a small amount of apparently normal LCAT owing to the high variability in LCAT mass and activity measurements according to the methods and the substrates used, standardised assays need to be developed along the lines of those already reported [60-62]. It is especially important to describe whether exogenous or endogenous substrates should be used.

As none of the molecular defects uncovered to date involves one of the "hot spots" of the enzyme (catalytic and glycosylation sites) it may be speculated that plasma LCAT deficiency may be caused by a defect in post translational processing of the enzyme or by the alteration of binding sites participating in the interaction between LCAT, the lipid substrates and activator proteins on the surface of lipoprotein particles. Further advances depend on more detailed knowledge of the secondary and tertiary structures of LCAT.

Additional studies are required to further elucidate the functions of the different domains of LCAT and its detailed role in cholesterol efflux and in plasma lipoprotein metabolism. The complete understanding of the pathophysiology of the clinical features associated with LCAT deficiency and the genetic and phenotypic heterogeneity of the different types of LCAT deficiency are dependant on such advances. For example, corneal arcus and opacification of the cornea which is the major and constant feature of primary LCAT deficiencies is only partially understood ; the defect in cholesterol efflux and the deposition of modified plasma lipoproteins have both been invoked to explain the electron microscopic data [63-65].

REFERENCES

[1]   Glomset J.A. and Norum K.R. The metabolic role of lecithin : choleste-
       rol acyltransferase. Adv. Lipid Res., 2, 1973, 1-65.
[2]   Hill J.S., Karmin O., Wang X., Paranjape S., Dimitrijevich D., Lacko A.G.
       and Pritchard P.H. Expression and characterization of recombinant lecithin :
       cholesterol acyl transferase. J. Lipid Res., 34, 1993, 1245-1251.
[3]   Fielding C.J. and Collet X. Phopholipase activity of lecithin : cholesterol
       acyltransferase. in : Methods in enzymology, Acad. Press., 197, 1991, 426-433.
[4]   Subbaiah P.V., Liu M., Bolan P.J. and Paltauf F. Altered positional specificity
       of human plasma lecithin : cholesterol acyltransferase in the presence of sn-2
       arachidonoryl phosphatidyl cholines. Mechanism of formation of saturated choles-
       teryl esters. Biochim. Biophys. Acta, 1128, 1992, 83-92.
[5]   Parks J.S., Thuren T.Y. and Schmitt J.D. Inhibition of lecithin : cholesterol acyl
       acyltransferase activity by synthetic phosphatidylcholine species containing eico-
       sapentaenoic acid or docosahexaenoic acid in the sn-2 position. J. Lipid Res.,
       33, 1992, 879-887.
[6]   Thomas M.S., Babiak J. and Rude L.L. Lecithin : cholesterol acyltransferase
       (LCAT) catalyzes transacylation of intact cholesteryl esters. J. Biol. Chem.,
       265, 1990, 2665-2670.
[7]   Carlson L.A. and Holmquist L. Evidence for deficiency of high density lipoprotein
       lecithin : cholesterol acyltransferase activity (alpha LCAT) in Fish Eye Disease.
       Acta Med. Scand., 218, 1985, 189-196.
[8]   Brown M.L., Hesler C. and Tall R. Plasma enzymes and transfer proteins in
       cholesterol metabolism. Curr. Opinion Lipidol., 1, 1990, 122-127.
[9]   Francone O.L., Gurakar A. and Fielding C. Distribution and functions of lecithin :
       cholesterol acyltransferase and cholesteryl ester transfer protein in plasma lipo-
       proteins. J. Biol. Chem., 264, 1989, 7066-7072.
[10]   Norum K.R. and Gjone E. Familial plasma lecithin : cholesterol acyltransferase
       deficiency. Biochemical study of a new inborn error of metabolism.
       Scand. J. Clin. Lab. Invest., 20, 1967, 231-243.
[11]   Carlson L.A. and Philipson B. Fish-Eye Disease : A new familial condition with
       massive corneal opacities and dyslipoproteinemia. Lancet, 2, 1979, 922-924.
[12]   Frolich J., Hoag G., Mc Leod R., Hayden M., Godin D.V., Wadsworth L.D.,
       Critchley J.D. and Pritchard P.N. Hypoalphalipoproteinemia resembling Fish Eye
       Disease. Acta Med. Scand., 221, 1987, 291-298.
[13]   Assmann G., Von Eckardstein A. and Funke H. Lecithin : cholesterol acyltrans-
       ferase and Fish Eye Disease. Curr. Opinion Lipid., 2, 1991, 110-117.
[14]   Assmann G., Schmitz A., Funke H. and Von Eckardstein A. Apolipoprotein A-I
       and HDL deficiency. Curr. Opinion Lipid., 1, 1990, 110-115.
[15]   Matsuzawa Y., Yamashita S., Kameda K., Kubo M., Tarin S and Hara I.
       Marked hyper-HDL2-cholesterolemia associated with premature corneal opacity.
       Atherosclerosis, 53, 1984, 207-212.
[16]   Mc Lean J., Fielding C., Drayna D., Dieplinger H., Baer B., Kohr W., Heuzel W.
       and Lawn R. Cloning and expression of human lecithin : cholesterol acyltransfera-
       se cDNA. Proc. Natl. Acad. Sci. USA, 83, 1986, 2335-2339.
[17]   Chung K.S., Jahnani M., Hara S. and Lacko A.B. Human plasma LCAT amino
       acid sequence. Can. J. Biochem. Cell. Biol., 61, 1983, 875-881.
[18]   Teisberg P. and Gjone E. A probable linkage of LCAT in man to the alpha hapto-
       globine locus on chromosome 16. Nature, 249, 1974, 550-551.
[19]   Azoulay M., Henry I., Tata F., Weil D., Grzeschik K.H., Mc Intyre N., Williamson
       R., Humphries S.E. and Juniere C. The structural gene for human lecithin : choles-
       terol acyltransferase maps to 16 q 22. Cytogenet. Cell. Genet., 40, 1985, 573.

[20] Meroni G., Malgaretti N., Pontoglio M., Ottolenghi S. and Tarannelli R. Functional analysis of the human LCAT gene promoter. B.B.R.C., 180, 1991, 1469-1475.

[21] Yang C.Y., Manaogian D., Pao Q., Fee F.S., Knapp R.D., Gotta A.M. and Pownall H.J. Lecithin : cholesterol acyltransferase : functionnal regions and a structural model of the enzyme. J. Biochem., 262, 1987, 3086-3091.

[22] Jauhiainen M., Stevenson K.J. and Dolphin P.J. Human lecithin : cholesterol acyltransferase. The vicinal nature of cysteine 31 and cysteine 184 in the catalytic site. J. Biol. Chem., 263, 1988, 6525-6533.

[23] Francone O.L. and Fielding C.J. Structure - Function relationships in human lecithin : cholesterol acyltransferase site - directed mutagenesis at serine residues 181 and 216. Biochemistry, 30, 1991, 10074-10077.

[24] Francone O.L. and Fielding C.J. Effects of site -directed mutagenesis at residues of cysteine-31 and cysteine-184 on lecithin = cholesterol acyltransferase activity. Proc. Natl. Acad. Sci. USA, 88, 1991, 1716-1720.

[25] Jonas A. Lecithin : cholesterol acyltransferase in the metabolism of high density lipoproteins. Biochim. Biophys. Acta, 1084, 1991, 205-220.

[26] Subbaiah P.V. Lysolecithin acyltransferase of human plasma : assay and characterization of enzyme activity. in : Methods in Enzymology, Acad. Press, 129, 1986, 790-797.

[27] Parthasarathy S. and Barnett J. Phospholipase A2 activity of low density lipoprotein : evidence for an intrinsic phospholipase A2 activity of apolipoprotein B-100. Proc. Natl. Acad. Sci. USA, 87, 1990, 9741-9745.

[28] Collet X. and Fielding C.J. Effects of inhibitors of N-linked oligo saccharide processing on the secretion, stability, an activity of LCAT. Biochemistry, 30, 1991, 3228-3234.

[29] Francone O.L., Evangelista L. and Fielding C.J. Lecithin-cholesterol acyltransferase : effects of mutagenesis at N-linket oligo saccharide attachment sites on acyl acceptor specificity. Biochim. Biophys. Acta, 1166, 1993, 301-304.

[30] Anantharamaiah G.M., Venkatachalapathi Y.V., Brouillette C.G. and Segrest J.P. Use of synthetic peptide analogues to localize lecithin : cholesterol acyl transferase activating domain in apolipoprotein A-I. Arteriosclerosis, 10, 1990, 95-105.

[31] Minnich A., Collet X., Roghani A., Cladaras C., Hamilton R.L., Fielding C. and Zannis V.I. Site directed mutagenesis and structure - function analysis of the human apo A-I. Relation between LCAT activation and lipid binding. J. Biol. Chem., 267, 1992, 16553-16560.

[32] Banka C.L., Bonnet D.J., Black A.S., Smith R.S. and Curtiss L.K. Localization of an apolipoprotein A-I epitope critical for activation of LCAT. J. Biol. Chem., 266, 1991, 23886-23892.

[33] Wong L., Curtiss L.K., Huang J., Mann C.J., Maldonado B. and Roheim P.S. Altered epitope expression of human interstitial fluid apolipoprotein A-I reduces its ability to activate LCAT. J. Clin. Invest., 90, 1992, 2370, 2375.

[34] Calvo C., Ulloa N., Del Pozo R and Verdugo C. Decreased activation of LCAT by glucated apolipoprotein A-I. Eur. J. Clin. Chem. Clin. Biochem., 31, 1993, 217-220.

[35] Taranelli R., Pontoglio M., Candiani G., Ottolenghi S., Dieplinger H., Catapeno A., Albers J., Vergani C. and Mc Lean J. Lecithin cholesterol acyl transferase deficiency : molecular analysis of a mutated allele. Hum. Genet., 85, 1990, 195-199.

[36] Gotoda T., Yamada N., Murase T., Sakuma M., Murayama N., Shimano H., Kozaki K., Albers J., Yazaki Y. and Adanuma Y. Differential phenotypic expression by three mutant alleles in familial LCAT deficiency. Lancet, 338, 1991, 778-781.

[37] Maeda E., Naka Y., Matozaki T., Sakuma M., Akanuma Y., Yoshino G. and Kasuga M. LCAT deficiency with a missense mutation in exon 6 of the LCAT gene. B.B.R.C., 178, 1991, 460-466.

[38] Skretting G., Blomhoff J.P., Solheim J. and Prydz H. The genetic defect of the original Norwelfian lecithin : cholesterol and transferase deficiency families. F.E.B.S., 309, 1992, 307-310.

[39] Funke H., Von Eckardstein A., Pritchard P.H., Hornby A.E., Wiebusch H., Motti C., Hayden M.R., Dachet C., Jacotot B., Gerdes U., Faergeman O., Albers J.J., Colleoni N., Catapano A., Frohlich J. and Assmann G. Genetic and phenotypic heterogeneity in familial lecithin : cholesterol acyltransferase (LCAT) deficiency. J. Clin. Invest., 91, 1993, 677-683.

[40] Klein H.G., Lohse P., Duverger N., Albers J.J., Rader D.J., Zech L.A., Fojo S.S. and Brewer H.B. Two different allelic mutations in the LCAT gene resulting in classic LCAT deficiency : LCAT (TYR 83 → STOP) and LCAT (TYR 156 → As N). J.Lipid Res., 34, 1993, 49-58.

[41] Dumon M.F., Berard A., Dabadie H., Rougier M.B. and Clerc M. Observation biochimique d'un nouveau cas de déficit classique en lécithine : cholesterol acyltransférase. Ann. Biol. Clin. (submitted).

[42] Frolich J., Hoag G., Mc Leod R., Hayden M., Godin D.V., Wadsworth L.D. Critchley J.D., Pritchard P.H. Hypoalphalipoproteinemia resembling Fish Eye Disease. Acat Med. Scand., 221, 1987, 297-298.

[43] Funke H., Von Eckardstein A., Pritchard P.H., Albers J.J., Kastelein J.J.P., Drost C. and Assmann G. A molecular defect causing Fish Eye Disease : an amino acid exchange in LCAT leads to the selective loss of alpha-LCAT activity. Proc. Natl. Acad. Sci. USA, 88, 1991, 4855-4859.

[44] Klein H.G., Lohse P., Pritchard P.H., Bojanovski D., Schmidt H. and Brewer H.B. Two different allelic mutations in the LCAT gene associated with the Fish Eye syndrome. J. Clin. Invest., 89, 1992, 499-506.

[45] Skretting G. and Prydz H. An amino acid exchange in exon I of the human LCAT gene is associated with Fish Eye Disease. B.B.R.C., 182, 1992, 583-587.

[46] Clerc M., Dumon M.F., Sess D., Freneix-Clerc M., Mackness M. and Conri C. A "Fish-Eye disease" familial condition with massive corneal opacities and hypoalphalipoproteinemia : clinical, biochemical and genetic features. Europ. J. Clin. Invest., 21, 1991, 616-624.

[47] Klein H.G., Fojo S.S., Duverger N., Clerc M., Dumon M.F., Albers J.J., Marcovina S. and Brewer H.B. Fish-Eye syndrome : a molecular defect in LCAT gene associated with normal alpha-LCAT specific activity. Implications for classification and diagnosis. J. Clin. Invest., 92, 1993, 379-485.

[48] Rader D.J., Ikewaki K., Duverger N., Schmidt H., Pritchard H., Frolich J., Clerc M., Dumon M.F., Fairwell T., Loren Z., Fojo S.S. and Brewer H.B. Rapid catabolism of apolipoprotein A-II and high density lipoproteins containing apo A-II in classic LCAT deficiency and Fish-Eye disease. J. Clin. Invest. (in press).

[49] Kastelein J.J.P., Pritchard P.H., Erkelens D.W., Kuivenhoven J.A., Albers J.J. and Frolich J.J. Familial high density lipoprotein deficiency causing corneal opacities (Fish Eye disease) in a familial dutch descent. J. Intern. Med., 231, 1992, 413-419.

[50] Assmann G., Schmitz G. and Brewer H.B. Familial high density lipoprotein deficiency : Tangier disease. in : The metabolic basis of inherited disease. Scriver C.R., Beaudet A.L., Sly W.S. and Valle D. Ed. Mc Graw-Hill Inc, N.Y., 1989, 1167-1282.

[51] Dumon M.F., Freneix-Clerc M., Maviel M.J. and Clerc M. Familial hypocholes-
terolemia and HDL deficiency. Adv. Exp. Med., 285,1991, 161-171.

[52] Norum R.A., Lakier J.B., Goldstein S., Angel A., Goldberg R.B., Block W.D.,
Nofze D., Dolphin P.J., Edelglass J., Bogorad DD and Alaupovic P. Familial
deficiency of apolipoproteins A-I and C-III and precocious coronary-artery
disease. N. Engl. J. Med., 306, 1982, 1513-1519.

[53] Oraovas J.M., Cassidy D.K., Civeira F., Bisgaier C.L. and Schaeffer E.J. Familial
apo A-I, C-III and A-IV deficiency and premature atherosclerosis due to a deletion
of a gene complex on chromosome 11. J. Biol. Chem., 264, 1989, 16339-16342.

[54] Funke H.A., Von Eckardstein A., Pritchard P.H., Karas M., Albers J.J. and
Assmann G. A frameshift mutation in the apolipoprotein A-I gene causes high density lipoprotein deficiency, partial lecithin : cholesterol acyltransferase deficiency
and corneal opacities. J. Clin. Invest., 87, 1991, 371-376.

[55] Schmitz G. and Williamson E. High-density lipoprotein metabolism, reverse cholesterol transport and membrane protection. Curr. Opin. Lipidol., 2, 1991, 177-189.

[56] Neary R., Bhatnagar D., Durrington P., Ishola M., Arrol S. and Mackness M.
An investigation of the role of LCAT and triglyceride-rich lipoprotein in the metabolism of pre-beta high density lipoproteins. Atherosclerosis, 89, 1991 35-48.

[57] Mackness M.I., Walker C.H. and Carlson L.A. Low A-Esterase activity in serum
of patients with Fish-Eye disease. Clin. Chem., 33, 1987, 587-588.

[58] Mackness M.I., Peuchant E., Dumon M.F., Walker C.H. and Clerc M.
Absence of "A"-Esterase activity in the serum of a patient with Tangier disease.
Clin. Biochem., 22, 1989, 475-478.

[59] Albers J.J., Bergelin R.O. and Aldolphson J.C. Population-based reference values
for LCAT. Atherosclerosis, 43, 1982, 369-379.

[60] Stokke K.T. and Norum K.R. Determination of LCAT in human blood plasma.
Scand. J. Clin. Lab. Invest., 27, 1971, 21-27.

[61] Channon K.M., Clegg R.J., Bathnagar D., Ishola M., Arrol S. and Durrington P.N.
Investigation of lipid transfer in human serum leading to the development of an iso-
topic method for the determination of endogenous cholesterol esterification and
transfer. Atherosclerosis, 80, 1990, 217-226.

[62] Frohlich J. and Mc Leod R. LCAT deficiency syndromes. Adv. Exp. Med. Biol.,
201, 1987, 181-194.

[63] Barchiesi B.J., Eckel R.H. and Ellis P.P. The cornea and disorders of lipid metabolism. Survey. Ophtalm., 36, 1991, 1-22.

[64] Koster H., Savoldelli M., Dumon M.F., Dubourg L., Clerc M. and Pouliquen Y.J.M.
A Fish-Eye disease like familial condition with massive corneal clouding and dyslipoproteinemia. Report of clinical, histologic, electron microscopic and biochemical
features. Cornea, 11, 1992, 452-464.

[65] Clerc M. and Pouliquen Y.J.M. L'Arcus juvenilis et les fonctions de la lecithine :
cholesterol acyltransferase. A propos d'un cas familial de Fish-Eye disease.
Bull. Acad. Natl. Med. Paris, 1993 (in press).

DEVELOPMENT OF A SPECIFIC ASSAY FOR PANCREATIC LIPASE ACTIVITY FOR DIAGNOSTIC PURPOSES

Georges Férard,[1] Jean Marc Lessinger,[1] Panteleimon Arzoglou,[2] Atanase Visvikis,[3] Wolfgang Junge[4]

[1]Laboratoire de Biochimie appliquée, Faculté de Pharmacie, Université Louis Pasteur de Strasbourg, F 67400 Illkirch, France
[2]Laboratory of Biochemistry, Department of Chemistry, University of Thessaloniki, 540 06 Thessaloniki, Greece
[3]Centre du Médicament, 30 rue Lionnois, F 54000 Nancy, France
[4]Friedrich-Ebert Krankenhaus, Neumunster, Friesenstrasse 11, DW 2350 Neumunster, Germany

INTRODUCTION

Determination of pancreatic lipase (E.C. 3.1.1.3) in serum is frequently required for diagnosing pancreatitis for at least two reasons. First, an increase in levels of pancreatic enzymes in blood or in urine is a part of the definition of acute pancreatitis [1]. Among the pancreatic enzymes, serum lipase activity is presently considered as the most efficient marker of this disease as shown in many comparative studies. As an example, data from 15 comparative studies published since 1985 indicate that the mean values of sensitivity of serum amylase, pancreatic isoamylase and lipase activity for the diagnosis of acute pancreatitis are 0.88 0.85 and 0.91 and those of specificity are 0.74 0.81 and 0.88, respectively. Second, several assays recently developed are easy to perform in an automated manner. Our goal is to analyze the characteristics of contemporary methods for lipase assay and to compare the results obtained by these methods for plasma samples from patients suffering from acute pancreatitis. The interassay agreement will be discussed and some recommendations will be formulated to improve the specificity and the accuracy of lipase assays for diagnostic use.

COMPARISON OF CONTEMPORARY ASSAYS FOR LIPASE

The characteristics of four currently used methods have been compared in order to determine their common features and pecularities. One is a turbidimetric assay [2] from Boehringer Mannheim, the second is performed by reflectometry [3] and is proposed by Eastman Kodak, the third is a colorimetric test [4] manufactured by Sigma and the last one is an UV spectrophotometric assay [5] from Wako. All the assays were performed on the same apparatus (Cobas Fara) at 30 °C, except for the reflectometric technique which was performed at 37 °C on a Ektachem 700 (Eastman Kodak). In fact, although different modes of measurement have been employed, all methods use an emulsified tri- or a diglyceride as substrate. The four assays have in common the use of a supramicellar concentration of one or two detergents, most often constituted by bile salt(s). Such a condition is regarded as inhibitory for all enzymes active on the ester bonds of a substrate forming an interface. All reaction media contain colipase in order to reverse specifically the inhibition of pancreatic lipase as well as one or two bivalent cations. Adaptations on an automatic analyser were described for all the procedures which thus can be performed in a few minutes with an acceptable interserial reproducibility.

Nevertheless, procedures differ by the nature of the substrate : Triglyceride with long chains such as triolein, triglyceride with one long chain at C1 and 2 acetyls or a 1,2-diglyceride do not yield the same interface and, perhaps, do not ensure the same analytical specificity towards esterase activities. Moreover, the nature and the concentration of detergent(s) vary with the procedure. More particularly, the potency of dodecylbenzene sulfate as a lipase inhibitor is not well known and the concentration of bile salts in some assays seems low. Furthermore, it is not possible to know the actual concentration of each reagent in the dry chemistry assay. Colipase and cation concentrations are not discussed here, because their concentrations have to be optimized with regards to the nature, the dispersion and the concentration of the substrate as well as the nature and the concentration of detergent. In other words, different combinations of optimized concentrations of the reagents are possible and have been observed in practice. Such conditions are only optimized in the sense that they allow the expression of a maximal lipase activity. In fact, a complete optimization study must include an evaluation of the analytical specificity in order to retain the most appropriate conditions for lipase activity measurement.

Calibration of lipase assays is often particular since most assays need a calibrator to transform arbitrary signals into catalytic activities. Lipase standards are proposed by manufacturers, but the catalytic properties as well as the modes of titration of the products vary [6,7]. Some standards are from human origin, others from porcine origin. It is of prime importance to ensure a similarity of catalytic properties between the calibrator and the enzyme which is intended to be determined in patient specimens i.e. the "pancreatitis" lipase mentioned by different authors [8,9].

INTERASSAY AGREEMENT OF ROUTINE METHODS

Twenty eight plasma samples from different patients suffering from acute pancreatitis have been assayed by the aforementioned procedures. For each patient, only the sample containing maximal lipase activity has been retained for further experiments.

Results of lipase activity in patients' plasma are highly method dependent since mean values differed by a factor up to 14. To take into account this variability, the range of usual values for the particular method of measurement is indicated after the result. The mean values for each method of this ratio unfortunately varied by a factor up to 13. This signifies that data of plasma lipase activities obtained by different procedures cannot be today utilized by the clinicians in a logical manner. Different cutoff values should be retained according to the methods used. Our observation is confirmed by published studies since authors have proposed cutoff values ranging from one the nine times the upper limit of usual values for diagnosing acute pancreatitis [10]. This interassay discrepancy may be due to at least two factors: the calibrators including their titration and the analytical specificity. In order to evaluate the magnitude of the first factor all methods were calibrated by the same enzyme preparation i. e. purified and stabilized human pancreatic lipase [11]. All the results were expressed in relation to the reaction rate observed with this preparation. After such treatment, mean values were calculated and a considerable improvement was observed for interassay agreement since mean values varied by a factor of 1.0 to 1.4. In other words, it becomes possible to transmit comprehensive information to the clinicians. This information is much less arbitrary since the method dependancy of the lipase assay is dramatically reduced. To achieve this goal, a reference material has to be developed. One feasibility study starting from human pancreatic juice has been performed. The purified preparation has been stabilized in an albumin matrix and exhibits catalytic properties similar to those of the corresponding enzyme in the plasma of patients suffering from acute pancreatitis [11]. Another approach is the use of cDNA of human pancreatic lipase. This product has been subcloned in an eucaryotic expression vector and used to transfect V79 Chinese hamster lung cells. One cell line produced lipase which was purified. Two immunoreactive polypeptides, one of the two being glycosylated have been demonstrated. Thus, it is possible to purify and stabilise human pancreatic lipase without significant alteration of catalytic properties.

RECOMMENDATIONS FOR A REFERENCE METHOD

Several variables have been studied sequentially, concerning the choice and concentrations of emulsifier, substrate, detergent as well as the optimal concentrations of colipase and bivalent cations. Hydroxypropylmethylcellulose at a concentration of 12 g/l is proposed because it is not contaminated by bivalent cations in contrast to gum arabic. Keltrol® was not retained because it creates some difficulties during the emulsification process. Refined olive oil (Sigma) was chosen in this study because it gave in our conditions results similar to triolein for purified human pancreatic lipase and for plasma pools of patients. A 100 ml/l concentration ensures a maximal lipase activity. Olive oil may be dispersed by an ultrasonic homogenizer. This apparatus was preferred to a blender because the micelle size was more homogenous and because it was possible to prepare easily in a reproducible manner, 500 ml lots. The emulsion was found to be usable for at least 10 days when stored at +4 °C, since a constant lipase activity was observed with a stabilized lipase preparation during this delay. Another critical step is to define both the type of bile salt to be employed and the most appropriate concentration in order to inhibit as completely as possible any hydrolytic activity towards the emulsified substrate. It was previously shown that all the types of bile salts at a concentration of 70 mmol/l inhibit completely pancreatic lipase activity [12]. In the presence of 100 ml olive oil and 70 mmol deoxycholate per liter, lipase activity was found to be maximal with 0.2 mg colipase and 0.5 mmol $CaCl_2$ per liter when released fatty acids were determined at 37 °C and at pH 9.0 using a pH-Stat method with an automatic titrator. This observation was obtained both with purified human pancreatic lipase and with plasma pools of patients suffering of acute pancreatitis.

ANALYTICAL SPECIFICITY: PROBLEMS AND SOLUTIONS

Data on interassay agreement show that a part of interassay discrepancy cannot be corrected by the use of a common and adequate calibrator. It is suggested that some hydrolases may interfere in a variable manner according to the method of measurement. thus, under these optimized conditions, the effects of potentially interfering enzymes such as lipoprotein and hepatic lipases, of acid lipases, and of carboxylesterases remain to be evaluated. In particular, the interference of an activity of human pancreatic origin should be tested [13]. Carboxylesterases are readily inactivated by diisopropylfluorophosphate at low concentrations and sodium dodecylsulfate inhibits hepatic lipase [14]. Such inhibitors should be added to improve the specificity of pancreatic lipase assays. In order to improve analytical specificity of a lipase assay, a substrate specific for lipase remains to be developed. Figure 1 illustrates that when S_1 is a triglyceride with long chains, the product formed 1,2-diglyceride becomes a substrate for pancreatic lipase and, probably, for esterases. This sequence indicates that a substrate with an unique bound hydrolysable by pancreatic lipase would be preferable and that specific inhibitors of esterase remain to be identified.

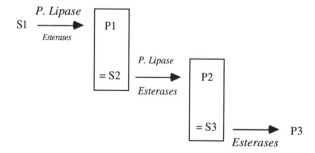

Figure 1.Scheme of action of pancreatic lipase (P. Lipase) and of esterases. In this example S1 has three ester bounds which may be hydrolysed at different rates by P. Lipase and esterases (S =substrate; P = product)

Contemporary lipase assays have in common the use of micellar substrates, detergents, colipase and bivalent cations. Although they assays are reproducible and easy to perform, they exhibit considerable interassay discrepancy whose most part can be corrected by the use of a common calibrator. In fact, a reference method and a reference material are necessary to standardize lipase assay and to improve the usefulness of lipase assay for the diagnosis of acute pancreatitis. Specific substrate as well as specific esterase inhibitors remains to be developed.

REFERENCES

1. Sarles H., Adler G., Dani R., Frey C., Gullo L., Harada H., Martin E., Norohna M., and Scuro L.A., Classifications of pancreatitis and definition of pancreatic diseases. *Digestion* 43: 1989; 234- 236.
2. Ziegenhorn J., Neumann U., Knitsch K.W., and Zwez W., Determination of serum lipase. *Clin. Chem.* 25: 1979; 1067.
3. Mauck J.C., Weaver M.S., and Stanton C., Development of a Kodak Ektachem clinical chemistry slide for serum lipase. *Clin. Chem.* 30: 1984; 1058-1059.
4. Imamura S., and Misaki H., An enzymatic method using 1,2-diglyceride for pancreatic lipase test in serum. *Clin. Chem.* 35: 1989; 1126.
5. Imamura S., and Misaki H. (1984) A sensitive method for assay of lipase activity by coupling with β-oxydation enzymes of fatty acid, *in* : Selected Topics in Clinical Enzymology, vol 12 pp. 73-77, Goldberg D.M., Werner M., eds, Walter de Gruyter, Berlin-New York.
6. Rick W., Kinetischer Test zur Bestimmung der Serumlipaseaktivität. *Z. Klin. Chem. Klin. Biochem.* 7: 1969; 530-539.
7. Tietz N.W., and Repique E.V., Proposed standard method for measuring lipase activity in serum by a continuous sampling technique. *Clin. Chem.* 19: 1973; 1268-1275.
8. Lessinger J.M., Arzoglou P.L., and Férard G., Evidence for multiple forms of pancreatic lipase in human plasma. *Adv. Clin. Enzymol.* 3: 1986; 139-150.
9. Lott J.A., and Lu C.L., Lipase isoforms and amylase isoenzymes: Assays and application in the diagnosis of acute pancreatitis. *Clin. Chem.* 37: 1991; 361-368.
10. Clavien P.A., Burgan S., and Moossa A.R., Serum enzymes and other laboratory tests in acute pancreatitis. *Br. J. Surg.* 76: 1989; 1234-1243.
11. Lessinger J.M.,Tavridou A., Arzoglou P.L., and Férard G., Interest of using a purified, stable and commutable preparation of human pancreatic lipase in indirect assays. *Anal. Lett.* 25: 1992; 1453-1468.
12. Tavridou A., Avranas A., and Arzoglou P.L., A mathematical approach to lipolysis based on the interrelationship of physicochemical and biochemical data. *Bioch. Biophys. Res. Comms* 186: 1992; 746-752 .
13. Junge W., Leybold K., and Philipp B., Identification of a non-specific carboxylesterase in human pancreas using vinyl 8-phenyloctanoate as a substrate. *Clin. Chim. Acta* 94: 1979; 109-114.
14. Demanet C., Goedhuys W., Haentjens M., Huyghens L., Blaton V., and Gorus F., Two automated fully enzymatic assays for lipase activity in serum compared: positive interference from post-heparin lipase activity. *Clin. Chem.* 38: 1992; 288-292.

POST-HEPARIN LIPOLYTIC ACTIVITIES IN CHRONIC RENAL FAILURE PATIENTS IN RELATION TO PLASMA TRIGLYCERIDES

Ali Abdallah, Maria Pascual de Zulueta, Pascale Richard,
Denise Higueret, André Cassaigne, and Albert Iron

Département de Biochimie Médicale et Biologie Moléculaire
Université de Bordeaux 2
33076 Bordeaux Cédex, France

INTRODUCTION

Chronic renal failure (CRF) is responsible for several abnormalities of lipoprotein metabolism and represents an increased risk of atherosclerosis [1,2]. Hypertriglyceridemia is very often observed in patients suffering from terminal CRF, treated by hemodialysis or not, while plasma cholesterol is either normal or slightly elevated. Among the numerous risk factors presented by uremic patients, it is not easy to know precisely the role of hypertriglyceridemia. The general increase in all the triglyceride-rich lipoproteins is a consequence of their impaired metabolism.

Plasma lipoproteins are metabolized and their concentration regulated by lipolytic enzymes, mainly lipoprotein lipase (LPL), responsible for the hydrolysis of chylomicron and VLDL triglycerides [3], and hepatic triglyceride lipase (HTGL), implicated in chylomicron remnant degradation, in IDL conversion to LDL and in HDL catabolism [4].

Previous studies have suggested that, in the post-heparinized plasma of uremic patients, LPL activity was reduced or normal while HTGL activity was markedly decreased [5,6]. Among the endogenous factors that might explain the fall in lipolytic activities, we have to consider a possible deficiency in apolipoprotein C-II (apo C-II), an activating cofactor, and the level of apolipoprotein C-III (apo C-III), an inhibitory modulator. Moreover, all the abnormalities of lipoprotein metabolism (potentialized by CRF itself) can undoubtedly explain the high cardiovascular risk in uremic patients.

Esterases, Lipases and Phospholipases, Edited by M.I. Mackness
and M. Clerc, Plenum Press, New York, 1994

The purpose of this article is : firstly to compare normotriglyceridemic and hypertriglyceridemic CRF patients by analysis of triglyceride metabolism including the measurement of lipase activities in the post-heparinized plasma and of apolipoprotein and lipoparticle cofactors ; on the other hand and secondly, to investigate the cardiovascular risk by studying atherosclerosis markers and protective factors.

PATIENTS AND METHODS

Thirty six CRF patients (13 men and 23 women) were included. The levels of plasma triglycerides (TG) allowed us to take into account two groups of CRF subjects : 16 normotriglyceridemics (TG < 1.8 mmol/l) and 20 hypertriglyceridemics (TG ≥ 1.8 mmol/l). Plasma total cholesterol and triglycerides were determined by enzymatic techniques. Apolipoproteins A-I and B were measured by immunonephelometry, apo C-II and apo C-III by radial immunodiffusion. Electroimmunodiffusion allowed the evaluation of lipoparticles LP A-I, LP C-III:B, LP C-III:nonB [7]. Post-heparin lipoprotein lipase (LPL) and hepatic triglyceride lipase (HTGL) were assayed in EDTA blood samples [8] after intravenous injection of 100 U of heparin per kg body weight. Statistical analysis was done by using Student's t-test and linear regression analysis.

RESULTS

The comparison of CRF patients selected according to whether they had normotriglyceridemia (NTG) or hypertriglyceridemia (HTG) indicated some significant variations of plasma lipid parameters (Figure 1) : (i) a higher cholesterolemia in HTG patients; (ii) a decrease of the hepatic triglyceride lipase activity in HTG patients ; (iii) an increase in apo C-II and apo C-III, in LP C-III:B and LP C-III:non B in HTG subjects ; (iv) an increase in apo B and a concomitant decrease of the apo A-I/apo B ratio in HTG subjects.

CONCLUSION

In relation to TG metabolism, there were negative correlations between, HTGL activity and TG level, and between HTGL activity and apo C-III. Conversely, there was a positive correlation between HTGL activity and the apo A-I/apo B ratio.

We noted, in relation to the cardiovascular risk, a positive correlation (which has already been reported) between the apo A-I/apo B ratio and the concentration of LP A-I particles which are considered a protection factor, and a negative correlation between the apo A-I/apo B ratio and LP CIII:B particles which represent a risk factor.

Therefore, we have confirmed that HTG uremics are more exposed to CV risk than NTG uremics. Further investigations taking into account the interaction of nutritional and

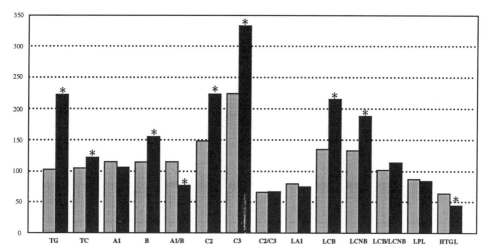

Figure 1. Comparison between normotriglyceridemic (▩) and hypertriglyceridemic (■) CRF patients. Abscissæ : plasma lipidic parameters, TG : triglycerides; TC : total cholesterol ; A1 : apo A-I ; B : apo B ; A1/B : apo A-I/apo B ratio ; C2 : apo C-II ; C-3 : apo C-III ; C2/C3: apo C-II/apo C-III ratio ; LA1 : LP A-I ; LCB : LP C-III:B ; LCNB : LP C-III:nonB ; LCB/LCNB : LP C-III:B/LP C-III:nonB ratio ; LPL : lipoprotein lipase activity ; HTGL : hepatic triglyceride lipase activity. Ordinates : values expressed as percentiles. (*) : significant (p < 0.05).

genetic factors in CRF patients could be necessary for a better understanding of the occurrence of hypertriglyceridemia and to appreciate the level of their atherogenic risk.

ACKNOWLEDGEMENTS

We are greatful to Pr M. Aparicio (Néphrologie, Centre Hospitalier et Universitaire de Bordeaux, France) and Dr N. Ammar (Néphrologie, Centre Hospitalier Général de Libourne, France) for providing blood samples of CRF patients.

REFERENCES

1. Appel, G., Lipid abnormalities in renal disease, *Kidney Int.* 39 (1991) 169-183 .

2. Attman, P., and Alaupovic, P., Lipid and apolipoprotein profiles of uremic dyslipoproteinemia : relation to renal function and dialysis, *Nephron* 57 (1991) 401-410 .

3. Taskinen, M.R., Lipoprotein lipase in hypertriglyceridemias, *in* : "Lipoprotein lipase," J. Borensztajn, ed., Evener Publishers, Chicago (1987) pp 201-227.

4. Simard, G., and Perret, B., La triglycéride lipase hépatique, *Ann. Biol. Clin.* 48 (1990) 61-76.

5. Attman, P., Gustafon, A., and Alaupovic, P., Lipid metabolism in patients with chronic renal failure in the predialytic phase, *Contr. Nephrol.* 65 (1988) 24-32.

6. Shoji, T., Nishizawa, Y., Nishitani, H., et al., Impaired metabolism of high density lipoprotein in uremic patients, *Kidney Int.* 41 (1992) 653-1661.

7. Fruchart, J.C., Du cholestérol aux particules lipoprotéiques marqueurs et/ou facteurs de risque, *Bull. Acad. Natle. Méd.* 175 (1991) 51-65.

8. Nilsson-Ehle, P., Measurements of lipoprotein lipase activity, *in* : "Lipoprotein lipase," J. Borensztajn, ed., Evener Publishers, Chicago (1987) pp 59-77.

RABBIT PANCREATIC LIPASE: PURIFICATION AND SOME NOVEL CATALYTIC PROPERTIES

A. Lykidis,[1] A. Avranas,[2] V. Mougios,[3] P. Arzoglou[1]

[1] Laboratory of Biochemistry, School of Chemistry
[2] Laboratory of Physical Chemistry, School of Chemistry
[3] Laboratory of Exercise Biochemistry, Department of Physical Education and Sport Science
Aristotle University of Thessaloniki, 54006 Thessaloniki, Greece

INTRODUCTION

One of the main features of lipolysis is that it occurs as a two-step reaction. In the first step, triglycerides (TG) are hydrolyzed producing 1,2-diglycerides (DG) which are subsequently hydrolyzed to 2-monoglycerides (MG). Therefore, DG constitute both the first product of lipolysis and the substrate for the second reaction. In order to compare these reactions, lipolysis has been considered to be a common chemical sequential reaction. Following the determination of the relative concentrations of the component parts, the rate constants of the two reactions could be calculated. These rate constants are the kinetic parameters we use in order to compare the two sequential reactions.

It has been shown in a previous publication that the partition of bile salts between the lipidic and the aqueous phase was of importance in the expression of lipase activity [1]. This partition may be affected by lecithin [2]. We thus decided to investigate both the effect of lecithin on the reaction rate in the presence of various bile salt concentrations as well as the effect of adding lecithin on the zeta potential of the substrate droplets and, subsequently, on the partition of bile salts.

The present studies were carried out with rabbit pancreatic lipase which was purified for the first time by a procedure described herein.

MATERIALS AND METHODS

Purification. The lipolytic activity was extracted from 1 g of pancreas acetone powder by continuous stirring in 30 ml of a 20 mM sodium acetate buffer, pH 4.0, for 60 min at 4°C. After centrifugation at 10000 g for 10 min the pH of the supernatant was adjusted to 5.0 with NaOH. The preparation was loaded onto an 8x2.5 cm CM52 column previously equilibrated at pH 5.0. Lipase activity was eluted from the column with a NaCl gradient (0-0.3 M) and precipitated with $(NH_4)_2SO_4$; the 20-50 % pellet (which contained the lipolytic activity) was

Esterases, Lipases and Phospholipases, Edited by M.I. Mackness
and M. Clerc, Plenum Press, New York, 1994

Table 1. Main purification steps of rabbit pancreas lipase

Step	Units	mg protein	U/mg	Purification	Recovery (%)
Extraction	814	54	15	1	100
CM-52	592	12	49	3	72
Ultrogel	493	1.6	308	20	60
Phenyl-Seph	358	0.7	477	32	44
Octyl-Seph	255	0.4	637	42	31
DEAE-Trisacryl	200	0.3	666	44	24

collected and resuspended in 1 ml of a 30 mM Tris/HCl buffer, pH 9.0 (buffer A). All subsequent steps were carried out using this buffer. The enzyme-containing preparation was loaded onto a 55x1 cm gel filtration column (Ultrogel AcA 44). The eluate was subjected to two steps of hydrophobic chromatography : a 4x1.4 cm Phenyl-Sepharose column, in which the lipolytic activity was not retained, and 3x1.4 cm Octyl-Sepharose column from which the enzyme activity was eluted with a 4 mM taurodeoxycholate in buffer A. Finally, an anion exchange chromatography (5x1.4 cm, DEAE-Trisacryl) column was employed, previously equilibrated with a solution of 4 mM taurodeoxycholate in buffer A, and the lipolytic activity was eluted with a NaCl gradient (0-0.3 M).

Assays. The standard assay procedure followed is described in [3]. The rate of production of diglycerides, monoglycerides and fatty acids was determined at different time intervals. This was achieved following extraction of lipids from the reaction mixture, separation of lipid classes by TLC, (trans)methylation and quantitation of fatty acid methyl esters by capillary gas chromatography.

RESULTS AND DISCUSSION

We purified lipase from rabbit pancreas according to the original procedure described in Materials and Methods. The results are shown in Table 1.

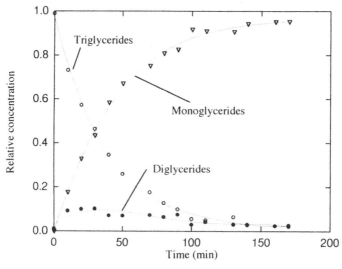

Figure 1. Time course of lipolysis in the presence of 2.5 mM triolein, 35 mM deoxycholate, 0.5 mg/l colipase, pH 9.0 at 37^0 C

The enzyme preparation appears on SDS-PAGE as one band with an approximate molecular mass of 45 kD. The isoelectric point is around 6.4.

The profile of relative concentrations of TG, DG, MG vs time is depicted in Figure 1. As is the case for other pancreatic lipases the final products are MG and free fatty acids.

The profile of Fig.1 is indicative of first order sequential reactions. The kinetic equations for such reactions are:

$$[TG] = [TG]_0 \, e^{-k_1 t}$$

$$[DG] = [TG]_0 (e^{-k_1 t} - e^{-k_1 t}) / (k_2 - k_1)$$

$$[MG] = [TG]_0 ((1-(k_2 e^{-k_1 t} - k_1 \, e^{-k_1 t})) / (k_2 - k_1))$$

where $[TG]_0$ is the initial concentration of TG and k_1 and k_2 are the rate constants for TG and DG hydrolysis, respectively.

Using the above equations, and with the aid of Sigma Plot software k_1 and k_2 are calculated to be

$k_1 = 0.0111 \text{ min}^{-1}$ (0.00018 Std error of estimate)
$k_2 = 0.0363 \text{ min}^{-1}$ (0.00206 Std error of estimate)

So, k_2 appears to be greater than k_1 indicating a faster hydrolysis of DG than TG.

We also attempted to investigate the effect of dipalmitoylphosphatidyl-choline (DPPC) and deoxycholate (DOC) in the presence of different combinations of concentrations. The first observation was that DPPC is not inhibitory at all concentrations; the presence of 8 mM DPPC together with 35 mM DOC leads to activation in the order of 60%. Increasing DPPC concentration leads to inhibition which may be total in the simultaneous presence of high DOC concentrations. Via the determination of the zeta potential [4] we observed that the addition of DPPC decreases the charge of the interface perhaps by enhancing the removal of adsorbed DOC molecules. Most probably, this can be ascribed to the formation of mixed micelles between DPPC and DOC in the bulk. It is known from earlier studies that increasing the lecithin to bile salt ratio leads to mixed micelles of a larger size [5].

Low DPPC concentrations may enhance the enzyme activity, probably by removing some DOC molecules between the lipase and colipase via the formation of mixed micelles and the subsequent formation of a high-affinity complex between the enzyme and its cofactor [6]. At higher DPPC concentrations, large mixed micelles may bind colipase molecules and therefore act as a competitive inhibitor. Under these conditions lipase activity is only partially restored with the aid of colipase concentrations 300-fold higher than those usually employed.

CONCLUSIONS

The main conclusions to be drawn from these results are the following:
-Lipolysis can be considered as a sequential reaction, where both hydrolysis steps obey first order kinetics. Furthermore diglycerides are hydrolyzed faster than triglycerides.
-Dipalmitoylphosphatidylcholine is not an absolute inhibitor of lipase activity; there are optimal combinations of concentrations of deoxycholate and DPPC that ensure maximal lipase activity.

REFERENCES

1. A. Tavridou, A. Avranas, and P. Arzoglou, A mathematical approach to lipolysis based on the interrelationship of physicochemical and biochemical data, *Biochem. Biophys. Res. Commun.* 186: (1992) 746-752
2. D. Lairon, G. Nalbone, H. Lafont, J. Leonardi, N. Domingo, J.C. Hauton, R. Verger, Possible roles of bile lipids and colipase in lipase adsorption, *Biochemistry.* 17. (1978) 5263-5269
3. P. Arzoglou, A. Tavridou, JM. Lessinger, G. Tzimas, and G. Férard, Spectro photometric determination of lipase activity in the presence of increased triolein concentration, *Ann. Biol. Clin.* 50:(1992)155-160
4. A. Avranas, G. Stalidis, and G. Ritzoulis, Demulsification rate and zeta potential of O/W emulsions, *Colloid Polym. Sci.* 266:(1988) 937-940
5. N. Mazer, G. Benedek, and M. Carey, Quasielastic light-scattering studies of aqueous biliary lipid systems. Mixed micelle formation in bile salt-lecithin solutions, *Biochemistry.* 19:(1980) 601-615
6. C. Erlanson-Albertsson, The interaction between pancreatic lipase and colipase: a protein-protein interaction regulated by a lipid, *FEBS Lett.* 162:(1983) 225-229

<u>PHOSPHOLIPASE A2</u>

PHOSPHOLIPASES A₂ AND PROSTAGLANDIN FORMATION IN RAT GLOMERULAR MESANGIAL CELLS

Henk van den Bosch[1], Margriet J.B.M. Vervoordeldonk[1], Rosa M. Sanchéz[1], Josef Pfeilschifter[2], and Casper G. Schalkwijk[1]

[1] Center for Biomembranes and Lipid Enzymology
Utrecht University
Padualaan 8, NL-3584 CH Utrecht
The Netherlands

[2] Department of Pharmacology
Biocenter, University of Basel
Klingelbergstrasse 70, CH-4056 Basel
Switzerland

INTRODUCTION AND SUMMARY

Phospholipases A₂ are believed to play important roles in the production of bioactive mediators by regulating the release of precursors for these compounds from structural membrane phosphoglycerides. It has become clear during the last years that several and distinct cellular phospholipases A₂ exist. These can be distinguished in low molecular weight 14 kDa group I (pancreatic-type) and group II (non-pancreatic-type) enzymes and in high molecular weight 85 kDa or cytosolic phospholipase A₂. These require Ca^{2+} for activity, either in their catalytic mechanism as in the case of 14 kDa phospholipases or for translocation from the cytosol to membranes as in the case of cytosolic phospholipase A₂. In addition, Ca^{2+}-independent enzymes exist. Each of these phospholipase A₂ activities appears to be subject to multiple regulation mechanisms, both at the transcriptional and post-translational level, depending on the cell type and the cell stimulators. This paper summarizes our data on the regulation of phospholipase A₂ activities in relation to prostaglandin formation in rat glomerular mesangial cells.

Initial experiments indicated that pro-inflammatory cytokines, such as interleukin-1β (IL-1β) and tumor necrosis factor, induced the secretion of phospholipase A₂ activity from these cells in parallel to enhanced prostaglandin E₂ (PGE₂) formation. Using monoclonal antibodies the secreted enzyme was identified as 14 kDa group II phospholipase A₂ (PLA₂)

Esterases, Lipases and Phospholipases, Edited by M.I. Mackness
and M. Clerc, Plenum Press, New York, 1994

and this was confirmed by sequence analysis. The enzyme is not secreted from a pre-existing cellular pool (as is present in *e.g.* platelets) but appears to be *de novo* synthesized upon cytokine treatment prior to secretion. Immunofluorescence localization studies are in line with this interpretation. Attempts to study the contribution of this enzyme to arachidonate release for prostaglandin formation by increasing its cellular level after IL-1 induction by admission of brefeldin A, to dissociate the Golgi structure, were unsuccessful due to inhibition of group II PLA$_2$ synthesis by this drug.

The cytokine-induced synthesis of prostaglandins can be prevented by dexamethasone. In parallel, PLA$_2$ activity becomes diminished. Western blot analysis indicated that this is not due to inhibition of the enzyme, but is caused by suppression of the cytokine-induced synthesis of group II PLA$_2$. These results provide a novel mechanism for the anti-inflammatory action of glucocorticosteroids not involving lipocortin induction, *i.e.* direct suppression of group II PLA$_2$ gene expression. Independent experiments confirm that lipocortins are not additionally involved in the observed decrease in PLA$_2$ activity upon dexamethasone treatment of mesangial cells. Using cDNA for group II PLA$_2$, generated from rat liver by PCR technology, indicated that the suppression of interleukin-induced PLA$_2$ synthesis by dexamethasone is mainly at the post-transcriptional level. By contrast, the suppression of forskolin-induced PLA$_2$ synthesis by dexamethasone appears to be caused mainly by inhibition of mRNA synthesis.

The interleukin- and forskolin-induced synthesis of group II PLA$_2$ can also be completely suppressed by transforming growth factor-β2 (TGF-β2). In both cases the suppression is due to decreased mRNA levels. This suppression leads also to an inhibition of interleukin-induced prostaglandin E$_2$ formation, but in contrast to dexamethasone, the addition of TGF-β2 does not completely block prostaglandin formation. This can be explained by the observation that TGF-β2 at one hand suppresses group II PLA$_2$ synthesis but at the same time increases the activity of a high molecular weight cytosolic phospholipase A$_2$.

The possible contribution of the secreted group II PLA$_2$ to arachidonate release for prostaglandin formation was investigated either by stimulation of cells in the presence of a neutralizing antibody or by addition of immunopurified enzyme to control cells and will be briefly discussed.

RESULTS AND DISCUSSION

Identification of Secreted Phospholipase A$_2$

In initial studies with cultured rat mesangial cells it was observed that treatment with IL-1β and tumor necrosis factor α induced the release of a PLA$_2$ activity into the culture medium in parallel with a highly enhanced PGE$_2$ synthesis [1,2]. However, the type of PLA$_2$ responsible for the secreted enzymatic activity remained to be characterized. The only type of PLA$_2$ reported to be present in mesangial cells was a hormonally-regulated cytosolic PLA$_2$ (cPLA$_2$). [3,4]. Using monoclonal antibodies against rat liver group II PLA$_2$, that showed no cross-reactivity with rat group I 14 kDa PLA$_2$ [5], all PLA$_2$ activity secreted from rat mesangial cells in response to IL-1β, tumor necrosis factor or forskolin could be immunoprecipitated. Western blotting and immunostaining detected only a single band of a

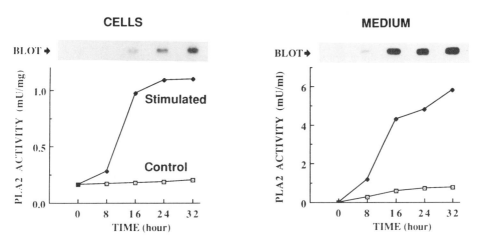

Figure 1. Induction and secretion of PLA$_2$ activity and PLA$_2$ protein after stimulation of mesangial cells with IL-1β plus forskolin. Reproduced with permission from [6].

14 kDa PLA$_2$ [6]. These experiments showed that the high molecular weight cPLA$_2$ does not become secreted and that the secreted PLA$_2$ is immunologically related to rat group II PLA$_2$. Immunoaffinity purification of this enzyme and determination of the sequence of the first 40 N-terminal amino acids [7] confirmed that this sequence is identical to that of rat platelet [8], rat spleen [9] and rat liver mitochondrial [10] group II PLA$_2$.

Induction and Secretion of Phospholipases A$_2$

It has been observed that group II PLA$_2$ in platelets is largely secreted upon thrombin stimulation [11,12], suggesting a granular localization. The presence of a membrane-associated form of PLA$_2$, with an N-terminal sequence identical to that of the secreted enzyme, left the possibility that the enzyme occurred in a membrane-associated form and became released upon thrombin activation. However, immunogold labeling experiments detected a large part of the gold particles over α-granules, directly demonstrating that the enzyme is present in a soluble form in the matrix of these granules [13,14]. To investigate whether the enhanced levels of PLA$_2$ activity and protein in the culture medium of IL-1β-stimulated mesangial cells were also caused by an enhanced secretion of enzyme from a pre-existing cellular pool, which then would become depleted upon stimulation, the cellular levels of PLA$_2$ were studied as a function of time after IL-1β stimulation. As shown in figure 1, PLA$_2$ protein cannot be detected by Western blotting of control cells. After a lag period of approximately 8 h a sharp increase in cellular PLA$_2$ activity is observed. This is not due to activation of pre-existing cellular PLA$_2$ but, as shown in the blots of stimulated cells, is caused by new synthesis of PLA$_2$, which now becomes detectable by immunostaining. No increase of PLA$_2$ activity or PLA$_2$ protein in blots of control cells was observed (data not shown). On the same time scale both PLA$_2$ enzymatic activity and PLA$_2$ protein becomes secreted into the culture medium. It can be calculated from the activity measurements that at all time points after the lag period about 85 to 90% of the PLA$_2$ activity appeared in the medium indicating that the majority of newly synthesized enzyme becomes secreted [6]. The enzyme is further denoted as sPLA$_2$.

Effect of Brefeldin A on IL-1β-Induced sPLA₂

The localization of the sPLA₂ remaining in mesangial cells after IL-1β stimulation was studied with immunofluorescence. In full accord with the results of figure 1 the enzyme could not be detected in unstimulated cells. In IL-1β-stimulated cells the major part of the fluorescence was detected in the Golgi area and in punctate fluorescent spots throughout the cytoplasm, presumably representing secretory vesicles [15,16]. In view of this localization, an attempt was made to accumulate sPLA₂ in IL-1β-stimulated cells by blocking its secretion, in order to study the possible contribution of this enzyme in intracellular arachidonate release for prostaglandin synthesis. For that purpose we used brefeldin A, a drug known to block protein secretion by causing dismantling of the Golgi cisternae in many cultured cells [17]. As expected, brefeldin A blocked the secretion of sPLA₂ from mesangial cells. Surprisingly, however, immunoblots from the treated cells (figure 2) indicated that brefeldin A did not cause an accumulation of sPLA₂ in the cells. Rather, the drug prevented the *de novo* synthesis of sPLA₂ protein as induced by IL-1β. These results were confirmed by immunofluorescence experiments. We then studied the effect of brefeldin A on [³H]leucine incorporation into total cellular proteins and confirmed a recent report [18] that this drug inhibits protein synthesis in general. However, the IL-1β-induced synthesis of sPLA₂ was more sensitive to brefeldin A, both with respect to the dose-dependency and the magnitude of the inhibition, than that of protein synthesis in general [16].

Dexamethasone Inhibits Induced sPLA₂ Activity by Suppression of Enzyme Synthesis

It has long been know that corticosteroids inhibit prostaglandin formation in many cells and animal models of inflammation. Initially, this inhibition was ascribed to induced synthesis of PLA₂ inhibitory proteins termed lipocortins or annexins [19,20]. In view of the fact that IL-1β-stimulated formation of PGE₂ in rat mesangial cells was also subject to inhibition by dexamethasone [2], we studied the effect of dexamethasone on the IL-1β-induced synthesis of sPLA₂ [21]. As shown in figure 3, PLA₂ activity increased sharply upon stimulation with either IL-1β or IL-1β + forskolin and a gradual decrease to basal

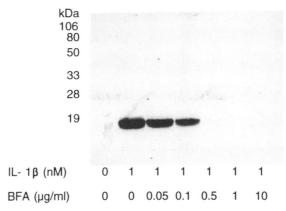

Figure 2. Dose-dependent suppression of IL-1β-induced sPLA₂ synthesis by brefeldin A. Reproduced with permission from [16].

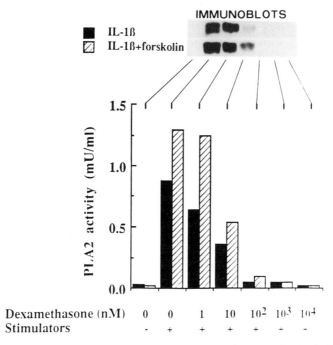

Figure 3. Dose-dependency of the inhibition of IL-1β- or IL-1β + forskolin-induced synthesis of sPLA2 by dexamethasone. Reproduced with permission from [21].

levels was observed when the cells were pretreated with increasing concentrations of dexamethasone. As shown in the blots, this lower PLA2 activity was not due to inhibition of the enzyme but caused by suppression of the synthesis of the enzyme. It has long been recognized that the regulation of PLA2 activities is complex and may involve superimposable mechanisms. Although the inhibition of sPLA2 synthesis by dexamethasone can fully explain the decrease in enzyme activity, this does not rule out *a priori* that other mechanisms are not operative. We, therefore, investigated whether dexamethasone-induced synthesis of annexins could potentially contribute to the observed decrease in sPLA2 activity. A prerequisite for this is that dexamethasone affects the levels of annexins in mesangial cells or changes their subcellular localization from a non-inhibitory form in the cytoplasm to an inhibitory, membrane-bound, form that would be able to sequester phospholipid substrate making it unavailable for PLA2 attack. As can be seen in figure 4, rat mesangial cells contain annexins I/II, detected by a single cross-reactive antibody, and annexin V. The levels of neither of them changes after IL-1β, IL-1β + dexamethasone- or dexamethasone treatment. Further experiments showed that the subcellular localization of these annexins, as assessed by cell fractionation and Western blotting or by immunofluorescence, did not change after dexamethasone treatment. Similar observations were made for annexins III, IV and VI [15]. We conclude from these experiments that the dexamethasone-induced inhibition of sPLA2 activity in rat mesangial is neither mediated by induced synthesis nor by induced translocation of annexins and can be fully accounted for by the dexamethasone-induced suppression of sPLA2 protein synthesis as shown in figure 3.

We then addressed the question whether the induced sPLA2 synthesis was suppressed by dexamethasone at the transcriptional or translational level. Previous

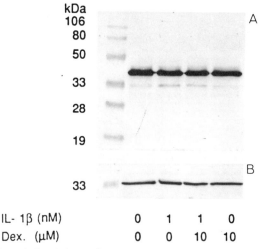

Figure 4. Annexin levels in IL-1β- and dexamethasone-treated mesangial cells. A, annexin I/II; B, annexin V.

experiments [22] had shown that combined stimulation of mesangial cells with IL-1β + forskolin resulted in increased sPLA$_2$ mRNA levels which became only partially suppressed by dexamethasone under conditions where sPLA$_2$ synthesis was totally inhibited. By using a recently cloned cDNA for group II PLA$_2$ [10] Vervoordeldonk (unpublished experiments) was able to show that the level at which dexamethasone suppressed sPLA$_2$ protein synthesis depended on the type of stimulation used. Forskolin induced the synthesis of sPLA$_2$ mRNA and the levels thereof were suppressed by dexamethasone (data not shown). By contrast, the IL-1β-induced synthesis of sPLA$_2$ mRNA was hardly affected by dexamethasone treatment, implicating post-transcriptional inhibition of sPLA$_2$ synthesis (figure 5). These results in mesangial cells confirm the previous results of Nakano *et al.* [23] in rat smooth muscle cells. These authors demonstrated that dexamethasone suppressed the forskolin-induced synthesis of sPLA$_2$ by blocking the accumulation of sPLA$_2$ coding mRNA. However, in the case of

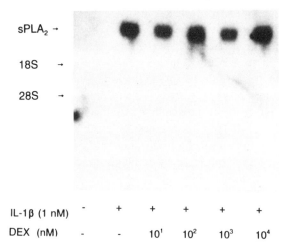

Figure 5. Northern blot experiments demonstrating that IL-1β-induced sPLA$_2$ mRNA synthesis is not affected by dexamethasone.

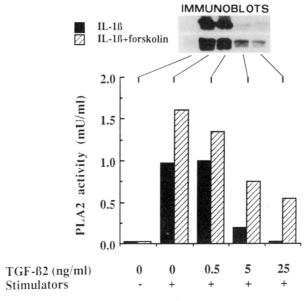

Figure 6. Dose-dependency effect of TGF-β2 on induced secretion of sPLA2. Reproduced with permission from [7].

tumor necrosis factor-induced enzyme synthesis the mRNA levels were less affected by dexamethasone and an inhibition of post-transcriptional expression of group II PLA2 was inferred. It now appears that similar events as observed for tumor necrosis factor are also involved in the suppression of IL-1β-induced sPLA2 synthesis by dexamethasone.

Transforming Growth Factor-β2 Suppresses sPLA2 Gene Expression and Increases cPLA2 Activity

TGF-β2 exerts many diverse activities on a variety of cells and in many cases counteracts the effects of pro-inflammatory cytokines [24]. Having observed the profound effect of IL-1β on sPLA2 in mesangial cells and knowing that these cells contain high affinity receptors for TGF-β it became highly appropriate to study whether TGF-β2 affected sPLA2. The results indicated (figure 6) that TGF-β2 inhibited the IL-1β, and to a lesser extent the IL-1β + forskolin, induced secretion of sPLA2 in a dose-dependent manner. Further experiments [7] showed also that the forskolin-induced sPLA2 secretion was subject to TGF-β2 inhibition and that the attenuation of secretion was due to suppression of sPLA2 synthesis. In this respect the effects of TGF-β2 on sPLA2 resemble those of dexamethasone. With respect to PGE2 formation an important difference was noted, however. Whereas the 100-fold increase of PGE2 formation after IL-1β treatment was completely blocked by dexamethasone [21], it became only partially inhibited by TGF-β2 [7], leaving a PGE2 production that was still 20-fold that in control cells. This appeared to be explainable by the fact that TGF-β2 had additional effects on the high molecular weight cytosolic cPLA2 [25]. As can be seen in figure 7, TGF-β2 increased cPLA2 activity 2.5-fold while leaving sPLA2 activity unaffected. A similar increase in cPLA2 activity was induced by IL-1β, while this cytokine induced sPLA2 activity 20 to 30-fold. These results implicate the involvement of cPLA$_2$ activation in the TGF-β2-induced PGE$_2$ synthesis. The time-dependency of cPLA$_2$ activation by both IL-1β and TGF-β2 suggested that protein synthesis was involved in this phenomenon and this was confirmed by the observation that the enhancement of cPLA$_2$ activity was abolished by both actinomycin D and cycloheximide [25]. Additional experiments (Schalkwijk, unpublished) then showed that this protein synthesis involved cPLA$_2$ itself. IL-1β induced the synthesis of both cPLA$_2$ coding mRNA and cPLA$_2$ protein. The IL-1β-induced cPLA$_2$ synthesis was completely abolished by dexamethasone. These observations provide an explanation for the complete inhibition of IL-1β-induced PGE$_2$ formation.

CONCLUDING REMARKS

The results described in this paper can be summarized as shown in figure 8. TGF-β2 overrules the induction of sPLA$_2$ by IL-1β and completely suppresses the synthesis and secretion of sPLA$_2$. By stimulation of mesangial cells with IL-1β in the presence of a neutralizing antibody in one series of experiments and by addition of immunopurified sPLA$_2$ to cells in another series of experiments we recently provided evidence that this secreted PLA$_2$ in its extracellular form can contribute to arachidonate release for prostaglandin formation [26]. Thus, by abolishing the IL-1β-induced sPLA$_2$ secretion TGF-β2 blocks the propagation of inflammatory reactions and the possible destructive action of this sPLA$_2$. The formation of PGE$_2$ as induced by IL-1β is indeed inhibited by TGF-β2 [7] but is not completely attenuated. Under conditions where TGF-β2 fully suppresses sPLA$_2$ it still enhances PGE$_2$ formation in comparison to control cells. This effect can be explained because TGF-β2, as does IL-1β, causes an increase in cPLA$_2$ activity. These increases are

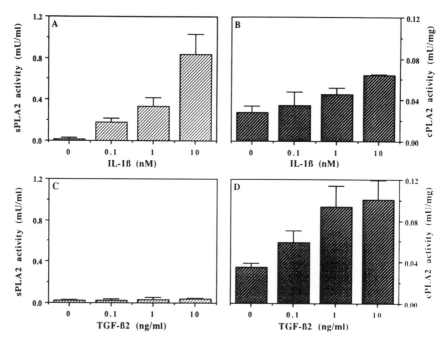

Figure 7. Dose-dependency effects of IL-1β and TGF-β2 on sPLA$_2$ and cPLA$_2$ activity in mesangial cells. Reproduced with permission from [25]

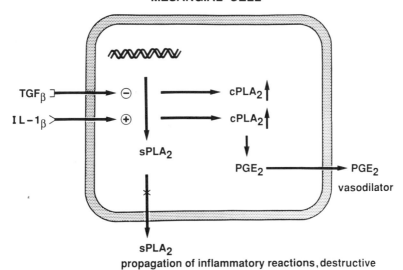

Figure 8. Schematic representation of cytokine effects on PLA_2 activities and PGE_2 formation in rat mesangial cells.

only 2 to 3-fold but it has to be realized that this represents the activation factor that is measurable in the cell-free cytosolic fraction due presumably to IL-1β-induced *de novo* synthesis of the enzyme. The actual activity increase in the cells may be considerably higher due to additional increases of enzymatic activity resulting from translocation of the $cPLA_2$ from cytoplasm to membrane. The PGE_2 formed in response to $cPLA_2$ activation may, through its vasodilatory action, serve to maintain glomerular filtration rate and plasma flow during inflammation and the repair phases thereof.

REFERENCES

1. Pfeilschifter, J., Pignat, W., Vosbeck, K. and Märki, F., Biochem. Biophys. Res. Commun. 159 (1989) 385-394.
2. Pfeilschifter, J., Pignat, W., Vosbeck, K., Märki, F. and Wiesenberg, I., Biochem. Soc. Trans. 17 (1989) 916-917.
3. Gronich, J.H., Bonventre, J.V. and Nemenoff, R.A., J. Biol. Chem. 263 (1988) 16645-16651.
4. Gronich, J.H., Bonventre, J.V. and Nemenoff, R.A., Biochem. J. 271 (1990) 37-43.
5. De Jong, J.G.N., Amesz, H., Aarsman, A.J., Lenting, H.B.M. and van den Bosch, H., Eur. J. Biochem. 164 (1987) 129-135.
6. Schalkwijk, C., Pfeilschifter, J., Märki, F. and van den Bosch, H., Biochem. Biophys. Res. Commun. 174 (1991) 268-275.
7. Schalkwijk, C., Pfeilschifter, J., Märki, F. and van den Bosch, H., J. Biol. Chem. 267 (1992) 8846-8851.
8. Hayakawa, M., Kudo, I., Tomita, M., Nojima, S. and Inoue, K., J. Biochem. 104 (1988) 767-772.
9. Ono, T., Tojo, H., Kuramitsu, S., Kagamiyama, H. and Okamoto, M., J. Biol. Chem. 263 (1988) 5732-5738.
10. Van Schaik, R.H.N., Verhoeven, N.M., Neys, F.W., Aarsman, A.J. and van den Bosch, H. Biochim. Biophys. Acta 1169 (1993) 1-11.
11. Mizushima, H., Kudo, I., Horigome, K., Murakami, M., Hayakawa, M., Kim, D.-K., Koudo, E., Tomita, M. and Inoue, K., J. Biochem. 105 (1989) 520-525.

12. Kramer, R.M., Hession, C., Johansen, B., Hayes, G., McGray, P., Pinchong Chow, E., Tizard, R. and Pepinsky, R.B., J. Biol. Chem. 264 (1989) 5768-5775.
13. Aarsman, A.J., Leunissen-Bijvelt, J., van den Koedijk, C.D.M.A., Neys, F.W., Verkleij, A.J. and van den Bosch, H., J. Lipid Med. 1 (1989) 49-61.
14. Van den Bosch, H., Aarsman, A.J., van Schaik, R.H.N., Schalkwijk, C.G., Neys, F.W. and Sturk, A., Biochem. Soc. Trans. 18 (1990) 781-785.
15. Vervoordeldonk, M.J.B.M., Schalkwijk, C.G., Römisch, J., Aarsman, A.J. and van den Bosch, H., Biochim. Biophys. Acta, submitted.
16. Sánchez, R.M., Vervoordeldonk, M.J.B.M., Schalkwijk, C.G. and van den Bosch, H., FEBS Lett., in press.
17. Lippincott-Schwarz, J., Yuan, L., Tipper, C., Amherdt, M., Orci, L. and Klausner, R.D., Cell 67 (1991) 601-616.
18. Fishman, P.H. and Curran, P.K., FEBS Lett. 314 (1992) 371-374.
19. Flower, R.J., Br. J. Pharmacol. 94 (1988) 987-1015.
20. Davidson, F.F. and Dennis, E.A., Biochem. Pharmacol. 38 (1989) 3645-3651.
21. Schalkwijk, C., Vervoordeldonk, M., Pfeilschifter, J., Märki, F. and van den Bosch, H., Biochem. Biophys. Res. Commun. 180 (1991) 46-52.
22. Mühl, H. Geiger, T., Pignat, W., Märki, F., van den Bosch, H., Cerletti, N., Cox, D., McMaster, G., Vosbeck, K. and Pfeilschifter, J., FEBS Lett. 301 (1992) 190-194.
23. Nakano, T., Ohara, O., Teraoka, H. and Arita, H., J. Biol. Chem. 264 (1990) 12745-12748.
24. Massagué, J., Annu. Rev. Cell Biol. 6 (1990) 597-641.
25. Schalkwijk, C.G., de Vet, E., Pfeilschifter, J. and van den Bosch, H., Eur. J. Biochem. 210 (1992) 169-176.
26. Pfeilschifter, J., Schalkwijk, C., Briner, V.A. and van den Bosch, H., J. Clin. Invest., in press.

PHOSPHOLIPASE A2 AND THE PATHOGENESIS OF MULTISYSTEM ORGAN FAILURE IN EXPERIMENTAL AND CLINICAL ENDOTOXIN SHOCK

Peter Vadas, Jeffrey Edelson, and Waldemar Pruzanski

Divisions of Immunology and Respirology
Department of Medicine
University of Toronto
Toronto, Ontario, Canada M4Y 1J3

INTRODUCTION

Synthesis and activation of phospholipases A2 have been recognized to be an integral part of the host's response to inflammatory and infectious agents. Several functionally and structurally distinct phospholipases A2 are present in humans. Their respective roles have not been fully elucidated. A growing body of evidence has implicated the secretory nonpancreatic phospholipase A2 (sPLA2) in the pathogenesis of the major manifestations of endotoxin shock. Herein, we will review the evidence implicating sPLA2 in the pathogenesis of septic shock in the context of the biochemical characterization, sources of synthesis in systemic inflammatory reactions such as endotoxin shock, endogenous regulation of synthesis and signal transduction pathways, endogenous modulation of activity of secreted PLA2 and experimental data which further substantiates the involvement of sPLA2 in the initiation and propagation of local and systemic inflammatory reactions.

Biochemical Characterization of Phospholipases A2

Phospholipases A2 (PLA2s) are lipolytic enzymes which cleave phospholipid substrate at the sn-2 position to yield 2 products in stoichiometric amounts, lysophosphatides and free fatty acids. Considerable attention has been brought to bear on the activity of this class of enzymes because liberation of these 2 products is generally regarded as rate-limiting in the synthesis of major classes of bioactive lipid mediators including eicosanoids and platelet activating factor (PAF). While the bioactivity of arachidonic acid and its metabolites has been well-described, relatively little attention has been paid to lysophosphatides which may have equally important biologic activities.

Esterases, Lipases and Phospholipases, Edited by M.I. Mackness
and M. Clerc, Plenum Press, New York, 1994

Generally, the biologic activities of PLA2s have been attributed to the products of phospholipid hydrolysis but recent evidence suggests that some of these activities may be independent of enzymatic activity, but rather be the result of binding of PLA2 to specific cell-surface receptors [1].

At least three mammalian PLA2s are recognized: pancreatic PLA2 (pPLA2), cytosolic PLA2 (cPLA2) and sPLA2. The pPLA2 is primarily a digestive enzyme although pPLA2 mRNA is also found in lung, spleen, stomach and intestine. The cPLA2 is an intracellular enzyme which participates in the rapid mobilization of arachidonic acid [2]. In contrast, sPLA2 does not exhibit preference for a specific fatty acid system in the 2-acyl position but does have a unique headgroup preference [2]. The respective biochemical characteristics of the cytosolic and secretory PLA2s are shown in table 1.

Table 1. Comparison of $cPLA_2$ and $sPLA_2$

	$cPLA_2$	$sPLA_2$
MW	85-110 kD	14 kD
Ca^{2+} optimum	μM	mM
pH optimum	8-10	7.5
Disulphide bridges	None	7
Dithiothreitol	Resistant	Sensitive
Polar headgroup preference	None	PG > PE > PC
sn-2 fatty acid preference	Arachidonyl	None
Chromosomal localization	?	1 q

Sources of Synthesis

Intravenous administration of endotoxin in experimental animals leads to markedly increased circulating levels of sPLA2 within 30-60 minutes [3]. Subsequent changes in serum sPLA2 activity correlate with specific manifestations of organ dysfunction as described below. Similar phenomena have been documented in patients with gram-negative and gram-positive septic shock [4]. The cellular sources of endotoxin-induced synthesis and secretion of sPLA2 have not been firmly established but the major source appears to be vascular smooth muscle cells and perhaps hepatocytes. SPLA2 mRNA is found in blood vessels and liver of endotoxemic rats [5]. Transcript levels in these tissues parallel sPLA2 activity in blood. Immunohistochemical studies of tissues taken at autopsy of patients who died of septic shock exhibited intense staining for sPLA2 exclusively in vascular smooth muscle of all tissues examined [6]. In contrast, human liver parenchyma did not stain for sPLA2 using monoclonal antibodies to recombinant human sPLA2. Whereas the vasculature is clearly an important source of systemic release of circulating sPLA2, the extent of the contribution by liver and other organs remains to be defined.

Table 2. Regulation of synthesis of sPLA2 and acute phase proteins by Hep G2 cells [1]

Agent(s)	sPLA2	Haptoglobin	α_1-Antichymotrypsin
IL-1β	↑	↑	↑
TNFα	↑	↑	↑
IL-6	↑	↑	↑
Oncostatin M	↑	↑	↑
CNTF[2]	↔	↔	↔
LIF	↔	↔	↔
IL-6 + Dex	↔	↔	↔
8-Br-cAMP	↑	ND	ND
Bt$_2$-cAMP	↑	ND	ND
IBMX	↔	ND	ND

[1]Vadas, P., Pruzanski, W., and Gauldie, J. Unpublished.

[2]Abbreviations:

CNTF	-	ciliary neurotrophic growth factor
LIF	-	leukemia inhibitory factor
Br-cAMP	-	8-Bromo-cAMP
Bt$_2$-cAMP	-	dibutyryl cAMP
IBMX	-	Isobutylmethylxanthine
ND	-	not determined

These data are further corroborated by studies of vascular smooth muscle cells and human hepatoma cells in culture. Hep G2 cells, a human hepatoma line, synthesize and secrete sPLA2 in response to interleukin 1 (IL-1) tumour necrosis factor (TNF), IL-6, and oncostatin M but not the α-1 helical cytokines CNTF and LIF (Vadas, P. Gauldie, J., Pruzanski, W. unpublished data). Co-addition of IL-6 to IL-1 or TNF has a synergistic effect. Glucocorticoids down-regulate the cytokine induced synthesis of sPLA2. The expression of sPLA2 by Hep G2 cells parallels that of two recognized acute phase proteins, haptoglobin and α_1 antichymotrypsin (summarized in Table 2).

SPLA2 is co-expressed in vivo with C-reactive protein (CRP) in patients with septic shock. In longitudinal studies of these patients, both sPLA2 and CRP appeared in increasing concentrations in blood shortly after peak IL-6 levels. These data suggest that sPLA2 and acute phase proteins such as CRP, haptoglobin, and α_1 antichymotrypsin share common regulatory mechanism. Data presented below show that acute phase proteins such as CRP may act as endogenous modulators of sPLA2 activity in endotoxin shock.

Regulation of Synthesis

A number of cell types derived from embryonic mesenchyme express sPLA2 in response to defined stimuli. The response to some stimuli is cell-type specific. For example, endotoxin-induced sPLA2 expression is seen in astrocytes, osteoblasts, smooth muscle cells but not Hep G2 cells [6]. IL-6 and oncostatin M-induced PLA2 expression is observed only in Hep G2 cells, but not other mesenchymal cells. In Hep G2 cells, intracellular signal transduction pathways of cytokine-induced sPLA2 expression are only partly defined. Increased intracellular cAMP levels tend to augment sPLA2 expression. Activation of adenylate cyclase by forskolin or the addition of cell-permeable cAMP analogues such as 8-bromo-cAMP or dibutyryl-cAMP leads to potentiation of sPLA2 synthesis (Table 3). However, the cAMP-protein kinase A pathway is not the primary pathway of signal transduction. Inhibition of adenylate cyclase by dideoxyadenosine (DDA) does not inhibit cytokine-induced sPLA2 synthesis. Partial inhibition is seen with the tyrosine kinase inhibitors genistein and tyrophostine-25. The protein kinase C activators, PMA and OAG enhanced sPLA2 synthesis by three-fold.

SPLA2 expression involves a cholera toxin-sensitive G protein. However, pertussis toxin was without effect. These observations in Hep G2 cells are consistent with findings in renal mesangial cells and vascular smooth muscle cells [7, 8].

Modulation of sPLA2 Activity

As described above, sPLA2 expression is induced by IL-1, TNF and oncostatin M as well as endotoxin. Glucocorticoids exert down-regulatory transcriptional and post-transcriptional control. Several growth factors including transforming growth factor-β (TGFβ), platelet-derived growth factor (PDGF), epidermal growth factor (EGF) and fibroblast growth factor (FGF) also exert inhibitory control at transcriptional and translational levels [6]. Thus, in the complex environment of inflammatory or infectious foci, control of sPLA2 synthesis is the product of the interplay between a variety of cytokines and growth factors.

However, it is clear that situations arise in which extraordinarily high levels of sPLA2 are synthesized and rapidly secreted into the extracellular environment. In rheumatoid arthritis, high levels of sPLA2 are sequestered into the articular compartment and in endotoxin shock equally high levels of sPLA2 accumulate in the intravascular compartment. In the absence of demonstrable circulating inhibitors of sPLA2, this enzyme would have potentially unlimited access to phospholipid substrate in cell membrane structures, lipoproteins and chylomicrons. The notion of regulation of sPLA2 activity by sequestration of substrate provides a plausible mechanism for control of the activity of this lipolytic enzyme.

The requirements for modulatory control by substrate sequestration include co-expression of sPLA2 and the putative modulator, attainment of high concentrations of the modulator in the compartment in which sPLA2 resides and demonstrable inhibition of sPLA2 by the mechanism of substrate depletion by using the putative modulator in vivo. The observation that sPLA2 and CRP are co-regulated and co-expressed in patients with septic shock led us to

Table 3. Intracellular signal transduction pathways of sPLA2 expression in Hep G2 cells

Probe	Concentration	sPLA2 activity (units/ml)
control	-	61 ± 6
8-Br-cAMP[1]	4 mM	2336 ± 314
Bt$_2$-cAMP	2 mM	3046 ± 170
IBMX	4 mM	18 ± 8
PMA	10 nM	158 ± 5
Cholera Toxin	50 ng/ml	275 ± 2
Pertussis Toxin	500 ng/ml	24 ± 12
Oncostatin M	1 ng/ml	590 ± 21
Oncostatin M cholera Toxin		4227 ± 357

[1]Abbreviations:

8-Br-cAMP	-	8-bromo-cAMP
Bt$_2$-cAMP	-	dibutyryl cAMP
IBMX	-	isobutylmethylxanthine
PMA	-	phobol myristate acetate

investigate the possibility that CRP may fulfil the criteria of inhibitor of sPLA2 by the mechanism of substrate depletion.

CRP is a calcium-dependent, phospholipid binding protein synthesized and secreted during infectious and inflammatory stimuli. It is a prototypic acute phase protein whose expression is upregulated as much as 1000-fold in serum primarily by IL-6. IL-1, TNF and glucocorticoids also exert regulatory control [9]. In studies in vitro using PC: lyso-PC mixed micelles, CRP dose-dependently inhibited sPLA2-induced substrate hydrolysis. Inhibition was reversed by co-addition of phosphorylcholine. Since CRP preferentially binds to PC and especially lyso-PC with high affinity, these data are consistent with the potential for CRP to act as an endogenous modulator of sPLA2 activity in vivo. Of 2 other structurally related pentraxins studied to date, the human acute phase protein serum amyloid P (SAP) also inhibited sPLA2 activity by substrate sequestration whereas limulin, a pentraxin from horseshoe crab, with homology to CRP but known weak binding to PC, did not inhibit sPLA2 activity in this substrate system (Vadas, P. Stefanski, E., Pruzanski, W., unpublished data).

Thus, sPLA2 synthesis and secretion are regulated by the dynamic interplay of several cytokines and growth factors as well as glucocorticoids. Once secreted extracellularly, sPLA2 activity may well be modulated in vivo by the acute phase proteins, CRP and SAP. Indeed, CRP and SAP fulfil the requirements for co-expression, co-regulation, attainment of high serum levels as well as co-localization during infection and inflammation. In vitro, both CRP and SAP inhibit sPLA2 hydrolysis of phospholipid by kinetics consistent with substrate depletion.

Table 4. Potentiation of sPLA2-induced lung injury by LPS and C5a

Priming stimulus	Route	Secondary stimulus	Route	Cell count (x 10^6)	Protein concentration (mg/ml)	p
LPS 50 ng	IT*	vehicle	IT	3.81 ± 0.23	0.22 ± 0.04	ns*
LPS 50 ng	IT	sPLA2 150,000	IT	3.59 ± 0.60	0.25 ± 0.03	ns
LPS 3 ng	iv	vehicle	IT	2.42 ± 0.28	0.10 ± 0.07	<.005
LPS 3 ng	iv	sPLA2	IT	3.35 ± 0.30	0.22 ± 0.08	
C5a 1 ng	IT	vehicle	IT	2.72 ± 0.39	0.11 ± 0.02	<.005
C5a 1 ng	IT	sPLA2	IT	5.20 ± 0.15	0.26 ± 0.13	

*IT - intratracheal
ns - not significant

Involvement of sPLA2 in the Pathogenesis of Multisystem Organ Dysfunction in Endotoxin Shock

A number of systemic inflammatory reactions such as endotoxin shock, pancreatitis and salicylate intoxication, share common manifestations and, in fact, may be virtually indistinguishable [6, 10]. These clinical syndromes as well as that induced by snake bite, share the common pathogenic determinant of high levels of circulating sPLA2. Markedly high circulating PLA2 activity in conjunction with generalized capillary leak allows this enzyme access to the interstitium and parenchyma of major organ systems such as lung, kidney, liver and myocardium. Structural perturbation of cell membranes by sPLA2 may lead to loss of functional integrity with resultant organ system dysfunction [6]. Indeed, in patients with endotoxin shock, pancreatitis and salicylate intoxication, we have shown that sPLA2 activity correlates with the duration and magnitude of organ system dysfunction [4]. Clearly, sPLA2 is not the only mediator responsible for cell and tissue injury. There are other mediators which are potentially injurious to host tissue and there is a very clear potential for interaction and synergy amongst these mediators and sPLA2.

Acute lung injury is a situation in which the local accumulation of PLA2 may be linked directly to organ dysfunction. Pancreatitis and gram negative sepsis, conditions characterized by high serum levels of PLA2 are both associated with an increased risk of developing acute lung injury [11, 12]. In an experimental model of sepsis-induced lung injury, Von Wichert observed five fold elevation in BAL levels of PLA2 and lysophosphatidylcholine [13]. High levels of PLA2 and its hydrolysis products, free fatty acids and, lysoPC have been documented in bronchoalveolar lavage fluid from patients with ARDS.

To directly address the possibility that PLA2 is a mediator of inflammation in the lower respiratory tract we have developed an injury model that follows intratracheal instillation of a sublethal amount of PLA2 . An acute lung injury syndrome follows with many similarities to ARDS including intrapulmonary accumulation of inflammatory cells, the development of hyaline membranes, increased wet-dry lung weights, transepithelial leak of fluid and protein, impaired gas exchange and a mortality of 30% [14,15].

One mechanism of injury in the PLA2-induced lung injury model may well be enzyme mediated hydrolysis of saturated phosphatidylcholine (and other surfactant phospholipids). It has been documented that PLA2 alters and biophysically inactivates surfactant preparations in vitro [16]. However the lysophospholipid and protein present in the alveolus in the context of acute lung injury also lead to indirect inactivation of surfactant biophysical activity [17]. In addition, PLA2 is cytotoxic to alveolar epithelial cells [18] and damage to alveolar epithelial cells may lead to impairment of surfactant synthesis and secretion. Thus, although our understanding of this injury model is far from complete, it does appear that surfactant is directly degraded by PLA2, inactivated by the presence of lysoPC and protein and its synthesis and secretion may be impaired as the result of epithelial cell damage.

Further studies in experimental animals and man have shed light on regulatory aspects of endotoxin-cytokine-sPLA2 interactions. Experimental intravenous administration of endotoxin in human volunteers induced a marked increase in circulating TNF levels followed by a 4.4 fold increase in circulating sPLA2 within 3 hrs after endotoxin challenge. A mean maximal increase of 14-fold above control was seen at 24 hrs after the challenge. Endotoxin challenge did not alter serum pPLA2 levels [19]. When endotoxin was repeatedly given intravenously to patients with various malignancies, an increase in circulating sPLA2 of 15-20 fold was observed after each challenge. The initial administration led to an almost immediate increase in circulating TNF levels followed by increase in IL-6, in turn followed by increase in sPLA2. Subsequent sequential challenges with iv endotoxin led to much less pronounced rise of TNF and IL-6, however the increases in circulating sPLA2 remained high [20]. Intravenous administration of recombinant human TNF to cancer patients was associated with increased circulating sPLA2 activity that started 3 hrs after rh-TNF infusion and reached a mean maximum increase of 16-fold in 18 hrs. A significant correlation was noted between the dose of infused rh-TNF and the maximum increase in sPLA2 activity [21]. By virtue of the fact that only minute doses of endotoxin or of TNF could be given to man, the infused individuals did not develop marked hypotension or manifestations of organ system dysfunction in the experimental setting. However, the above studies demonstrate a direct relationship between challenge with endotoxin or TNF and a subsequent marked rise in sPLA2.

In rabbits challenged with an LD_{50} dose of E. coli endotoxin, there was a clear inverse correlation between the increase in serum sPLA2 activity and the fall in mean arterial blood pressure [3]. Infusion of exogenous sPLA2 in normal rabbits reproduced the fall in blood pressure. Inactivation of sPLA2 by para-bromophenacyl bromide prior to infusion abrogated the fall in blood pressure, suggesting that sPLA2 is vasoactive and may mediate, in part, the circulatory collapse induced by endotoxin. Pre-treatment of rabbits with dexamethasone before infusion of endotoxin not only attenuated the fall in blood pressure but also prevented the increase in circulating sPLA2 activity [3].

In patients with gram-negative septic shock, serum sPLA2 activity correlated with both the increased risk of adult respiratory distress syndrome (ARDS) as well as with increased mortality [22]. In a prospective study of patients with septic shock, serum sPLA2 activity correlated with the magnitude and duration of circulatory collapse in all patients [23]. Furthermore, the extent of organ system dysfunction as assessed by APACHE II scores, paralleled changes in serum sPLA2 activity.

In the only intervention study reported to date, baboons were passively immunized with neutralizing monoclonal antibody to human sPLA2 prior to challenge with a lethal dose of gram-negative bacteria. In a cohort of only 3 experimental animals, there was a trend to reversal of hypotension [24], but the sample size was inadequate for statistical analysis.

Although the experimental and clinical data reviewed above provide compelling evidence for the role of sPLA2 in the pathogenesis of circulatory collapse and multiorgan system dysfunction in endotoxin shock, the specific and selective inhibition of sPLA2 activity and the accompanying reversal or abrogation of the manifestations of endotoxin shock will provide additional evidence to substantiate a significant pathogenic role for sPLA2.

Acknowledgments

The authors thank Ms. Michèle Lydon for her skilful secretarial assistance. These studies were supported by Physicians Services Incorporated and the Medical Research Council of Canada.

References

1. Tohkin, M, Kishino, J., Ishizaki, et al. Pancreatic-type phospholipase A2 stimulates prostaglandin synthesis in mouse osteoblastic cells (MC3T3-E1) via a specific binding site. J Biol Chem, 268:1993; 2865-2871

2. Leslie, C.C., and Channon, J.Y. Anionic phospholipids stimulate an arachidonoyl-hydrolyzing phospholipase A2 from macrophages and reduce the calcium requirement for activity. Biochim Biophys Acta, 1045:1990; 261-270.

3. Vadas, P, and Hay, J.B. Involvement of circulating phospholipase A2 in the pathogenesis of the hemodynamic changes in endotoxin shock. Can J Physiol Pharmacol, 61:1983; 561-566.

4. Vadas, P, and Pruzanski, W. Induction of group II phospholipase A2 expression and pathogenesis of the sepsis syndrome, Circ Shock, 39:1993; 160-167.

5. Nakano, T., and Arita, H. Enhanced expression of group II phospholipase A2 gene in the tissues of endotoxin shock rats and its suppression by glucocorticoid, FEBS Lett, 273:1990; 23-26

6. Vadas, P., Browning, J., Edelson, J., et al. Extracellular phospholipase A2 expression and inflammation:the relationship with associated disease states, J Lipid Mediators, 8:1993; 1-30.

7. Pfeilschifter, J., Leighton, J, Pignat, W, et al. Cyclic AMP mimics, but does not mediate, interleukin-1 and tumour-necrosis-factor-stimulated phospholipase A2 secretion from rat mesangial cells. Biochem J, 273:1991; 199-204

8. Nakano, T., Ohara, O, Teraoka, H., et al. Group II phospholipase A2 mRNA synthesis is stimulated by two distinct mechanisms in rat vascular smooth muscle cells, FEBS Lett, 261:1990; 171-174

9. Schultz, D.R., and Arnold, P.I. Properties of four acute phase proteins: C-reactive protein, serum amyloid A protein, α1-acid glycoprotein, and fibrinogen, Seminars in Arthritis and Rheumatism. 20:1990; 129-147

10. Vadas, P., Schouten, B.D., Stefanski, E., et al. Association of hyperphospholipasemia A2 with multiple system organ dysfunction due to salicylate intoxication. Crit Care Med, 21:1990; 1087-1091.

11. Hallman, M., Spragg, R., Harrell, J.H., et al. Evidence of lung surfactant abnormality in respiratory failure. Study of bronchoalveolar lavage phospholipids, surface activity, phospholipase activity and plasma myoinositol, J Clin Invest. 70:1982; 673-683

12. Gregory, T.J., Longmore, W.J., Moxley, M.A. et al. Surfactant chemical composition and biophysical activity in acute respiratory distress syndrome, J Clin Invest. 88:1991; 1976-1981

13. Von Wichert, P., Tennesfeld, M., and Meyer, W. Influence of septic shock upon phosphatidylcholine remodelling mechanism in rat lung, Biochim Biophys Acta 664:1981; 487-497.

14. Edelson, J.D., Vadas, P., Mullen, B, et al. Phospholipase A2 - induced acute lung injury in rats, Am Rev Respir Dis 143:1991;1102-1109.

15. Mullen, J.B.M., Pittman, J.A., Feinmesser, M.E., and Edelson, J.D. Phospholipase A2 - induced lung injury; morphometric analysis. Am Rev Respir Dis 145:1992; A606

16. Holm, B.A., Keicher, L., Liu, M.Y. et al. Inhibition of pulmonary surfactant function by phospholipases. J Appl Physiol, 71:1991; 317-321.

17. Cockshutt, A.M., and Possmayer, F. Lysophosphatidylcholine sensitizes lipid extracts of pulmonary surfactant to inhibition by serum proteins. Biochim Biophys Acta, 1086:1991;63-71.

18. Edelson, J.D., Vadas, P., Pruzanski, W., et al. Phospholipase A2 is cytotoxic to alveolar type II cells, Am Rev Respir Dis 145:1992; A360

19. Pruzanski, W., Wilmore, D.W., Suffredini, A. et al. Hyperphospholipasemia A2 in human volunteers challenged with intravenous endotoxin. Inflammation, 16:1992; 561-570.

20. Pruzanski, W., Mackensen, A, Engelhardt, R, et al. Induction of circulating phospholipase A2 activity by intravenous infusion of endotoxin in patients with neoplasia. J Immunotherapy, 12:1992; 242-246.

21. Pruzanski, W., Sherman, M.L., Kufe, D.W., et al. Induction of circulating phospholipase A2 by intravenous administration of recombinant human tumour necrosis factor. Mediators Inflam, 1:1992; 235-240.

22. Vadas, P. Elevated plasma phospholipase A2 levels: Correlation with the hemodynamic and pulmonary changes in gram-negative septic shock. J Lab Clin Med, 104:1984; 873-881.

23. Vadas, P., Pruzanski, P., Stefanski, E. et al. Pathogenesis of hypotension in septic shock: Correlation of circulating phospholipase A2 levels with circulatory collapse. Crit Care Med, 16:1988; 1-7.

24. Scott, K.F., Smith, G.M., Green, J.A., et al. Neutralising monoclonal antibody to human secretory phospholipase A2 is anti-hypotensive in a baboon E. coli induced septic shock model, in "Prostaglandins and Related Compounds",1992; A151.

TRANSCRIPTIONAL REGULATION OF SECRETED PHOSPHOLIPASE A$_2$ IN HUMAN HepG2 CELLS, RAT ASTROCYTES AND RABBIT ARTICULAR CHONDROCYTES

Mouloud Ziari[1], Francis Berenbaum[2], Qishi Fan[1], Colette Salvat[1], Marie Thérèse Corvol[2], Jean Luc Olivier[1] and Gilbert Bereziat[1]

[1]Laboratoire de Biochimie, CHU Saint-Antoine, URA CNRS 1283
27 rue Chaligny 75571 Paris Cedex 12
[2]INSERM U30, Hopital Necker, 149 rue de Sèvres 75743 Paris Cedex 15

SUMMARY

We studied the regulation of the release and transcriptional regulation of the secreted type II phospholipase A$_2$ (PLA$_2$) in rat astrocytes, rabbit articular chondrocytes and the human hepatoma cell line HepG2. The interleukins IL-1, IL-6, TNFα stimulate PLA$_2$ release in these cell lines. Okadaïc acid and TPA decrease the release of PLA$_2$ by HepG2 cells which is contrary to the effect of forskolin. Synergisms between IL-1 and forskolin in rat astrocytes, IL-6 and forskolin in HepG2 cells were observed in the secretion of PLA$_2$. The same synergisms were observed for transcriptional activity measured in CAT assays. Surprisingly IL-1 and TNF as well as okadaïc acid and TPA decrease the CAT activities induced in HepG2 cells. The transcription of the PLA$_2$ gene in rabbit chondrocytes and rat astrocytes appears to be regulated by positive regulatory DNA elements located in the first hundred base pairs upstream to the first exon. In rabbit chondrocytes, the region [-326; -264] was found to be critical for the transcriptional activity. The regulation of the promoter in HepG2 cells seems to be more complex and results from the combination of several positive and negative regulatory elements located in the region which spans from positions -1614 to -87.

INTRODUCTION

Phospholipases A$_2$ (PLA$_2$s) release the fatty acid at the sn2 position of glycerophospholipids. This reaction leads to the production of eicosanoids when the esterified chain is an arachidonic acid. Therefore these enzymes have been implicated in inflammatory reactions. Since 1989, two main group of proteins have been characterized in mammals. The first enzyme is a high molecular weight cytosolic protein (85 kDa) which was first purified and cloned from the U937 macrophage cell line [1, 2]. The second type of enzyme is a low molecular weight secreted protein (14 kDa) which is related to the snake venoms and pancreatic enzymes [3, 4]. Although the structures of the pancreatic and the inflammatory PLA$_2$s are roughly identical and differ only by the absence of one disulfide bond in the latter group (6 bonds instead of 7), these proteins are encoded by two different genes [5]. The inflammatory secreted and cytosolic PLA$_2$s are expressed in several cell types, some of which express both these enzymes, e.g. synoviocytes [6], chondrocytes [7] and mesangial cells [8]. Until now the cytosolic PLA$_2$ activity seemed to be mainly affected by post translational modification [9]. On the other hand, the expression of the secreted PLA$_2$ gene is stimulated by interleukins and other mediators such as forskolin and the phorbol ester TPA. Thus interleukin 1 (IL-1) and tumor necrosis factor α (TNFα) stimulate PLA$_2$ synthesis and secretion in rat chondrocytes [10], rat mesangial cells [11], rat vascular smooth cells [12], rat calvarial osteoblasts [13] and rat astrocytes [14]. Interleukin 6 (IL-6) alone or in combination with TNFα and IL-1 stimulates PLA$_2$ synthesis in the human hepatoma cell HepG2 [15]. The time required for the maximal activation of PLA$_2$ in many cell types is as long as 24-48h, the suppression of their effects by actinomycin D established by Oka and Arita in astrocytes [14], Lyons-Giordano et al in rabbit chondrocytes [16] and Crowl et al in HepG2 cells [15] strongly suggest that interleukins regulate PLA$_2$ at the transcriptional level. Furthermore different transduction pathways might be involved. For instance, Oka and Arita [14] indicated that the LPS and TNF stimulation of PLA$_2$ in astrocytes are differently affected by agents which increase intracellular cAMP concentration on one hand and those which modify the protein kinase C activity on the other hand.

The aim of this study was to investigate the regulation of the PLA$_2$ promoter by several agents in different cell types. We have chosen three cell models which are known to secrete PLA$_2$ and which can play a role in the local or general inflammatory responses i. e. rabbit articular chondrocytes, rat astrocytes and human hepatoma cells HepG2. Secreted PLA$_2$ has been purified from the synovial fluid of inflammed joints where it accumulates through secretion by synoviocytes or chondrocytes [3, 4]. Astrocytes are the major glial cells of the central nervous system. They are involved in the protection of neuronal cells and in the immune response in the brain [17]. The secretion of PLA$_2$ by HepG2 cells suggests that this protein might belong to the group of acute phase responsive proteins. The agents used in this study included IL-1, IL-6, TNFα as well as agents which can modulate their effects by affecting the PKA or PKC transduction pathways. We compared the effects of these mediators on both PLA$_2$ activity and promoter activity. We focused our work on the functional aspects of the transcriptional regulation and the definition of the DNA elements involved in the different cell types. This study constitutes a preliminary work which will be followed by the identification of the various transcription factors involved in the transcriptional regulation of PLA$_2$ in the different cell types and the mechanisms which modulate their function.

MATERIAL AND METHODS

Materials

Restriction enzymes, T4 kinase, proteinase K, polyacrylamide, TEMED, ammonium persulfate, and agarose were purchased from Appligene. MSL medium was provided by Eurobio. Taq polymerase was obtained from Cetus. Materials for cell culture were purchased from Gibco BRL (Dubelco's modified Eagle medium supplemented, fetal calf serum, HEPES, trypsin), or from Boehringer (hyaluronidase, collagenase) and from Falcon Inc. (flasks and petri dishes). We used b-galactosidase expression vector pSV40β-gal from Clontech, acetyl CoA and deoxynucleotides from Pharmacia. Radioactive materials were obtained from Amersham. All other chemicals were purchased from Fluka.

Synthetic oligonucleotides and Plasmid constructions

Oligonucleotides were synthesized by the solid-phase phosphide triester method using an automated oligonucleotide synthesizer (Applied Biosystems, Inc., Model 380-B). The 5' flanking fragment of the PLA$_2$ gene was generated by PCR using human genomic DNA as template. The primers were designed according to the sequence published by Kramer et al.[4] and corresponded respectively to the regions [12/40] and [2403/2433] in their sequence. This PCR product was reamplified using the derivated primers containing the Sal I and Hind III restriction sites, inserted into the polylinker site of the pUC-SH-CAT plasmid [18]. This product is referred to as [-1614; +806] in this paper with respect to the putative transcription starting site proposed by Seilhamer et al [3]. The 5' deleted fragments [-326/+20], [-264/+20], [-210/+20], [-159/+20], [-118/+20], [-87/+20] were amplified by PCR from the pUC-[-326/+20]-PLA$_2$-CAT construct and subcloned at the XbaI/Hind III sites into the pUC-SH-CAT plasmid [18]. The pUC-PLA$_2$-SH-CAT constructs were sequenced and found to be identical to the sequence already published [3, 4].

Cell cultures, transfections and CAT assays

For chondrocyte cultures, the shoulder, knees and femoral heads from four week-old rabbits were dissected under sterile conditions. The articular cartilage was removed and cut into small pieces and digested with 0.05 % testicular hyaluronidase in DMEM for 15 min, then by 0.25 % trypsin for 30 min and by 0.2 % clostridial collagenase for 90 min at 37°C; The resulting pure suspension of chondrocytes was split into 25 cm^2 flasks in DMEM supplemented with 10 % fetal calf serum, 10 UI/ml penicillin and 100 μg/ml streptomycin. The cells were maintained at 37°C in 5 % CO_2. The cells reached confluency within 6-7 days and were used for experiments.

Astrocytes were prepared from cerebral hemispheres of 7 day-old Sprague-Dawley rats. After removal of the meninges, the brains were minced and dissociated using 0.1 % trypsin. Cells were collected by centrifugation and resuspended in DMEM supplemented with 10 % fetal calf serum in 75 cm^2 flasks. They were incubated at 37°C in 5 % CO_2 and reached confluency in 7-8 days.

HepG2 cells were grown in DMEM supplemented with 10% fetal calf serum. The cultures were incubated with 5% CO2 at 37°C.

Sixteen-twenty hours prior to transfection, 75 cm^2 confluent flasks were trypsinized and cells were plated on 60 mm dishes at a density of 10^6 cells/dish. Plasmids used in transfections were purified through two Cesium Chloride gradients. Cells were transfected using the calcium phosphate DNA coprecipitation method [19]. Briefly, cells were incubated with the transfection mixture containing 12 μg pUC-SH-CAT constructs and 2.5 μg of a plasmid bearing the β-galactosidase gene [20] for 4 h, and then, shocked with HBS buffer (HEPES 21 mM, pH 7.1, 16 mM Dextrose, 0.8 mM Na_2HPO_4, 5 mM KCl and 137 mM NaCl), containing 15% glycerol for 30 s. Eighteen hours later, the medium was changed and the cells were grown for 30 additional hours in the presence or absence of the various effectors. The cells were harvested 48 h after the glycerol-shock, resuspended in 100 μl 0.25M Tris-HCl pH 7.5 and lysed by 3 cycles of freezing/thawing [21]. Cell debris was removed by low speed centrifugation at 4°C for 5 minutes in a microfuge and the supernatants stored at -80°C.

The CAT assays were performed as described by Gorman et al [22]. The concentration of the lysates was selected to assure linear conversion of the chloramphenicol into the acetylated forms (i.e < 30% acetylation). The products were extracted into ethyl acetate. The monoacetylated and diacetylated forms were separated by thin layer chromatography on silica acid resistant gel using chloroform/methanol 95 : 5 for development. The radioactive spots, detected by autoradiography were scraped from the thin layer plates and counted. The β-galactosidase activity of the cell lysates was determined as described [19] and the values were used to normalize variabilities in the efficiency of the transfections.

Phospholipase A$_2$ assays

Fluorescent substrate (1-palmitoyl 2-(10-pyrenyldecanoyl)-sn glycero 3-monomethyl phosphatidic acid) was furnished by Interchim (USA). The hydrolytic activity of PLA$_2$ was determined according to Radvanyi et al [23]. Typically, 10 to 50 μl of sample was incubated with 2 nmol of fluorescent substrate, in the presence of 0.1% delipidated BSA, in 1 ml 10 mM Tris/HCl pH 9 buffer containing 10 mM $CaCl_2$. The reaction kinetics were monitored, at room temperature, by measuring the increase of the emitted light at 397 nm, with an excitation wavelength of 345 nm.

RESULTS

Release of PLA$_2$ by rat astrocytes, rabbit chondrocytes and human HepG2 cells in the presence or absence of interleukins and other effectors

The three cell types, rat astrocytes, rabbit chondrocyte and human cells HepG2 exhibited low basal secretion of PLA$_2$ (Fig. 1). Only in the culture medium of HepG2 cells, PLA$_2$ secretion increased sharply after 40 hours incubation-time and reached noticeable concentrations at 48 hours. The kinetics of the interleukin- stimulated release of PLA$_2$ from HepG2 cells on one hand and from rat astrocytes on the other hand differed as shown by figure 1. IL-1 stimulation was mainly achieved after 24 hour incubation in rat astrocytes. It only began after this delay in HepG2 cells. The time-course of the stimulations of PLA$_2$ release by the other interleukins IL-6 and TNFα displayed the same cell-type dependency (data not shown).

Interleukins IL-1, IL-6 and TNFα enhanced PLA$_2$ secretion of rat astrocytes and of HepG2 cells by 3-4 fold after 48 hours incubation time. These interleukins displayed additive effects on the PLA$_2$ secretion by HepG2 cells (Fig. 2). For the rat astrocytes, the addition of IL-6 and IL-1 induced a stimulation slightly higher than the stimulation obtained with IL-6 or IL-1 alone (Fig. 2). On the other hand TNFα and IL-1 exhibited a synergistic effect at concentrations respectively of 100 ng/ml and 10 ng/ml. The agents which increase the intracellular cAMP concentration and stimulate PKA, enhanced the secretion of PLA$_2$. Thus, 10 μM forskolin increased the secretion of PLA$_2$ by 7 fold in HepG2 cells and 4 fold in rat astrocytes. The non metabolized cAMP analog, dibutyryl cAMP was a less potent activator. This may be due to a lower diffusion through the cell membrane compared to forskolin. Both forskolin and dibutyril cAMP synergized with IL-6 in HepG2 cells and forskolin synergized with IL-1 in rat astrocytes. On the other hand, the PKC activator TPA did not potentialize the effect of IL-1 on rat astrocytes although, used alone, it enhanced the PLA$_2$ secretion by these cells. In HepG2 cells TPA slightly lowered the secretion of PLA$_2$ and more obviously inhibited the stimulation by IL-6 (Fig. 2). Furthermore okadaic acid which inhibits phosphatases and might increase the cAMP concentration, suppressed both the unstimulated and the IL-6 induced secretion of PLA$_2$ by HepG2 cells (Fig. 2).

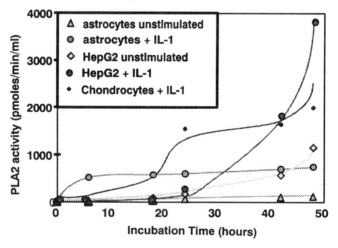

Figure 1. Time course of the unstimulated and IL-1-stimulated release of PLA$_2$ by rat astrocytes, rabbit chondrocytes and human HepG2.

Cells were incubated in P60 dishes with 10 ng/ml IL-1. PLA$_2$ activities were measured with the fluorescent substrate as indicated in material and methods. No PLA$_2$ activity could be detected in the supernatants of rabbit chondrocytes in absence of IL-1.

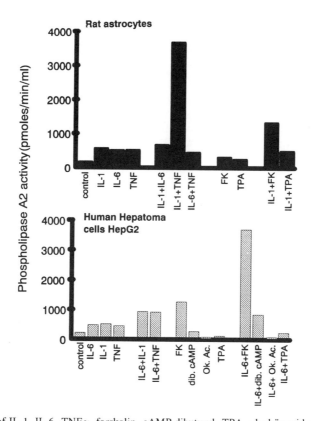

Figure 2. Effects of IL-1, IL-6, TNFα, forskolin, cAMP dibutyryl, TPA, okadaïc acid on the PLA$_2$ release by rat astrocytes, rabbit chondrocytes and human HepG2.

PLA$_2$ activities were measured after 48 hours incubation with the various effectors. Rat astrocytes were grown in p60 dishes, HepG2 in p35 multiwell plates. Concentrations were 10 ng/ml for interleukins IL-1, IL-6, TNF except when IL-1 was added to forskolin to the astrocytes cultures. In this case the IL-1 concentration was lowered to 1 ng/ml. Concentrations of the other effectors were 10^{-6} M for forskolin (FK), 10^{-7} M for TPA, cAMP dibutyryl (dib. cAMP), okadaïc acid (Ok. Ac.).

Transcriptional activity of various fragments of the 5' flanking region of the PLA$_2$ gene in rabbit chondrocytes, rat astrocytes and human HepG2 cells

The gene for the secreted PLA$_2$ contains five exons. The first is a short DNA fragment of 20 base pairs. We have cloned various fragments of the 5' flanking region upstream to the first exon spanning the positions -1614 to -87 and we have inserted them in front of a CAT reporter gene in the PUC-SH-CAT plasmid [18]. We transfected HepG2 cells, rabbit chondrocytes and rat astrocytes with these plasmids using the calcium phosphate method and measured the basal transcription activity driven by the various fragments by taking as reference the activity induced by the longest fragment [-1614;+20] in each cell type. The data are shown in figure 3; Rat astrocytes transfected by the fragments progressively deleted from -1614 to -118 displayed almost constant CAT activities. Only the last fragment [-87;+20] exhibited much lower activity (20% of the control) indicating that some elements required for the transcription have been lost. Therefore the DNA elements responsible of the transcription of the PLA$_2$ gene in rat astrocytes appeared to be located in the first hundred base pairs upstream to the first exon.

In rabbit chondrocytes, CAT activities dropped from 115% to 27.5% when the CAT construct containing the [-2264;+20] fragment was used instead of the plasmid including the [-326;+20] fragment. Thus, the region [-326;-264] seems to contain an important element which positively regulates the transcription in rabbit chondrocytes. The transfection experiments in both these cell types suggest a simple regulation of the promoter by a limited number of positive regulatory elements located in the first hundred base pairs upstream to the first exon. On the other hand, the regulation of the unstimulated transcription in HepG2 cells appears more complex. In this cell line, CAT activities increased 8 fold when the [-1614;-326] region was deleted and 10 fold again when a further deletion up to the position -210 was made. Deletion up to the position -87 induced a progressive decrease of CAT activities indicating that positive regulatory elements might be contained in the -210 to -87 region.

As the [-326;+20] fragment could drive the maximal non stimulated transcription activity in the three cell types used in these experiments and therefore might enclose the various tissue specific DNA elements required for the transcription, we studied the effects

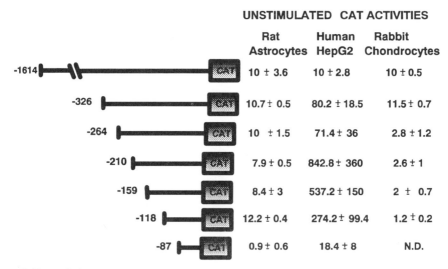

UNSTIMULATED CAT ACTIVITIES

	Rat Astrocytes	Human HepG2	Rabbit Chondrocytes
-1614	10 ± 3.6	10 ± 2.8	10 ± 0.5
-326	10.7 ± 0.5	80.2 ± 18.5	11.5 ± 0.7
-264	10 ± 1.5	71.4 ± 36	2.8 ± 1.2
-210	7.9 ± 0.5	842.8 ± 360	2.6 ± 1
-159	8.4 ± 3	537.2 ± 150	2 ± 0.7
-118	12.2 ± 0.4	274.2 ± 99.4	1.2 ± 0.2
-87	0.9 ± 0.6	18.4 ± 8	N.D.

Figure 3. Transcription activities of the various fragments of the 5' flanking region of the PLA$_2$ gene in rat astrocytes, rabbit chondrocytes and human HepG2.
CAT activities were measured 48 hours after transfections of the plasmids as explained in material and methods.

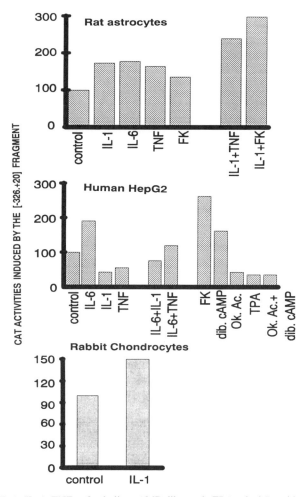

Figure 4. Effects of IL-1, IL-6, TNFα, forskolin, cAMP dibutyryl, TPA, okadaïc acid on the transcription activity of the [-326; +20] fragment of the PLA$_2$ promoter in rat astrocytes, rabbit chondrocytes and human HepG2.

The various effectors were added 18 hours after the transfections and the cells were incubated for 30 additional hours and then scraped for the measurement of CAT activities. Concentrations of the various effectors were as in figure 2.

of interleukins, phorbol ester and cAMP modulating agents on the CAT activities induced by this fragment. The data of these experiments are shown in figure 4. IL-6 increased CAT activities of transfected HepG2 cells two fold but surprisingly TNFα and IL-1 reduced them to 50% of the control. This peculiar effect of TNFα and IL-1 was confirmed by the inhibition of the IL-6 stimulation when transfected HepG2 cells were incubated with IL-1 or TNFα in addition to IL-6. Forskolin (10 μM) mimicked IL-6 stimulation (263% of the control). cAMP dibutyryl was a less potent activator of CAT activities than forskolin. A stimulation by 1.6 instead of 2.6 fold was obtained with this agent. Interestingly TPA lowered CAT activities to 39% of the control as well as okadaic acid which in addition inhibited the stimulation by cAMP dibutyryl. In rat astrocytes the interleukins IL-1, IL-6, TNFα and forskolin stimulated CAT activities by 1.5 fold. But the synergisms between IL-1 and TNFα on one hand and IL-1 and forskolin on the other hand were more obvious since IL-1 with TNFα and IL-1 with forkolin increased CAT activities respectively by 2.5 and 3 fold. Finally IL-1 enhanced CAT activities by 1.5 fold in rabbit chondrocytes.

DISCUSSION

Our data concerning the regulation of the secretion and synthesis of PLA_2 can be correlated with those previously described in the same cell types. We have shown that PLA_2 activities increase 3-6 fold in the supernatants of astrocytes and HepG2 cells treated with interleukins and forskolin. Oka and Arita [14] found a 600-fold enhancement of PLA_2 activities in presence of IL-1. Crowl et al [15] showed a 50 fold stimulation of the release of PLA_2 by HepG2 cells treated by IL-6 and $TNF\alpha$. The difference between our data and previous reports might be due to the use of a fluorescent substrate for measuring the PLA_2 activity instead of radiolabeled E. Coli phospholipids. In our view, the use of this fluorescent substrate leads to more precise and reliable estimations of the basal activities [23]. The three interleukins $TNF\alpha$, IL-1, IL-6 stimulate the release of PLA_2 in astrocytes and HepG2 cell supernatants in a similar range of values. We have observed a synergism between the effects of IL-1 and TNF and IL-6 and forskolin in rat astrocytes. This data can be related to the synergism between TNF and forskolin described by Oka and Arita [14]. Only IL-6 and forskolin cotreatment of HepG2 cells leads to a synergized release of PLA_2 by these cells.

The non-stimulated release of PLA_2 by the three cell types used in this study is low although it is noticeable in HepG2 cell supernatants at 48 hours incubation time. Its stimulation by interleukin and forsholin requires from 6 hours in astrocytes to more than 24 hours in HepG2 cells. These delays suggest the involvement of transcriptional regulatory mechanisms. Therefore the PLA_2 gene appears to be highly stimulable.

Recently extensive information has been accumulated in the role of regulatory DNA elements and protein factors which are involved in the induction of expression of the genes in various tissues, mainly in the liver. In a number of cases, the promoter elements which are crucial for the gene expression, are located in the first hundred base pairs upstream from the transcription start site. These elements are the binding sites of trans-acting factors. Tissue and effector-specific gene expressions appear to be the result of the combined effect of a set of factors which are bound to different regulatory elements.

Some putative binding sites can be recognized in the first 300 hundred base pairs upstream to the first exon of the PLA_2 gene : AP1 [-125;-86], NF-Y^m [-122;-102] NFκB [-208.-180] CREB [-224;-217] consensus sequences. In addition homologies with the binding sites of C/EBP factors can be detected in several sequences of this fragment.

Some transcription factors are more or less ubiquitous such as the members of C/EBP and CREB/ATF families, NFκB, AP1 although their expression can vary according to the environment and the differentiation state of the tissues. The expression and the transactivation properties of some transcription factors were demonstrated to be regulated and modified by various pharmacological effectors. For example, IL-1 and TPA were shown to regulate NFκB [24, 25]. TPA also affects the fos and jun dimers which constitute AP1 [26]. It is tempting to compare the inhibitory effects of IL-1 and TPA on the CAT activities in HepG2 cells and to correlate them to a putative binding of NFκB and AP1 to the promoter. But IL-1 does not inhibit the CAT activities of rat astrocytes although these factors might be present in glial cells as well as in HepG2 cells. IL-6 has been shown to induce the expression of 2 members of the C/EBP family, C/EBPβ and C/EBPδ [27]. In addition C/EBPβ can also be regulated by IL-6 at a post transcriptional level [28]. C/EBPδ is strongly expressed in brain and C/EBPβ in liver [29]. Furthermore IL-1 induces IL-6 gene expression in astrocytes [17]. These factors can be assumed to play a role in the regulation of the PLA_2 promoter by interleukins in astrocytes and HepG2 cells. Moreover it has been recently demonstrated that PKC dependant phosphorylation modulate the transactivation properties of C/EBPβ [30]. Forskolin and cAMP dibutyryl activate PKA which phosphorylates the members of the CREB/ATF family [31]. These factors might be responsible of the induction of the release of PLA_2 and CAT activities in HepG2 cells and in rat astrocytes.

The decrease of CAT activities in the presence of okadaïc acid is puzzling at first glance since this agent inhibits some phosphatases [31] and would potentialize the effects of forskolin and cAMP dibutyryl. But okadaïc acid has recently been shown to also modulate the transactivation function of AP1 and other transcription factors [32].

The analysis of the transcriptional regulation of the PLA$_2$ gene requires topographical studies which will identify the various DNA elements and characterize the factors bound to these elements in the various cell types. Such data would allow the clarification of the target(s) of interleukins and other agents whose effects lead to the secretion of PLA$_2$.

Acknowledgements : This work was partly supported by CRE INSERM n° 900210.

REFERENCES

1. Clark, J.D., Lin, L.-L., Kriz, R.W., Ramesha, C.S., Sultzman, L.A., Lin, A.Y., Milona, N. and Knopf, J.L. A novel arachidonic acid-selective PLA$_2$ contains a Ca^{2+}-dependent translocation domain with homology to PKC and GAP. *Cell* **65** (1991) 1043-1051.
2. Sharp, J.D., White, D.L., Chiou, X.G., Goodson, T., Gamboa, G.C., McClure, D., Burgett, S., Hoskins, J., Skatrud, P.L., Sportsman, J.R., Becker, G.W., Kang, L.H., Roberts, E.F. and Kramer, R. Molecular cloning and expression of human Ca^{2+}-sensitive cytosolic phospholipase A$_2$. *J. Biol. Chem.* **266** (1991) 14850-14853
3. Seilhamer, J.J., Pruzanski, W., Vadas, P., Plant, S., Miller, J.A., Kloss, J. and Johnson, L.K. Cloning and recombinant expression of phospholipase A$_2$ present in rheumatoid arthritic synovial fluid. *J. Biol. Chem.* **266** (1989) 5335-5338.
4. Kramer, RM.M., Hession, C., Johansen, B., Hayes, G., McGray, P., Chow, E.P., Tizard, R. and Pepinsky, R.B. Structure and properties of a human non-pancreatic phospholipase A$_2$. *J. Biol. Chem.* **264** (1989) 5768-5775.
5. Seilhamer, J.J., Randall, T.L., Yamanaka, M. and Johnson, L.K. Pancreatic Phospholipase A$_2$: Isolation of the human gene and cDNAs from porcine pancreas and human lung. *DNA* **5** (1986) 519-527.
6. Hulkower, K.I., Hope, W.C., Chen, T., Anderson, C.M., Coffey, J.W. and Morgan, D.W. Interleukin-1β stimulates cytosolic phospholipase A$_2$ in rheumatoid synovial fibroblasts. *Biochem. Biophys. Res. Commun.* **184** (1992) 712-718.
7. Berenbaum, F., Thomas, G., Poiraudeau, S., Bereziat, G., Corvol, M.T. and Masliah, J. Insulin like growth factor inhibit arachidonic acid release and type II phospholipase A$_2$ gene expression by interleukin-1β in articular rabbit chondrocytes. (1993) *submitted to publication*
8. Schalkwijk, C., Pfeilschifter, J., Märki, F. and van den Bosch, H. Interleukin-1β- and forskolin-induced synthesis and secretion of group II phospholipase A$_2$ and prostaglandin E$_2$ in rat mesangial cells is prevented by transforming growth factor-β$_2$. *J. Biol. Chem.* **267** (1992) 8846-8851
9. Lin, L.-L., Lin, A.Y. and Knopf, J. L. Cytosolic phospholipase A$_2$ is coupled to hormonally regulated release of arachidonic acid. *Proc. Natl. Acad. Sci. USA* **89** (1992) 6147-6151
10. Suffys, P., Roy, F.V. and Fiers, W. Tumor Necrosis Factor and Interleukin 1 Activate Phospholipase in Rat Chondrocytes *FEBS Lett.* **232** (1988) 24-28.
11. Pfeilschifter, J., Pignat, W., Vosbeck, K. and Märki, F. Interleukin 1 and tumor necrosis factor synergistically stimulate prostaglandin synthesis and phospholipase A$_2$ release from rat renal mesangial cells *Biochem. Biophys. Res. Commun.* **159** (1989) 385-394.
12. Nakano, T., Ohara, O., Teraoka, H. and Arita, H. Group II phospholipase A$_2$ mRNA synthesis is stimulated by two distinct mechanisms in rat vascular smooth muscle cells. *FEBS Lett* **261** (1990) 171-174.
13. Vadas, P., Pruzanski, W., Stefanski, E., Ellies, L.G., Aubin, J.E., Sos, A. and Melcher, A. Extracellular phospholipase A$_2$ secretion is a common effector pathway of interleukin-1 and tumor necrosis factor action. *Immunol. Lett.* **28** (1991) 187-194.
14. Oka, S. and Arita, H. Inflammatory factors stimulate expression of group II phospholipase A$_2$ in rat cultured astrocytes *J. Biol. Chem.* **266** (1991) 9956-9960.

15. Crowl, R.M., Stoller, T.J., Conroy, R.R. and Stoner, C.R. Induction of phospholipase A_2 gene expression in human hepatoma cells by mediators of the acute phase response. *J. Biol. Chem.* **266** (1991) 2647-2651.

16. Lyons-Giordano, B., Davis, G.L., Galbraith, W., Pratta, M.A. and Arner, E.G. Interleukin-1β stimulates phospholipase A_2 mRNA synthesis in rabbit articular chondrocytes. *Biochem. Biophys. Res. Commun.* **164** (1989) 488-495.

17. Benveniste, E.N., Sparacio, S.M., Norris, J.G., Grenett, H.E. and Fuller, G.M. Induction and regulation of interleukin-6 gene expression in rat astrocytes *J. Neuroimmunol.* **30** (1990) 201-212.

18. Ogami, K., Hadzopoulou-Cladaras, M., Cladaras, C., and Zannis, V.I. Promoter elements and factors required for hepatic and intestinal transcription of the human apoCIII gene. *J. Biol. Chem.* **265** (1990) 9808-9815.

19. Graham, F.L., and Van der Eb, A.J. A New Technique for the Assay of Infectivity of Human Adenovirus 5. *Virology* **52** (1973) 456-461.

20. Edlund, T., Walker, M.D., Barr, P.J., and Rutter, W.J. Cell-Specific Expression of the Rat Insulin Gene : Evidence for Role of Two Distinct 5' Flanking Elements. *Science* **230** (1985) 912-916.

21. Kumar, V; and Chambon, P. The Estrogen Receptor Binds Tightly to its Responsive Elements as a Ligand-Induced Homodimer. *Cell* **55** (1988) 145-156.

22. Gorman, C.M., Moffat, L.F., and Howard, B.H. Recombinant Genomes Which Express Chloramphenicol Acetyl-Transferase in Mammalian Cells. *Mol. Cell. Biol.* **2** (1982) 1044-1051.

23. Radvanyi, F., Jordan, L., Russo-Marie, F. and Bon, C. A Sensitive and continuous fluorometric assay for phospholipase A_2 using pyrene-labeled phospholipids in the presence of serum albumin. *Anal. Biochem.* **177** (1989) 103-109.

24. Sen, R. and Baltimore, D. Inducibility of κ immunoglobulin enhancer-binding protein NFκb by a posttranslational mechanism. *Cell* **47** (1986) 921-928.

25. Hohmann, H.P., Remy, R., Aigner, L., Brockhaus, M. and Loon, A.P.G.M. Protein kinases negatively affect nuclear factor-κB activation by tumor necrosis factor-α at two different stages in promyelocytic HL60 cells. *J. Biol. Chem.* **267** (1992) 2065-2072.

26. Konig, H., Ponta, H., Rahmsdorf, H. J., Herrlich, P. Interference between pathway-specific transcription factors : Glucocorticoids antagonize phorbol ester-induced AP-1 activity without altering AP-1 site occupation in vivo. *EMBO J.* **11** (1992) 2241-2246.

27. Akira, S., Isshiki, H., Sugita, T., Tanabe, O., Kinoshita, S., Nishio, Y., Nakajima, T., Hirano, T. and Kishimoto, T. A nuclear factor for IL-6 expression (NF-IL6) is a member of a C/EBP family *EMBO J* **9** (1990) 1897-1906.

28. Ranji, D.P., Vitelli, A., Tronche, F., Cortese, R. and Ciliberto, G. The two C/EBP isoforms, IL-6DBP/NF-IL6 and C/EBPδ/NF-IL6β, are induced by IL-6 to promote acute phase gene transcription via different mechanisms. *Nucleic Acids Res.* **21** (1993) 289-294.

29. Alam, T., An, M.R. and Papaconstantinou, J. Differential expression of three C/EBP isoforms in multiple tissues during the acute phase response. *J. Biol. Chem.* **267** (1992) 5021-5024.

30. Trautwein, C., Caelles, C., van der Geer, P., Hunter, T., Karin, M. and Chojkier, M. Transactivation by NF-IL6/LAP is enhanced by phosphorylation of its activation domain. *Nature* **364** (1993) 544-547.

31. Gomezmunoz, A., Hatch, G.M., Martin, A., Jamal, Z., Vance, D.E. and Brindley, D.N. Effects of okadaic acid on the activity of two distinct phosphatidate phosphohydrolases in rat hepatocytes. *FEBS Let.* **301** (1992) 103-106.

32. Park, K., Chung, M. and Kim, S.J. Inhibition of myogenesis by okadaic acid, and inhibitor of protein phosphatases, 1 and 2A, correlates with the induction of AP1. *J. Biol. Chem.* **267** (1992) 10810-10815.

PHOSPHOLIPASE A$_2$ IN INTENSIVE-CARE PATIENTS: THE GERMAN MULTICENTRE TRIAL

Waldemar H.Uhl and Markus W. Büchler and the German Phospholipase A2-Study Group

Department of Visceral and Transplantation Surgery
University of Berne (Inselspital)
Murtenstr. 35
CH-3010 Berne, Switzerland

INTRODUCTION

In animal experiments and clinical trials increased circulating catalytic activities of phospholipase A$_2$ were measured in various life-threatening diseases and conditions including septic shock [1-5], ARDS [6], multiple injuries [7], and diffuse peritonitis [8]. There seems to be no longer any question that phospholipases play a major role in the inflammatory responses associated with these diseases [9,10].

The mortality of patients in intensive care following surgery today is mainly determined by multiple organ failure (MOF syndrome) and its degree of severity. MOF syndrome can be characterized as a generalized autoaggressive inflammatory process [11,12]. An early and accurate prediction of the degree of severity of MOF syndrome can reduce the mortality of intensive care patients and help to save on the costs of treatment.

The objective of the present study was to evaluate the role of serum phospholipase A$_2$ in comparison with two known inflammatory parameters C-reactive protein and PMN elastase in four groups of surgical intensive-care patients. The study especially aimed to determine whether phospholipase A$_2$ activity can help to predict a lethal MOF syndrome early in the intensive-care treatment.

Esterases, Lipases and Phospholipases, Edited by M.I. Mackness
and M. Clerc, Plenum Press, New York, 1994

PATIENTS

Patient recruitment for this prospective multicentre study was carried out at seven German University Hospitals (H.G.Beger, M.Büchler and W.Uhl, University of Ulm; E.Hanisch, University of Frankfurt; A.Schild, University of Erlangen; Ch.Waydhas, University of Munich; E.Entholzner, Klinikum Rechts der Isar of Munich; K.Müller, University of Tübingen; W.Kellermann, Klinikum Munich-Großhadern and G.Hoffmann, Munich-Bogenhausen). Following approval of the Ethics Commission of the University of Ulm, a total of 223 intensive-care patients were admitted to the study over a period of 17 months (9/89 to 1/91) after having given their informed written consent. Excluded were patients under 18 years of age, patients who were enrolled in another clinical therapeutic trial at the same time, and those with pancreatic disease. Four different groups of surgical intensive-care patients were evaluated (Table 1):

Table 1. Patients characteristics in the four surgical intensive-care patient-groups, multiple injuries, peritonitis, sepsis and controls (high risk)

	Multiple injuries n=73	Peritonitis n=46	Sepsis n=52	Controls (High risk) n=52
Age				
Median	33	66	56	60
Min.-Max.	18-88	18-96	22-83	28-79
Sex				
Male	78%	59%	62%	79%
Female	22%	41%	38%	21%
	ISS	MPI	SSS	TG n=21 (40%)
Median	34	21	17	DAR n=27 (33%)
Min.-Max.	18-75	10-39	9-28	ER n=14 (27%)

ISS: injury severity score; MPI: Mannheim Peritonitis Index; SSS: Sepsis severity score. TG: total gastrectomy; DAR: deep anterior rectal resection; ER: esophageal resection.

Group I: multiple injuries (n=73; MI): Inclusion criteria were relevant injuries to at least one body cavity with additional extremity fractures, injury to a body cavity plus head injury, or at least three extremity fractures alone. An injury severity score (ISS) of at least 18 points was mandatory. Date of entry into the study (day 0) was the day on which the multiple injuries occurred.

Group II: peritonitis (n=46; P): Patients with intraoperatively confirmed diffuse peritonitis were enrolled in the study. The date of entry (day 0) was the day of the laparotomy. Excluded were patients with local peritonitis associated with perforated appendicitis.

Group III: septic shock (n=52; S): The criterion for entry (day 0) was considered fulfilled when four of the five following conditions were satisfied:
1. leukocytosis > 12 G/l or leukopenia < 4 G/l
2. thrombocytes < 100 G/l or thrombocyte reduction of 30% within 24 hours
3. rectal temperature > 38.5°C
4. positive blood culture or known focus of infection
5. metabolic acidosis (base excess > - 4 mmol/l)

Group IV: Control group (high risk) (n=52; HR): Patients were included in this group who were at a considerably higher risk for postoperative sepsis after defined surgical interventions (esophageal resection, total gastrectomy, deep anterior rectal resection). Date of entry into the study was the day prior to the planned operation (day -1).

MONITORING AND METHODS

In addition to the routine blood samples taken in the mornings, additional samples were taken for the three inflammatory parameters to be analyzed. These samples were taken during intensive-care treatment daily up to day 7 and on the day of discharge. Analysis were done after uninterupted refrigeration centrally and in a blinded fashion at the Institute for Clinical Chemistry, Städtisches Krankenhaus Munich-Bogenhausen.

For the determination of **C-reactive protein (CRP)** the nephelometric method developed by Behring Werke (Marburg, Germany) was used. The upper normal value for CRP is 6 mg/l by this test system. **PMN elastase (PMN)** was measured by a photometric immunoassay (IMAC technique). For the PMN elastase the upper normal value is 56 μg/l. The reaction principle for **Phospholipase A_2 (PLA_2)** is based on the photometric fixed-time determination of fatty acids [13] released from a standardized phospholipid emulsion. The upper limit of the reference range for pospholipase A_2 is 10 U/l.

MOF groups: The patients were divided into three groups according to their MOF scores (Table 2):

nMOF = no MOF: patients without multiple organ failure
rMOF = reversible MOF: patients who survived a multiple organ failure. To be assigned to this group a MOF score of more than 4 points on at least 2 successive days after the second day of patient's enrollement was required.
lMOF = lethal MOF: patients who died of multiple organ failure.

Table 2. Multiple organ failure (MOF) - subgroups in the four surgical intensive-care patient-groups, multiple injuries, peritonitis, sepsis and controls (high risk)

	Multiple injuries n=73	Peritonitis n=46	Sepsis n=52	Controls (High risk) n=52	Total n=223
nMOF	22 (30%)	22 (48%)	4 (8%)	49 (94%)	97 (43%)
rMOF	38 (52%)	15 (33%)	22 (42%)	2 (4%)	77 (35%)
lMOF	13 (18%)	9 (19%)	26 (50%)	1 (2%)	49 (22%)
MOF-incidence	70%	52%	92%	6%	57%
Mortality	18%	19%	50%	2%	22%

nMOF: no MOF; rMOF: reversible MOF; lMOF: lethal MOF

STATISTICS

Medians and quartiles were calculated after being checked for normal distribution. The Mann-Whitney U Test for independent random samples was used for the statistical significance test. The ROC curves were plotted to determine the optimal cut-off. To characterize a diagnostic test, sensitivity, specificity, efficiency, positive (PPV) and negative (NPV) predictive values of a test result were determined at the optimal cut-off.

RESULTS

The median curves for C-reactive protein, PMN elastase, and phospholipase A_2 in the four patient groups multiple

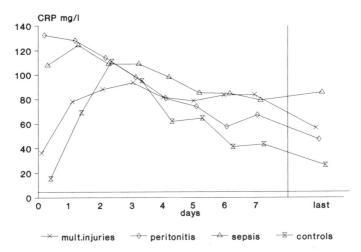

Figure 1a. Median curves for C-reactive protein in the four intensive-care patient-groups. Solid line = upper normal limit; last = last blood sample.

injuries (MI), peritonitis (P), sepsis (S) and controls (HR: high risk) are shown in Fig 1a-c.

C-reactive protein

For C-reactive protein (Fig. 1a) an elevated serum level far above the normal range was measured in the median in all four groups analyzed from day 0 (medians and quartiles in mg/l: MI: 36, 12-68; P: 128; 98-148; S: 108, 93-134; HR: 15, 7-33) to the last sample day (day 7: MI: 56, 26-98; P: 47, 24-77; S: 86, 26-116; HR: 26, 15-54). The CRP levels in

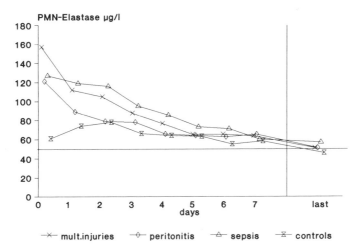

Figure 1b. Median curves for PMN elastase in the four intensive-care patient-groups. Solid line = upper normal limit; last = last blood sample.

patients with peritonitis and sepsis were the highest, already at the beginning. The maximum levels in the multiple injury and control (high risk) groups followed on days 2 and 3. The fastest drop was seen in the high-risk group.

PMN elastase

All of the median levels for PMN elastase (Fig. 1b) were above the normal range in the first week. In contrast to CRP, the highest PMN elastase levels were not found in the groups with sepsis and peritonitis, but among those with multiple injuries (medians and quartiles on day 0 in $\mu g/l$: MI: 157, 106-324; P: 121, 66-196; S: 127, 94-188; HR: 61, 39-88). During the further course, the PMN elastase levels were highest among the multiple injuries and sepsis groups, however, the median curves had dropped to normal range at the last blood-sampling.

Phospholipase A$_2$ activity

In the multiple injuries and high risk groups, the median of the PLA$_2$ activities was in the normal range throughout the entire observation period (Fig. 1c). The highest levels were measured in the groups with sepsis and peritonitis (medians and quartiles on day 0 in U/l: MI: 2, 0-6.5; P: 25, 9-72; S: 35, 12 63; HR: 1, 0-4), which also returned to the normal range in the first week.

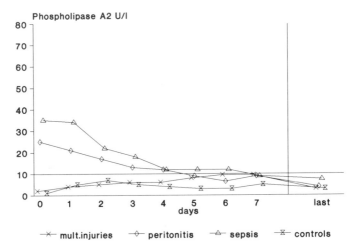

Figure 1c. Median curves for Serum-Phospholipase A2 activity in the four intensive-care patient-groups. Solid line = upper normal limit; last = last blood sample.

Table 3. Prediction of lethal MOF (lMOF) in patients suffering from multiple injuries and peritonitis by means of PMN elastase and phospholipase A_2 activity

	Multiple injuries			Peritonitis	
	day 0	day 1		day 0	day 1
PMN-Wert $\mu g/l$	170	205	PLA$_2$ U/l	80	60
Prevalence	16	17		21	19
Sensitivity	78	67		71	68
Specificity	64	88		88	57
Efficiency	**66**	**84**		**85**	**58**
PPV	29	53		63	25
NPV	93	93		92	87

(PPV) positive and (NPR) negative predictive value.

PREDICTION OF A LETHAL MOF (LMOF)

From a statistical point of view the two patient groups sepsis and controls (high risk) were not suited for further analysis in this study (Table 2). First of all, the frequency of multiple organ failure in the sepsis group was 92% (48/52), making the a-priori-probability of sustaining a MOF very high. Secondly, the mortality in the control group was only 2% (1/52) and the frequency of reversible MOF was also only 4% (2/52). Therefore, only the groups multiple injuries and peritonitis could be included in the special statistical analysis for predicting a lMOF (lMOF 18% and 19%, respectively).

The results of the diagnostic tests based on the optimized cut-off levels in the groups multiple injuries and peritonitis are given in Table 3. With a cut-off of 170 $\mu g/l$ and 205 $\mu g/l$, the efficiency of the PMN elastase in the multiple injuries group for the prediction of a lethal MOF was 66% on day 0 and 84% on day 1. The MOF score with the optimal cut-off calculated at 5 points was only 69% on day 1. In the peritonitis group the efficiency of the catalytic phospholipase A_2 activity for an lMOF on day 0 and 1 with a cut-off of 80 U/l and 60 U/l was 85% and 58%, respectively. On the day of laparotomy (day 0), the efficiency for the MOF score with 4 points was at the same level as the single parameter phospholipase A_2 activity (88% and 85%, respectively).

DISCUSSION

The analysis of three inflammatory parameters CRP, PMN elastase, and phospholipase A_2 showed very different kinetics among the four groups of patients analyzed. The patients defined as controls, who carried a far higher risk for developing postoperative sepsis, had the highest CRP values on the 2nd and 3rd days after surgery and the median values were still above the reference range at the end of the first week. This observation is known from CRP and is consistent with results given in the literature [14]. The PMN elastase increased only moderately in these patients. In contrast, the phospholipase A_2 remained in the reference range, which indicates that this parameter is not influenced by the postaggression metabolism to the extent that CRP is. Among the septic patients extremely high levels of all three biochemical markers were measured.

In the patients with multiple injuries the efficiency of the PMN elastase on day 1 for predicting a lethal MOF at a cut-off of 205 $\mu g/l$ was 84% (sensitivity 67%, specificity 88%); this was much better than the multifactorial MOF score with the best cut-off of 5 points with an efficiency of 69%.

The median survival of the patients with multiple injuries was 7 days. This finding can be used as a so-called "diagnostic or therapeutic window" that provides information to enable the physician to initiate early treatment measures in these patients. The serum phospholipase A_2 activity in patients with diffuse peritonitis, however, has an extremely high prognostic value, since an 85% efficiency for lMOF can already be measured at laparotomy (cut-off 80 U/l; sensitivity 71%, specificity 88%). In this respect, the other parameters measured were not relevant in this group.

Measurement of the serum-CRP does not contribute to an early estimation of the prognosis in either the group with multiple injuries or with peritonitis.

In summary, the PMN elastase in patients with multiple injuries and the phospholipase A_2 activity in patients with diffuse peritonitis are of high prognostic relevance very early in the treatment phase. The measurement of these inflammatory mediators can contribute greatly to estimating the individual risk for the occurrence of a MOF syndrome. The finding that individual laboratory parameters can have the same prognostic value as a multifactorial score system, which includes the essential organ and vital functions, is surprising and induces us to undertake further prospective studies.

The exact cellular source of phospholipase A_2 is largely unknown. A possible source is the liver, since it has been shown that inflammatory mediators such as interleukin-6 and tumor necrosis factor increase the release of phospholipase A_2 from liver cell cultures. Therefore, this phospholipase A_2 might represent an acute phase protein of the liver [15,16].

REFERENCES

1. Schmidt H., Creutzfeld W., The possible role of phospholipase A in the pathogenesis of acute pancreatitis, *Scand J Gastroenterol* 4: 1969, 39-48.

2. Büchler M., Malfertheiner P., Schädlich H. et al., Role of phospholipase A_2 in human acute pancreatitis, *Gastroenterology* 97: 1989, 1521-1526.

3. Vadas P. and Hay J.B., Involvement of circulation phospholipase A_2 in the pathogenesis of the hemodynamic changes in endotoxin shock, *Can J Physiol Pharmacol* 61: 1983, 561-566.

4. Vadas P., Elevated plasma phospholipases A_2 levels: Correlation with the hemodynamic and pulmonary changes in gram-negative septic shock. *J Lab Clin Med* 104: 1984, 873-881.

5. Schild A, Pscheidl.E. and v.Hintzenstern U., Phospholipase A - a parameter of sepsis? *Klin Wochenschr* 67: 1987, 207-211.

6. Kellermann W., Frentzel-Beyme R, Welte W., et al., Phospholipase A in acute lung injury after trauma and sepsis: its relation to the inflammatory mediators PMN elastase, C3a, and neopterin. *Klin Wochenschr* 67: 1989, 190-195.

7. Uhl W., Büchler M., Nevalainen T.J., et al., Serum phospholipase A_2 in patients with multiple injuries. *J Trauma* 30: 1990, 1285-1290.

8. Büchler M., Deller A., Malfertheiner P., et al., Serum phospholipase A_2 in intensive care patients with peritonitis, multiple injury, and necrotizing pancreatitis. *Klin Wochenschr* 67: 1989, 217-221.

9. Vadas P. and Pruzanski W., Biology of disease role of secretory phospholipase A_2 in the pathobiology of disease. *Lab Invest* 55: 1986, 391-404.

10. Kaiser E., Chiba P., and Zaky K., Phospholipases in biology and medicine. *Clin Biochem* 23: 1990, 349-370.

11. Goris R.J.A., Boekhorst T.P.A., Nuytinck J.K.S., et al., Multiple organ failure. *Arch Surg* 120: 1985, 1109-1115.

12. Waydhas C., Nast-Kolb D., Jochum M., et al., Inflammatory mediators, infection, sepsis, and multiple organ failure after severe trauma. *Arch Surg* 127:1992,460-467.

13. Hoffmann G.E., Schmidt D., Bastian B., et al., Photo-metric determination of phospholipase A. *J Clin Chem Clin Biochem* 24: 1986, 871-875.

14. Mustard R.A., Bohnen J.M.A., Haseeb S., et al., C-reactive protein levels predict postoperative septic complications. *Arch Surg* 122: 1987, 69-73.

15. Clark M.A., Chen M.J., Crooke S.T., et al., Tumor necrosis factor (cachectin) induces phospholipase A_2 activity and synthesis of a phospholipase A_2-activating protein in endothelial cells. *Biochem J* 250:1988,125-132.

16. Crowl R.M., Stoller T.J., Conroy R.R., et al., Induction of phospholipase A_2 gene expression in human hepatoma cells by mediators of the acute phase response. *J Bio Chem* 266: 1991, 2647-2651.

PHOSPHOLIPASES

STUDY OF SOME MECHANISMS THAT MAY CONTRIBUTE TO THE PRESENCE OF HIGH
LEVELS OF PHOSPHOLIPASE A_2 ACTIVITY IN PLASMA OF PATIENTS WITH SEPTICEMIA

Miguel Angel Gijón[1], Carolina García[1], César Pérez[2], Fernando
López·Díez[3], and Mariano Sánchez-Crespo[1]

[1]Departamento de Bioquímica y Fisiologia, Facultad de
Medicina, 47005-Valladolid, Spain
[2]Fundación Jiménez Díaz, Av. Reyes Católicos 2, 28040-Madrid
Spain
[3]Hospital de la Princesa, Diego de León 62, 28006-Madrid
Spain

INTRODUCTION

Phospholipase A_2 (EC 3.1.1.4) is a family of phosphatide 2-
acylhydrolases that play an important role in the metabolism of
phospholipids and membrane homeostasis [1-4]. These functions of
phospholipase A_2 seem relevant to the pathogenesis of various clinical
conditions, since the presence of high concentrations of phospholipase
A_2 activity in plasma and inflammatory exudates of patients suffering
from inflammatory arthritis, peritonitis and septic shock has been
reported. Administration of exogenous extracellular phospholipase into
experimental animals causes inflammatory hyperemia [5], and a correlation
of serum levels of phospholipase A_2 with the magnitude of hypotension has
been shown in endotoxin shock [6,7]. The possible pathogenetic role of
soluble phospholipase A_2 in endotoxin shock in connection to the cytokine
network has been emphasized by several reports showing the secretion of
type II phospholipase A_2 by rat mesangial cells [8,9], liver cells [10-
11] and human synovial cells [12,13] in response to tumor necrosis factor
and interleukin 1ß. A recent report from this laboratory has shown the
presence of high amounts of PAF associated with the platelets of patients
with septicemia [14] and this raises the question as to whether this
finding might be related to the secretion of phospholipase A_2 activity
and an ensuing formation of lipid mediators.
However, there are a number of difficulties to be resolved before
assigning a clearcut pathogenetic role in endotoxin shock to this enzyme
activity. For instance, Heath and coworkers [15] have stressed the
absence of correlations in an experimental model of peritonitis between
plasma phospholipase A_2 activity and the clinical course of the disease,
whereas Wright and coworkers [16] have identified the phospholipase A_2
accumulated in rabbit inflammatory ascites with the soluble enzyme
detected in serum under physiological conditions, and proposed a role for
this enzyme activity among the mechanisms of defense of the host, since

Esterases, Lipases and Phospholipases, Edited by M.I. Mackness
and M. Clerc, Plenum Press, New York, 1994

it favors the lysis of bacteria. Parthasarathy and Barnett [17] have detected a phospholipase A_2 activity associated with human apolipoprotein B-100, which only acts on oxidized phosphatidylcholine, and seems to play a primary role on the elimination fatty acids undergoing oxidation in the plasma.

In this study we have attempted the measurement and characterization of phospholipase A_2 activity in the plasma from patients suffering from septic shock and other serious conditions. We have also studied the mechanism of secretion of phospholipase A_2 from leukocytes in order to detect the origin of the plasma enzyme and its possible induction by cytokines.

MATERIALS & METHODS

Patients and Materials

Plasma from both normal volunteers and patients was obtained from venous blood anticoagulated with heparin. Plasma was stored at -70°C, and could be kept for up to four months without loss of the phospholipase A_2 activity. Sepsis syndrome was defined according to the criteria of Bone and coworkers [18]. APACHE II score was used for disease classification severity [19]. Essentially fatty acid-free BSA was obtained from Miles Laboratories. p-Bromophenacyl bromide was from Aldrich-Europe, Beerse, Belgium. Interleukin 1α, interleukin 1β and tumor necrosis factor-α were from Genzyme Inc. Boston, MA. *E. coli* endotoxin (serotype No. 0127:B8) was from Difco Labs, Detroit, MI. Reagents for the measurement of proteins according to the method of Bradford [20] were purchased from Bio-Rad Labs, Richmond, CA. Rabbit polyclonal antibody against hog pancreatic-type phospholipase A_2 was kindly provided by Dr. Pedro Esponda and was purified by precipitation with sodium sulfate followed by dialysis and affinity chromatography with Protein A coupled to Sepharose CL-4B (Sigma Chemical Co.).

Assay of phospholipase A_2 activity

The assay was carried out with 2 μl of patients' plasma in a volume of 0.5 ml. The procedure [21] consisted of the incubation of the sample with [^{14}C]oleate-labeled autoclaved *E. coli* of a K12 strain. The amount of phospholipid contained in the *E. coli* sample was 60-120 nmol, as assessed by the assay of phospholipid-associated phosphorus. The assay medium contained 0.1 M Tris/HCl, 1 mg/ml fatty acid-free BSA, and 1 mM $CaCl_2$, pH 7.5. The reaction proceeded for 30 min and was stopped by addition of 0.2 ml of ice-cold 2 N HCl and 80 μl of a solution containing 100 mg/ml of essentially fatty acid-free BSA, followed by centrifugation for 5 min in an Eppendorf microcentrifuge at 13,000 rpm. The radioactivity released into the supernatant was assayed by scintillation counting and expressed in arbitrary units. One unit was defined as 1% hydrolysis of the substrate per 30 min. To determine the positional specificity of the phospholipase activity, experiments were carried out with [^{14}C]palmitate-labeled *E. coli* since >90% of this fatty acid residue incorporates in the 1-ester position. In the case of experiments intended to define calcium-dependency, plasma was diluted in a Ca^{2+}/EGTA buffer, and buffer free Ca^{2+} concentration was calculated from the association constant for the Ca^{2+}-EGTA complex. For the study of the pH-dependency of phospholipase A_2 activity, a 0.1 M citrate/phosphate buffer was used for the range of pH 4.5-7, and 0.1 M Tris/HCl for the range 7.5-9.

Isolation of human leukocytes

Polymorphonuclear leukocytes were obtained from peripheral blood of normal volunteers anticoagulated with heparin. Red cells were removed by Dextran T-500 (Pharmacia LKB) sedimentation, and polymorphonuclear leukocytes were separated from mononuclear cells by centrifugation on Ficoll-Hypaque cushions. Peripheral blood monocytes were obtained by adherence to plastic dishes of mononuclear leukocytes obtained at the interface of Ficoll-Hypaque. These cells were maintained in a humidified atmosphere at a concentration of $1-2x10^6$ monocytes in tissue culture dishes (Sterilin Limited, Feltham, England), in RPMI 1640 supplemented with 2 mM L-glutamine, containing 100 U/ml penicillin and streptomycin. Contamination by platelets of the leukocyte preparation was usually less than one platelet per cell. The isolated leukocyte population was used for stimulation with different secretagogues and phospholipase A_2 activity was measured in both supernatants and cells after scraping and sonication. Samples to be used for enzyme assay were supplemented with 1 mM EDTA and 1 mM phenylmethylsulfonyl fluoride to avoid proteolytic degradation of the enzyme.

RESULTS

Phospholipase A_2 activity in plasma from sepsis patients

Plasma from 21 patients with septicemia, as assessed by the presence of at least one positive blood culture and clinical signs, contained high levels of phospholipase A activity (Table 1).

Table 1. Clinical data of sepsis patients

Microorganism	APACHE II Score	Platelet count[1]	Phospholipase A_2[2]	n
Fungi[3]	23±2,96	36333±15387	194±4	3
Combination	29±1,98	94625±34386	148±16	8
Gram positive	25±2,80	127000±31391	130±2	5
Gram negative	24±2,28	122600±30135	140±9	5

[1]Platelets/μl. [2]Units/mg protein. [3]This group includes two patients with a previous diagnosis of aplastic anemia.

Disease severity was high, although similar in the different etiological groups, as judged from the high APACHE II score and the overall mortality rate which reached 76.12%. The positional specificity of this activity was determined as type A_2, in view of its preferential effect on [14C]oleate-labeled *E. coli*, rather than on [14C]palmitate-labeled bacteria (not shown). p-Bromophenacyl bromide, a known inhibitor of phospholipase A_2 also inhibited the enzyme detected in patients' plasma in the range of concentrations 0.1-1 mM (not shown). In contrast, the enzyme activity was not influenced by incubation with purified polyclonal rabbit antiserum against pancreatic phospholipase A_2, which strongly suggests that this enzyme is not of pancreatic origin, in view of the very wide sequence homology that exists between human and hog pancreatic phospholipase A_2 [22,23].

Phospholipase A_2 activity did not correlate with renal function as measured by the clearance of creatinine. This can be a relevant finding since most reports show low molecular weights for soluble phospholipase A_2, a feature that would facilitate a rapid renal clearance provided there is adequate renal function. Three of the sepsis patients had received a bolus dose of methylprednisolone prior to the blood sample being taken, but this did not apparently interfere with the phospholipase A_2 activity assayed in plasma. Phospholipase A_2 plasma levels were not influenced by previous platelet count, since 4 patients with aplastic anemia showing less than 20,000 platelets per μl and 1,000 leukocytes per μl had levels of phospholipase A_2 indistinguishable from those detected in the remaining patients. Two of these patients showed fungal infection. Patients with either Gram-positive or Gram-negative bacteria showed no remarkable difference in the amount of phospholipase A_2 assayed in plasma (Table 1). We were unable to find any difference in the assay of phospholipase A_2 due to the use of either serum or plasma, which is in keeping with a recent report [15]. The enzyme activity detected in patients plasma was calcium-dependent, the optimal activity being observed with concentrations of calcium ions above 0.1 mM (Fig. 1A). Optimal enzyme activity was observed at alkaline pH.

Phospholipase A_2 activity in plasma of non-septic patients

The assay showed undetectable levels in a group of ten human volunteers (less than 10 U/mg plasma protein), and in eight patients diagnosed with different conditions and admitted to the intensive therapy unit because of either cardiogenic shock or organ failure. This group

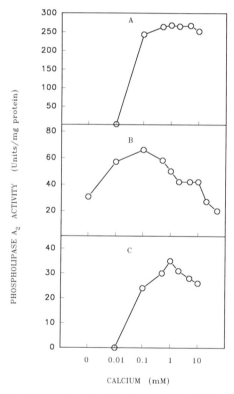

Figure 1. Effect of the concentration of calcium ions on phospholipase A_2 activity from plasma of a sepsis patient (A), supernatant of human polymorphonuclear leukocytes stimulated with opsonized zymosan particles (B), and supernatant of adhered monocytes stimulated with opsonized zymosan particles (C). Data represent mean values of two experiments in duplicate.

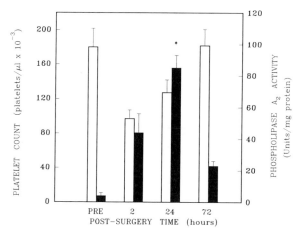

Figure 2. Phospholipase A$_2$ activity levels and platelet counts in a group of patients undergoing open-heart surgery, before and at different times after completion of surgery. Open columns indicate platelet count and black columns indicate phospholipase A$_2$ activity. Results are expressed as mean±S.E.M. of 10 individuals and the samples were assayed in duplicate. *Indicates a P<0.01 (two tailed t-test) as compared to pre-surgery levels.

also included patients with local infection and a patient with the diagnosis of malaria treated with chloroquine. However, high levels of phospholipase A$_2$ activity were detected in plasma of 8 out of ten patients that had undergone open-heart surgery because of either valvular substitution or coronary bypass surgery. These patients also presented thrombocytopenia most likely explained by the occurrence of platelet activation in the cardio-pulmonary by-pass. Phospholipase A$_2$ activity increased less markedly in these patients than in septicemia patients, peaked at 24-hours after completion of surgery, and returned to basal levels by 48-72 hours (Fig. 2). In an attempt to correlate this phospholipase A$_2$ rise with the activation of platelets, serial blood platelet counts were carried out, but platelet count reduction following surgery, did not correlate with the increase of enzyme activity, since maximal platelet count drop occurred two hours after completion of surgery and phospholipase A$_2$ peaked 24-hours thereafter (Fig. 3). Peak plasma phospholipase A$_2$ levels were higher in those patients that underwent longer periods in by-pass there being a correlation coefficient between both parameters of 0.60.

Release of phospholipase A$_2$ activity from human polymorphonuclear leukocytes

Since PMN have been reported to release phospholipase A$_2$ [15,20], these cells were considered as a possible source of the enzyme activity detected in plasma from sepsis patients. In order to substantiate this hypothesis, isolated human PMN were stimulated with complement-coated zymosan particles, and phospholipase A$_2$ activity assayed in supernatants. Enzyme activity was detected in supernatants from stimulated PMN, reached maximal activity after 30-60 min, and was present in higher amounts (2-3-fold) in supernatants from cells stimulated with zymosan particles (Fig. 4A), as compared to cells stimulated with ionophore A23187 (data not shown). Catalytic properties of the enzyme released from PMN showed a profile of calcium-dependency rather different from that displayed by the plasma enzyme, since the enzyme could be detected in the absence of calcium ions and concentrations above 5 mM calcium ions resulted in inhibition (Fig 1B). PMN phospholipase A_2 was best assayed when [^{14}C]oleate labeled autoclaved *E. coli* were used as substrate, having either lower or undetectable activity in the sonicated phospholipid assay (data not shown). Moreover, analysis of pH-dependency showed maximal activity at pH 5 (not shown). These features are consistent with a lysosomal origin of the enzyme released by PMN, and are rather different from the characteristics of the enzyme activity detected in the plasma of septic shock patients.

Release of phospholipase A_2 by human monocytes in culture

Incubation of adherent monocytes with opsonized-zymosan particles showed a time-dependent release of phospholipase A_2 activity that increased up to one hour (Fig. 3B). This profile seems rather different from that observed with human PMN, which peaked before completion of the incubation period. In other experiments, monocytes were maintained in culture for up to 48 hours and stimulated with IL-1, TNF, PAF and *E. coli* endotoxin.

As shown in Fig. 4B&D, endotoxin was the most potent stimulus for phospholipase A_2 release, whereas PAF was less active (Fig 4A&C). The phospholipase A_2 activity increased both in supernatants and within the

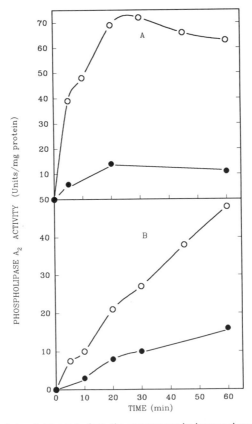

Figure 3. Release of phospholipase A_2 into the supernatant by human polymorphonuclear leukocytes (A) and monocytes (B) stimulated with complement-coated zymosan particles. At the times indicated the supernatant was removed for phospholipase A_2 assay and protein quantitation. This is a typical experiment of three showing an identical pattern.

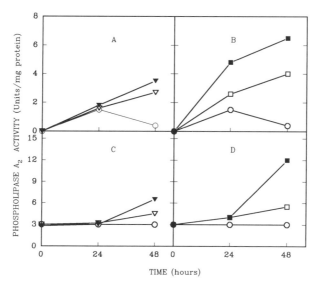

Figure 4. Release of phospholipase A$_2$ activity by adherent human monocytes in culture. Panels A and B represent supernatants and panels C and D represent cell-associated enzyme activity. Circles represent control monocytes. Triangles represent cells stimulated with PAF at the dose of 0.1 μM (open inverted triangles) and 1 μM (▼). Squares represent monocytes stimulated with bacterial endotoxin at the doses of 100 μg/ml (open squares) and 500 μg/ml (■).

cells, and could be optimally detected after 48 hours in culture. The enzyme activity obtained by stimulation of monocytes adhered to plastic dishes showed more analogies to the enzyme detected in patients' plasma samples than the enzyme obtained from polymorphonuclear leukocytes, since it was best assayed in the presence of calcium ions (Fig. 1C) and at alkaline pH.

IL-1α, IL-1ß and TNF-α were also tested on monocytes adhered to plastic dishes. IL-1α was unable to produce both secretion or increase in cell-associated phospholipase A$_2$ when tested in a range of doses of up to 10 U/ml (data not shown). However, TNF-α and a combination of TNF-α and IL-1ß enhanced the phospholipase A$_2$ activity associated with the cells after 16 hours in culture (Fig. 5B). In the case of the combination of both cytokines, an increase of phospholipase A$_2$ activity was also detected in the supernatants after 48 hours in culture (Fig. 5A).

DISCUSSION

This study confirms previous reports on the presence of high amounts of phospholipase A$_2$ activity in the blood of patients suffering from septicemia [6,7] and provides some suggestions as regards the cellular origin, biochemical characterization, and the delineation of mechanisms that may lead to the accumulation of this enzyme activity in plasma.

High levels of the enzyme were also detected in plasma from patients undergoing open heart surgery. This was an unexpected finding since the study of these patients was planned as a part of the control group of patients without bacterial infection, but in partial agreement with the original hypothesis, phospholipase A$_2$ levels in these group of patients reached levels significantly lower than those detected in septicemia patients, even in cases of long lasting cardiopulmonary by-pass, i.e., above 100 min. This suggests that phospholipase A$_2$ activity can rise in plasma under different clinical circumstances, and provides

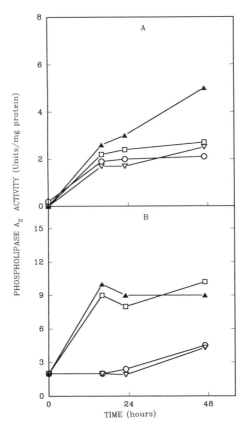

Figure 5. Release of phospholipase A_2 from human monocytes in culture. Cells were maintained in culture for the times indicated in the presence of no addition (O), IL-1ß 15 U/ml (inverted triangles), 1 nM TNF-α (open squares) and combination of both IL-1ß and TNF-α at the same doses (▲). Data represent a typical experiment out of four with identical patterns.

some indications about cellular origins and mechanism of production. Thus, the sequence of appearance of phospholipase A_2 activity in plasma, which peaks 24 hours after the occurrence of extracorporeal circulation, and the lack of correlation of plasma enzyme activity with thrombocytopenia suggest that a source of enzyme other than blood platelets and a mechanism distinct from the release of a preformed enzyme should be considered [24,25].

This is in keeping with results obtained in rabbit chondrocytes [26] and rat mesangial cells [8] which have shown the need for long periods of incubation to detect an increased synthesis of the enzyme, as judged from the inhibitory effect of actinomycin D and the presence of increased levels of phospholipase A_2 mRNA. Nevertheless, this does not exclude that there could exist structural and functional similarities between platelet phospholipase A_2 and phospholipase A_2 from other cellular sources [27-29].

Polymorphonuclear leukocytes are also an unlikely source of the enzyme since the activity released by these cells during phagocytosis shows remarkable differences as regards pH- and calcium-dependency with the plasma enzyme. Furthermore, the amount of enzyme detected in patients plasma seems to surpass the capacity of these cells to release phospholipase A_2. In keeping with this view is a report [15] showing that the phospholipase A_2 activity that accumulates in rabbit inflammatory exudates comes from serum rather than from infiltrating PMN.

Mononuclear phagocytes are the remaining blood cell type that should be considered as a source of the plasma phospholipase A_2 that can be assayed in septic shock patients. This was addressed by studying the effect of a variety of agonists on the release of this enzyme activity from adherent blood monocytes, and its functional characterization. We observed the release of a phospholipase A_2 activity showing identical features to the plasma enzyme as regards pH- and calcium-dependency from human monocytes in response to the phagocytosis of complement-coated zymosan particles and after incubation for longer periods of time with bacterial endotoxin and cytokines. This finding strongly points to mononuclear phagocytes as a type of cell involved in the secretion of the phospholipase A_2 detected in the plasma of patients with septicemia, and suggests the possible association of this phenomenon to phagocytosis of bacteria and stimulation of cells by bacterial-derived products and cytokines. However, the comparison between the levels assayed in patients' plasma and the amounts released by monocytes under these assay conditions indicate that the number of monocytes required to produce the levels detected in plasma are about 10-fold the average count of blood monocytes. This suggests that cells from different organs, e.g., mesangial cells [9], and Kuppfer cells and hepatocytes [10,11] could also contribute to the increased phospholipase A_2 activity detected in plasma from patients with septicemia.

ACKNOWLEDGMENTS
This study has been carried out with the help of a grant from DGICYT (Dirección General de Investigación Científica y Técnica). We thank Dr. F. Soriano for helping us to produce labeled *E. coli* for phospholipase assay, and Dr. T. Caparrós and the house staff from the Intensive Medicine Division by invaluable efforts to record clinical data and collect patients' blood samples.

REFERENCES
1. van den Bosch H. Biochim Biophys Acta 604, 1980:191-246.
2. Bereziat G, Etienne J, Kokkinidis M, Olivier JL, Pernas P. J Lipid Mediators 2, 1990:159-172.
3. Irvine RF. Biochem J 101, 1982:53-61.
4. Snyder F. Med Res Rev 5, 1985:107-140.
5. Vadas P, Wasi S, Movat H, Hay JH. Nature 293, 1981:583-585.
6. Vadas P. J Lab Clin Med. 104, 1984:873-881.
7. Vadas P, Pruzanski W, Stefanski E, Sternby B, Mustard R, Bohnen J, Fraser I, Farewell V, Bombardier C. Crit Care Med 16, 1988:1-7.
8. Schalkwijk C, Pfeilschifter J, Marki F, van den Bosch H. Biochem Biophys Res Commun 174, 1991:268-275.
9. Pfeilschifter J, Leighton J, Pignat W, Marki F, Vosbeck K. Biochem J 273, 1991:199-204.
10. Inada M, Tojo H, Kawata S, Seiichiro T, Okamoto M. Biochem Biophys Res Commun 174, 1991:1077-1083.
11. Crowl, R.M., T.J. Stoller, R.R. Conroy, and C.R. Stoner. 1991. J Biol Chem 266, 1991:2647-2651.
12. Chang, J, Gilman, SC, Lewis AJ. J. Immunol. 136, 1986:1283-1287.
13. Godfrey RW, Johnson WJ, Hoffstein ST. Biochem Biophys Res Commun 142, 1987:235-241.
14. López Díez F, Nieto ML, Fernández-Gallardo S, Gijón MA, Sánchez Crespo M. J Clin Invest 83, 1989:1733-1740.

15. Heath MF, Tighe D, Moss R, Bennet D. Crit Care Med 18, 1990:766-767.

16. Wright GW, Ooi CE, Weiss J, Elsbach P. J Biol Chem 265, 1990:6675-6681.

17. Parthasarathy S, Barnett J. Proc Natl Acad Sci USA 87, 1990:9741-9745.

18. Bone RC, Fisher CJ, Clemmer, TP, Slotman GJ, Metz CA, Balk RA. Crit Care Med 17, 1989:389-393.

19. Knauss WA, Draper EA, Wagner DP, Zimmerman JE. Crit Care Med. 13, 1985:818-819.

20. Bradford M. Anal Biochem 72,1976:248-254.

21. Elsbach P, Weiss J, Franson R, Beckerdite-Quagliata S, Schneider A, Harris L. J Biol Chem 254, 1979:11000-11009.

22. Puijk WC, Verheij HM, De Haas GH. Biochim Biophys Acta 492, 1977:254-259.

23. Verheij HM, Westerman J, Sternby B, De Haas GH. Biochim Biophys Acta 747, 1983:93-99.

24. Kramer RM, Checani GC, Deykin A, Pritzker CR, Deykin D. Biochim Biophys Acta 878, 1986:394-403.

25. Kramer RM, Hession C, Johansen B, Hayes G, McGray P, Pingchang Chow E, Tizard R, Blake Pepinsky R. J Biol Chem 264, 1989:5568-5575.

26. Lyons-Giordano B, Davis GL, Galbraith W, Pratta MA, Arner EC. Biochem Biophys Res Commun 164, 1989:488-495.

27. Lai C-Y, Wada K. Biochem Biophys Res Commun 157, 1988:488-493.

28. Kanda A, Ono T, Yoshida N, Tojo H, Okamoto M. Biochem Biophys Res Commun 163,1989:42-48.

29. Seilhamer JJ, Pruzanski W, Vadas P, Plant S, Miller JA, Kloss JY, Johnson LK. J Biol Chem 264, 1989:5335-5338.

PHOSPHOLIPASE ACTION IN THE GENERATION OF MODIFIED LDL

Sampath Parthasarathy,[1] and Joellen Barnett[2]

[1]Department of Gynecology and Obstetrics
Emory University
Atlanta, GA 30322, USA

[2]Department of Medicine
University of California, San Diego
La Jolla, CA 92093, USA

INTRODUCTION

A plethora of recent evidence suggests that some kind of oxidative process is involved in atherogenesis and that the oxidation of low density lipoprotein (LDL) may play a significant pathogenic role in atherosclerosis [1-3]. Based on the recommendations of the National Heart, Lung, and Blood Institute's expert panel [4], clinical trials to test the validity of the hypothesis in man are being undertaken using combinations of natural antioxidants .

A number of recent studies have suggested that antioxidants are effective against experimental atherosclerosis in animals [5-11]. Although such evidence is lacking in man, epidemiological studies have shown that the incidence of coronary artery disease (CHD) is inversely proportional to plasma antioxidant (vitamin E) levels [12]. The protective effect of β-carotene against CHD has also been attributed to its antioxidant effects. The precise mechanism(s) by which antioxidants may act to prevent atherosclerosis is not known. However, the oxidative modification of low density lipoprotein (LDL) has been suggested to be an important step in atherogenesis and the antioxidants are presumed to act by inhibiting such oxidation [1-3, 13].

The accumulation of lipids in the atherosclerotic lesion is predominantly intracellular. Both free and esterified cholesterol are present in the lesion, the latter as cytoplasmic lipid droplets and the former as crystalline cholesterol. LDL is the major source of cholesterol for the lesion. LDL is a spherical particle containing one apoprotein B_{100}, per particle. The surface of the lipoprotein is covered by a monolayer of phospholipid and the core is rich in cholesteryl esters. Phosphatidylcholine (PtdCho) is the predominant phospholipid. Phosphatidylethanolamine (PtdEtn), sphingomyelin and lysophosphatidylcholine (lyso PtdCho) are present in smaller amounts. Most of the cholesterol esters are linoleate derivatives. Very little, if any, triglycerides are present in LDL. Many lipid soluble antioxidants and hydrophobic proteins also associate with LDL in the plasma. The apoprotein B_{100} itself is a highly lipophilic protein and is tightly associated with the lipids. Large amounts of detergents are needed for its dissociation from the lipids. This protein is also very susceptible for proteolysis.

During the past 15 years, the cellular components associated with the early atherosclerotic lesion have been identified. Besides the cell types present in the normal artery macrophages and T-lymphocytes are also present in the atherosclerotic artery [14,15]. Macrophages, to a greater extent, and smooth muscle contribute to the foam cells of the fatty streak lesion. The paradoxical finding that macrophages do not take up native LDL in sufficient amounts to account for the accumulation of lipids had sparked an explosion of research activity on the mechanism(s) by which foam cells could be derived from LDL [1-3]. As a result, a number of modifications of LDL have been described that would result in their enhanced uptake by macrophages. A common feature of these modified lipoproteins is that they all bear negative charge and are presumed to be recognized, not by the LDL receptor, but by the macrophage scavenger receptors. One such modification, namely oxidative modification has received a great deal of attention during the past 10 years [16]. According to this modification, the oxidation of LDL-lipids is suggested to generate products (aldehydes) that would alter the lysine residues of apoprotein B_{100} by the formation of Schiff bases [16-18]. The increased negative charge, thus created, is presumed to be important in its recognition by the macrophage scavenger receptor although several other negatively charged molecules are not recognized by this receptor. Calculations, from the number of lysines that are oxidatively modified and the total available unsaturated fatty acids dictates that oxidation of the surface phospholipids alone insufficient to account for the modification. The fact that the protein is in a fluid state, weaving in and out of the particle into the core, would suggest that lysine residues, both surface exposed as well as those buried in the core might be exposed to products of lipid peroxidation and would be subject to modification. There are other oxidation-mediated changes in the apoprotein such as polypeptide breakdown, cross linking, amino acid alterations and structural modifications that may also be important in presenting the modified LDL to the macrophages. The oxidative modification of LDL can be accomplished by a variety of means. These conditions are shown in table 1.

Table 1. Systems that can induce the oxidation of LDL

----
1. Incubation with iron or copper, or chelated iron. Preexisting peroxides may be needed.
2. Incubation with specific cells in metal containing media.
3. Incubation with AAPH and other free radical generators. Considerable loss of apoprotein may result.
4. Incubation with enzymes such as lipoxygenase or cholesterol oxidase. Additional enzyme activity such as phospholipase A_2 may be needed.
5. Incubation with peroxidase in the presence of H_2O_2 or LOOH. Heme, hematin and hemoglobin may substitute for peroxidase activity.
6. Exposure to ionnizing radiation (UV, γ-rays, singlet oxygen) .and molecular oxygen
7. Incubation with thiols in the presence of metals or thiyl radical generators.
8. Incubation with peroxynitrite or simultaneous superoxide, nitric oxide generating systems.
9. Storage at low concentrations.

Table 2. Pro-atherogenic effects of lyso PtdCho

---

1. Lyso PtdCho is chemotactic to monocytes and T-lymphocytes which are important cellular components of the early atherosclerotic lesion.
2. Lyso PtdCho increases the uptake of native LDL and causes lipid accumulation in macrophages.
3. Lyso PtdCho induces the expression of adhesion molecules on the endothelial surface.
4. Lyso PtdCho affects the synthesis and secretion of endothelium dependent relaxing factor (EDRF) an important molecule that controls vascular tone.

It is obvious that some of these agents can freely diffuse into the core of the particle and initiate oxidation at random. On the other hand, enzymes or cell contact (if needed) must occur at the surface in order to initiate oxidation.

Of all these methods, incubation with cells or with copper has attracted wide spread attention. While the former condition may be understood, because of potential physiological relevance, the latter is hard to comprehend, except that this method offers a convenient means of generating a modified LDL. At the same time, any interpretations on the mechanism of oxidation based on the oxidation of LDL by copper should be approached with caution.

One of the earliest changes associated with the oxidation of LDL is the generation of lyso PtdCho [18, 19]. The formation of this lipid parallels the modification process. Free radical oxidation appeared to be essential for the hydrolysis of the phospholipid, as no hydrolysis was seen when oxidation was inhibited by antioxidants. Using PtdCho, labeled in the 2-position with linoleic acid, it was demonstrated that the fatty acid at the 2-position was cleaved during the oxidation. Radioactive fatty acid products were released in the reaction whereas no labeled lyso PtdCho could be seen. Since the fatty acid moiety was labeled at the carbon atom-1 [1-^{14}C-linoleic acid], the product represented a carboxyl esterase reaction and not the release of an oxidatively tailored fatty acid fragment. The following lines of evidence suggested that the product was indeed lyso PtdCho:

1. The product chromatographed with authentic lyso PtdCho on thin layer chromatography in different solvent systems.
2. The product could be acylated with fatty acid anhydrides in the presence of dimethylamino pyridine to yield PtdCho.
3. The product was cytolytic to red blood cells.

Furthermore, recent studies have identified several potent biological, pro-atherogenic properties that are identical to those shown by authentic lyso PtdCho. These are included in table 2.

The initial experiments were performed using endothelial cells as the oxidizing system. We speculated that cells might be responsible for this apparent phospholipase A_2 activity. However, when we used copper as the oxidant, similar results were obtained suggesting that the phospholipase A_2 activity might be associated with the lipoprotein. Moreover, LDL, pretreated with phospholipase A_2 inhibitors resisted oxidation *by cells*.

Most, if not all, phospholipase A_2 enzymes contain a histidine residue at the active site and are inhibited by histidine modifiers such as p-bromophenacyl bromide (pBPB). This inhibitor also inhibited the oxidation of LDL in a concentration dependent manner,

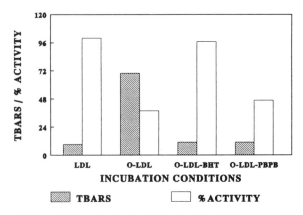

Figure 1. Inhibition of LDL oxidation by BHT but not by pBPB preserves PLA_2 activity. LDL (200 µg protein) was subjected to oxidation with 5 µM copper in the presence of 100 µM BHT or pBPB. PLA_2 activity and TBARS were measured as mentioned before (20).

completely inhibiting the modification at ~ 40 µM concentrations. However, at this concentration, oxidation, as measured by the formation of thiobarbituric acid products (TBARS), was also appreciably inhibited. This was not due to the antioxidant action of pBPB; two similar compounds that do not have a phospholipase A_2 inhibitory activity, namely, acetophenone and bromoacetophenone had no effect on the oxidation or modification of LDL. Further more, an aliphatic analog of pBPB, 1-bromooctan-2-one completely inhibited the modification of LDL with little or no effect on TBARS. Thus the effect of pBPB was not due to its antioxidant properties. PBPB also prevented several other changes related to the oxidation of LDL.

As a further demonstration, we incubated LDL under oxidizing conditions in the presence of BHT and 100 µM pBPB. At this concentration, pBPB completely inhibited the formation of TBARS. If BHT and pBPB both had acted as antioxidants, then complete activity must be restored at the end of the incubation. The results presented in fig.1 show that while BHT was able to prevent the oxidative loss of phospholipase A_2 activity, pBPB had very little protective effect.

The presence of a phospholipase A_2 activity was unequivocally demonstrated in the LDL protein by using soybean lipoxygenase oxidized PtdCho as the substrate. Unoxidized PtdCho was a poor substrate and the small amount of product released usually represented an oxidized fatty acid present in the PtdCho. Oxidized PtdCho was very efficiently cleaved by this enzyme. VLDL, another apolipoprotein containing particle showed powerful activity where as HDL that lacks the apoprotein B had no enzyme activity. The enhanced capacity of phospholipase A_2 to hydrolyze an oxidized phospholipid was known before; however, this was the first demonstration of a phospholipase A_2 activity that hydrolyzed only an oxidized phospholipid and spared unoxidized phospholipids. As antioxidants inhibited the availability of the substrate during the oxidation of LDL, we proposed that this phospholipase A_2 helped to propagate lipid peroxidation into the core of the particle. Accordingly, if an oxidizing system such as copper that can freely diffuse into the core is used, this enzyme would play less of a role.

Our previous studies had indicated that during the oxidation of LDL, there was over 30% loss of histidine residues. Since specific histidines were implicated in the catalysis of phospholipase A_2 reaction, we measured the phospholipase A_2 activity during modification.

Table 3. Comparison between PAF-hydrolase and LDL-phospholipase A_2 activities

---

1. The oxidized PtdCho hydrolyzing activity is preferentially enriched in LDL and VLDL, the apoprotein B containing lipoproteins, where as PAF-hydrolase is found more in association with HDL.

2. The LDL-phospholipase A_2 is inhibited by histidine modification and by pBPB. The plasma PAF-hydrolase was reported not to be affected by pBPB.

3. The PAF-hydrolase prefers short chain substituents in the 2-position and was reported not to hydrolyze long chain substituted PtdCho. The substrate used in the LDL-phospholipase A_2 studies are derived from purified lipoxygenase-treated PtdCho. We found no evidence to suggest that the substrate was degraded to shorter chain phospholipids.

4. Under the conditions the rose bengal inactivation destroyed more than 90% of LDL-phospholipase A_2 activity, very little inhibition was noticed in the PAF-hydrolase activity.

5. PAF-hydrolase was reported to be a small molecular weight (~40 kDa) protein. We could recover LDL-phospholipase A_2 activity in larger molecular weight fragments of apoprotein B.

As expected, the activity was lost gradually and the inclusion of antioxidants protected the loss of activity. In fact, preliminary results indicated that LDL from antioxidant treated animals showed higher phospholipase A_2 activity than that isolated from control animals.

We also subjected the LDL molecule to photo irradiation in the presence of rose bengal, a condition that generates singlet oxygen and destroys histidine residues specifically [20]. Even a 20 minute exposure to such conditions completely and specifically destroyed all the histidine residues with a total loss of phospholipase A_2 activity.

During the past few years, an enzyme reaction capable of hydrolyzing platelet activating factor (PAF) was described in the lipoprotein fractions [21-24]. This activity was also found in isolated LDL. Based on the ability of this enzyme to hydrolyze oxidatively degraded phospholipids, it is suggested that this enzyme might be responsible for the hydrolysis of oxidized PtdCho during the oxidative modification of LDL. This enzyme activity was purified from human LDL and was found to be of ~ 40-45 kDa size. However, no details of the amino acid sequence or structural properties are known. The differences between the two enzyme activities are shown in table 3.

The function of this phospholipase A_2 activity in LDL can only be speculated. We originally proposed that this enzyme would release a peroxidized fatty acid from the surface of the LDL, that would readily diffuse into the core of the particle and promote propagation. However, plasma contains peroxidase activities and the peroxides are very likely to be reduced. On the other hand, if the oxidation occurs, as suggested in the intima, indeed propagation may be facilitated. However, nature does not devise an enzyme system to promote a disease. We propose the function of this enzyme is to facilitate the repair process The damaged fatty acid component would be removed and the lipoprotein molecule would be restored to the natural state by the HDL-associated acyltransferase system.

Acknowledgment: This work was supported by National Heart, Lung and Blood Institute Grant HL-14197.

REFERENCES

1. Steinberg D, Parthasarathy S, Carew, TE, Khoo, JC, Witztum JL. Beyond Cholesterol. Modifications of low density lipoprotein that incresae its atherogenicity. N. Engl. J. Med. 320: 915-924 (1989).

2. Parthasarathy S, Steinberg D, Witztum,J.L. Annu. Rev. Med. 43: 219 225 (1992).

3 Parthasarathy, S and Rankin, S.M. Role of oxidized low density lipoprotein in atherogenesis. Progr. Lipid Res. 31: 127-143 (1992).

4. Steinberg D, Workshop Participants: Antioxidants in the prevention of human atherosclerosis. Summary of the proceedings of a National Heart, Lung and Blood Institute workshop. Circulation 85: 2337-2344 (1992).

5 Carew TE, Schwenke DC, Steinberg D: Antiatherogenic effect of probucol unrelated to its hypocholesterolemic effect: evidence that antioxidants in vivo can selectively inhibit low density lipoprotein degradation and slow the progression of atherosclerosis in Watanabe heritable hyperlipidemic rabbit. Proc. Natl. Acad. Sci. USA 84: 7725-7729 (1987).

6. Kita T, Nagano Y, Yokode M, Ishii K, Kume N, Ooshima A, Yoshida H, Kawai. Probucol prevents the progression of atherosclerosis in Watanabe heritable hyperlipidemic rabbit, an animal model for familial hypercholesterolemia. Proc. Natl. Acad. Sci. USA 84: 5928-5931 (1987).

7. Mao, S.J.T., Yates, M.T., Parker, R.A., Chi, E.M., and Jackson, R.L. Attenuation of atherosclerosis in a modified strain if hypercholesterolemic watanabe rabbit with use of a probucol analog (MDL 29, 311) that does not lower serum cholesterol. Arterios. Thromb. 11: 1266-1275 (1991).

8. Björkhem, I., Henriksson-Freyschuss, A., Breuer, O., Diczfalusy, U., Berglund, L., and Henriksson, P. The antioxidant butylated hydroxytoluene protects against atherosclerosisArterios. Thromb. 11: 15-22 (1991).

9. Sparrow, C.P., Doebber, T.W., Olszwski, J., Wu, M.S., Ventre, J., Stevens, K.A. and Chao, Y. Low density lipoprotein is protected from oxidation and the progression of atherosclerosis is slowed in cholesterol-fed rabbit by the antioxidant N,N'-diphenyphenylenediamine. J. Clin. Invest. 89: 1885-1891 (1992).

10 Verlangieri, A.J. and Bush, M.J. Effects of d-alpha-tocopherol supplementation on experimentally induced primate atherosclerosis. Am. Coll. Nutr. 11: 131-138 (1992).

11. Daugherty, A., Zweifel, B.S. and Schenfeld, G. The effects of probucol on the progression of atherosclerosis in mature Watanabe heritable hyperlipidaemic rabbits. Brit. J. Pharmacol. 103: 1013-1018 (1989).

12. Gey, K.F. and Puska, P. Plasma vitamin E and A inversely related to mortality from ischemic heart disease in cross-cultural epidemiology. Ann. Ny Acad. Sci. 570, 254-282 (1989)

13 Esterbauer, H., Schaur, R.J. and Zollner, H. The role of lipid peroxidation and antioxidants in oxidative modification of LDL. Free. Radic. Biol. Med. 11: 81-128 (1992).

14 Hansson, G.K. Seifert-P-S. Olsson-G. Bondjers-G. Immunohistochemical detection of macrophages and T lymphocytes in atherosclerotic lesion of cholesterol-fed rabbits. Arterioscler-Thromb. 11: 745-750 (1991).

15. Gerrity, R.G., Naito, H.K., Richardson, M. and Schwartz, C.J. Dietary induced atherogenesis in swine. Am. J. Pathol. 95: 775-792 (1979).

16. Steinbrecher .U. P, Witztum. J. L, Parthasarathy. S and Steinberg D. Decrease in reactive amino groups during oxidation or endothelial cell modification of LDL. Correlation with changes in receptor-mediated catabolism. Arteriosclerosis 7: 135-143 (1987).

17. Steinbrecher, U.P. Oxidation of human low density lipoprotein results in derivatization of lysine residues of apolipoprotein B by lipid peroxide decomposition products. J. Biol. Chem. 262: 3603-3608 (1987).

18. Steinbrecher U.P, Parthasarathy S, Leake D.S, Witztum J.L and Steinberg D. Modification of low density lipoprotein by endothelial cells involves lipid peroxidation and degradation of low densityl lipoprotein phospholipids. Proc. Natl. Acad. Sci. USA 81: 3883-3887 (1984).

19. Parthasarathy, S., Steinbrecher, U.P., Barnett, J., Witztum, J.L. and Steinberg, D. Essential role of phospholipase A_2 in endothelial cell-induced modification of low density lipoprotein. Proc. Natl. acad. Sci. USA. 82, 3000-3004 (1985).

20. Parthasarathy, S. and Barnett, J. Phospholipase A_2 activity of low density lipoprotein: Evidence for an intrinsic phospholipase A_2 activity of apoprotein B-100. Proc. Natl. Acad. Sci. USA. 87, 9741-9745 (1990).

21. Stafforini, D.M., McIntyre, T.M., Carter, M.E. and Prescott, S.M. Human plasma platelet-activating factor hydrolase. Association with lipoprotein particles and role in the degradation of platelet activating factor. J. Biol. Chem. 262, 4215-4222 (1987).

22. Stremler, K.E., Stafforini, D.M., Prescott, S.M., Zimmerman, G.A and McIntyre, T.M. An oxidized derivative of phosphatidylcholine is a substrate for the platelet activating factor acetylhydrolase from human plasma. J. Biol. Chem. 264, 5331-5334 (1989).

23. Steinbrecher, U.P. and Pritchard, H.P. Hydrolysis of phosphatidylcholine during LDL oxidation is mediated by platelet activating factor acetyl hydrolase. J. Lipid Res. 30, 305-315 (1989).

24. Stafforini, D.M., Carter, M.E., Zimmerman, G.A., McIntyre, T.M. and Prescott, S.M. Lipoproteins alter the catalytic behaviour of the platelet activating factor acetylhydrolase in human plasma. Proc. Natl. Acad. Sci. USA 86, 2393-2397 (1989).

DETECTION OF PHOSPHOLIPASES A2 IN SERUM IN ACUTE PANCREATITIS

Timo J. Nevalainen[1] and Juha M. Grönroos[2]

[1]Department of Pathology
[2]Department of Surgery
  University of Turku, Turku, Finland

INTRODUCTION

Phospholipase A2 (PLA2, E.C. 3.1.1.4) hydrolyzes phospholipids, e.g. phosphatidyl-choline, phosphatidylethanolamine and phosphatidylinositol into corresponding lysocom-pounds. PLA2 is the rate-limiting enzyme in the formation of eicosanoids, which are mediators of inflammation. It is assumed that PLA2 plays an important role in the pathology of various diseases, such as acute pancreatitis, septic shock and multiple injuries [1, 2, 3, 4, 5].

Acute pancreatitis is an autodigestive disease in which digestive proenzymes, synthesized in pancreatic acinar cells, are activated within the pancreas and injure the gland. Several studies suggest that PLA2 might play a key role in the pathology of acute pancreatitis [reviews: 1, 2, 6]. Recent advances in the methods of PLA2 determination have been essential in clarifying the involvement of PLA2 in acute pancreatitis. In the present paper we review recent developments in the field of PLA2 detection in serum in acute pancreatitis and illustrate the concepts by two case reports.

GROUP I AND GROUP II PHOSPHOLIPASES A2

Secretory 14 kDa PLA2s have been divided into two groups on the basis of the amino acid sequence of the enzyme protein [7]. Group I PLA2s contain a cysteine at position 11 forming a disulphide bridge with a cysteine at position 77. Group II PLA2s lack these cysteines and the corresponding disulphide bridge. Group I PLA2s are found in mammalian pancreas and group II PLA2 in many tissues e.g. in synovial fluid and platelets [8]. Since no commonly accepted terminology has been established for different forms of PLA2s, we call in this paper the 14 kDa group I PLA2 isolated from human pancreas "pancreatic PLA2 (pan-PLA2)" and the 14 kDa human non-pancreatic group II PLA2 "synovial-type PLA2 (syn-PLA2)".

Esterases, Lipases and Phospholipases, Edited by M.I. Mackness
and M. Clerc, Plenum Press, New York, 1994

PHOSPHOLIPASE A2 ASSAYS

The catalytic activity of PLA2 can be measured by titrimetric, fluorometric, colorimetric and radiometric methods [9]. A number of convenient and reliable methods are available for the measurement of the catalytic activity of PLA2. These methods are based on the quantification of radioactive fatty acid liberated by the enzyme from labelled *E. coli*-membranes [10] or synthetic phospholipid [11, 12].

Sensitive and specific methods have been developed for the determination of the concentrations of different PLA2s in body fluids including serum. These methods are based on immunoassays that use antibodies raised against PLA2s purified from various sources. The first immunoassays were developed in 1983 for the determination of human pan-PLA2 (radioimmunoassay, RIA, [13]; time-resolved fluoroimmunoassay, TR-FIA, [14]). These immunoassays have been used extensively in studies of pancreatic diseases [2]. Monoclonal anti-pan-PLA2 antibodies have been used in more recent immunoassays of pan-PLA2 [15, 16]. Recently, new immunoassays for the measurement of non-pancreatic syn-PLA2 have been developed [17, 18]. These methods are currently used in studies on the role of syn-PLA2 in various human diseases [19, 20, 21, 22]. Immunoassays are quite accurate in measuring concentrations of PLA2s in human serum despite the different measuring principles (time-resolved fluoroimmunometry, TR-FIA and radioimmunomeasurement, RIA). The mean concentration of human pan-PLA2 in sera of healthy individuals is 5.1 µg/l as measured by RIA [13] and 6.5 µg/l by TR-FIA [14]. Methods that measure syn-PLA2 also give comparable results: serum values for healthy individuals are 3.6 µg/l with RIA [17] and 10.8 µg/l with TR-FIA [18].

PHOSPHOLIPASE A2 IN SERUM IN ACUTE PANCREATITIS

Increased catalytic activity of PLA2 (cat-PLA2) in serum was found to correlate with the severity of the disease in acute pancreatitis [23]. Immunoreactive pan-PLA2 is found in the human in pancreatic acinar cells only [16, 24]. It was proposed that the catalytically active PLA2 in serum in experimental porcine pancreatitis was of pancreatic origin [25]. However, clinical studies on surgically treated patients [23] and immunoabsorption studies with a specific anti-pan-PLA2 antibody [26] supported the idea that the catalytically active PLA2 in acute pancreatitis might be of non-pancreatic origin.

The cat-PLA2 and the concentration of syn-PLA2 correlated statistically significantly with each other but not with the concentration of pan-PLA2 in acute pancreatitis [20]. This finding indicates that the catalytic activity of PLA2 in serum in acute pancreatitis is due to the presence of a non-pancreatic group II PLA2 and not of that of pancreatic group I PLA2. The cat-PLA2, syn-PLA2 and pan-PLA2 values increase in serum in the early stages of acute pancreatitis [20, 27].

The cellular source of the catalytically active group II PLA2 in acute pancreatitis is unknown. Syn-PLA2 is present e.g. in chondrocytes in human cartilage and in the Paneth cell secretory granules in the small intestine [24]. It has been postulated that syn-PLA2 found in serum in various inflammatory diseases represents an acute phase reactant [19, 28].

PHOSPHOLIPASE A2 AND PANCREATIC INJURY

In the beginning of this century it was hypothesized that trypsin causes the pancreatic tissue injury in acute pancreatitis [29]. However, during the last two decades it has been repeatedly proposed that PLA2 might be the enzyme responsible for tissue necrosis in acute pancreatitis [1, 2, 6].

Increased concentrations of lysolecithin were found in necrotic pancreatic tissue in human acute pancreatitis [30]. The lysolecithin content of pancreatic tissue also increased rapidly after the induction of acute pancreatitis in rats by an intraductal injection of sodium taurocholate [31]. Injections of PLA2, a solution containing PLA2 and bile acid, or lysolecithin into the pancreatic duct of the rat caused severe pancreatic necrosis [30, 32]. The lesions induced by intraductal injections of PLA2 and lysolecithin were identical [32].

The development of fat necrosis in acute pancreatitis is mediated by lipolytic enzymes including PLA2 [33]. It has been suggested that PLA2 activity could also account for pseudocyst formation after an attack of acute pancreatitis [34]. The degree of the increase in the cat-PLA2 in serum in acute pancreatitis separates the necrotizing from the oedematous form of the disease, while the measurement of immunoreactive pan-PLA2 is of no value in this respect [27].

PHOSPHOLIPASE A2 AND PULMONARY COMPLICATIONS

The measurement of the cat-PLA2 and the concentration of syn-PLA2 in serum provides a means of early assessment of pulmonary insufficiency in acute pancreatitis, whereas the degree of the increase in the concentration of pan-PLA2 in serum is not predictive in this respect [27, 35].

Circulating PLA2 has been shown to cause degradation [36] and inhibition of the function [37] of pulmonary surfactant resulting in the development of the acute respiratory distress syndrome (ARDS) in experimental animals. Rats with the highest PLA2 activities in serum had the lowest arterial pO_2 levels in experimental taurocholate-induced acute pancreatitis [38].

PHOSPHOLIPASE A2 AND RENAL COMPLICATIONS

The increase in the concentration of syn-PLA2 but not in pan-PLA2 in serum correlated to the increase in the serum creatinine levels in patients suffering from acute pancreatitis [35]. Immunohistochemical studies revealed that PLA2 was rapidly deposited in endocytotic vesicles and lysosomes in rat renal proximal tubular cells both after an intravenous injection of human pancreatic PLA2 and in experimental acute pancreatitis in the rat [39]. A positive correlation was also found between the early increase in cat-PLA2 in serum and the later development of renal tubular cell injury in human acute pancreatitis [40].

The mechanism of PLA2 toxicity has been difficult to elucidate. Tubular hypoxia due to hypovolemia may play a major role, since hypoxia has been found to increase PLA2 activity in proximal tubular cells resulting in increased production of toxic lysophospho-lipids and free fatty acids [41]. On the other hand, PLA2 has been shown to cause sustained hypotension and hypoxia by activating the lipooxygenase pathway [42]. PLA2 is also capable of damaging proximal tubular cells under mildly hypoxic conditions *in vitro* [43, 44]. The exact actions and interactions of PLA2 and hypoxia in renal injury in acute pancreatitis remain to be established.

CONCLUSIONS

The increase in the catalytic activity of PLA2 in serum correlates with the severity of acute pancreatitis. The catalytic activity correlates with the concentration of non-pancreatic synovial-type group II PLA2 in serum but not with the concentration of pancreatic group I PLA2. The cellular source of the catalytically active PLA2 in acute pancreatitis is unknown.

CASE REPORTS

Patient 1

The patient was a 38-year-old man who had earlier had two attacks of acute alcoholic pancreatitis. This time, after a few weeks of continuous alcohol consumption, he experienced severe, unrelenting epigastric pain and fullness for two days.

The physical examination revealed localized epigastric tenderness. Urine amylase excretion was increased to 24 060 U/l. There was profound dehydration and tachycardia, and pO_2 was 5.8 kPa (normal range 10-14 kPa) in the arterial blood gas analysis. The patient was hospitalized in the intensive care unit of the University Central Hospital of Turku. He required abundant fluid replacement, endotracheal intubation and mechanical ventilation. Peritoneal lavage was started, but his clinical condition deteriorated further. The concentration of C-reactive protein (CRP) in serum rose to 163 mg/l (the upper limit of the normal range is 10 mg/l). Haemorrhagic-necrotizing acute pancreatitis was suspected after the CT scan, and the chest x-ray revealed pleural effusions and pulmonary atelectasis.

Severe oedematous acute pancreatitis was found but no distinct necrosis at laparotomy on the sixth day of hospitalization. The pancreas was left intact, cholecystostomy was performed and drains were placed in the abdominal cavity. The postoperative recovery was uneventful.

The cat-PLA2, syn-PLA2 and pan-PLA2 values as measured daily in the intensive care unit are presented in Figure 1. The cat-PLA2 and syn-PLA2 values change in concert, whereas the pan-PLA2 values are increased at the early stages of the disease and then normalize rapidly.

The changes in cat-PLA2 and syn-PLA2 values reflected the clinical course of the patient, suggesting a role for non-pancreatic group II PLA2 in the pathophysiology of acute pancreatitis. The early increase in the concentration of pan-PLA2 in serum most probably reflects pancreatic tissue injury.

Patient 2

The patient was a 43-year-old man who had been treated conservatively for his first attack of acute alcoholic pancreatitis for five weeks before his admission to the intensive care unit of the University Central Hospital of Turku. At the onset of the disease, the patient presented with the classical clinical findings of acute pancreatitis (abdominal pain and tenderness, vomiting), and the urinary amylase excretion was 284 200 U/l. Although the CT scan indicated haemorrhagic-necrotizing acute pancreatitis and the concentration of CRP in serum was 206 mg/l, the patient was in a stable condition and managed without a respirator. Urinary amylase excretion decreased to 6590 U/l after the first week in the hospital. Although the symptoms were prolonged, the patient seemed to manage without operative treatment. However, during the fifth week in the hospital, the clinical state of the patient weakened, he had a new episode of unrelentig epigastric pain and the urinary amylase excretion rose to 98 000 U/l. There was a suspicion of a total necrosis of the pancreas in the CT scan. The concentration of CRP in serum rose to 290 mg/l and that of creatinine to 386 μmol/l, and the chest x-ray revealed pulmonary atelectasis. At laparotomy, the pancreas was almost totally black and necrotic. A next to total pancreatectomy and cholecystostomy were performed together with drainage of 500 ml of pus. The patient was treated postoperatively in the intensive care unit in a respirator for two weeks. *Staphylococcus aureus* was isolated from the blood culture on the first postoperative day and the intravenous antimicrobial drug was changed from ceftriaxone to imipenem. The patient recovered well, but had a postoperative wound infection and ventral hernia. The disease and its radical operative treatment were not followed by diabetes.

The daily changes in the cat-PLA2, syn-PLA2 and pan-PLA2 values are presented in Figure 2. The cat-PLA2 and syn-PLA2 values change in parallel. The pan-PLA2 values remain within the reference interval during the postoperative observation period.

Histological examination of the resected pancreas of patient 2 revealed total necrosis and loss of the exocrine and endocrine tissue components (Figure 3 A). In immunohisto-chemical specimens there was a total loss of pancreatic PLA2 from the tissue (Figure 3 B). For comparison, subnormal amounts of immunoreactive group I PLA2 are seen in the partially necrotic pancreatic tissue of another patient with acute pancreatitis (Figure 3 C and D), and normal PLA2 immunoreactivity in intact pancreatic tissue from a patient without acute pancreatitis (Figure 3 E and F).

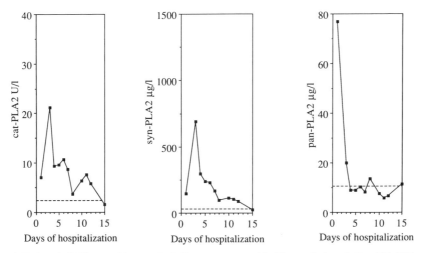

Figure 1. Patient 1. A 38-year-old man, alcoholic acute pancreatitis. The catalytic activity of PLA2 (cat-PLA2) and the concentration of synovial-type group II PLA2 (syn-PLA2) and pancreatic group I PLA2 (pan-PLA2) in serum during treatment in the intensive care unit. Laparotomy was performed on day six. The dashed horizontal lines indicate the upper limits of the reference intervals for cat-PLA2, syn-PLA2 and pan-PLA2.

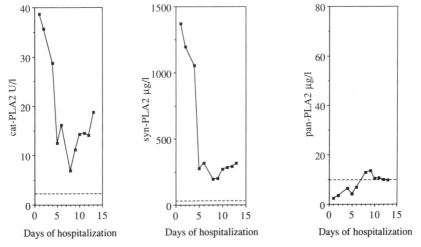

Figure 2. Patient 2. A 43-year-old man, alcoholic acute pancreatitis. The catalytic activity of PLA2 (cat-PLA2) and the concentration of synovial-type PLA2 (syn-PLA2) and pancreatic PLA2 (pan-PLA2) in serum during treatment in the intensive care unit. Next to total pancreatectomy was performed on day 0. The dashed horizontal lines indicate the upper limits of the reference intervals as in Figure 1.

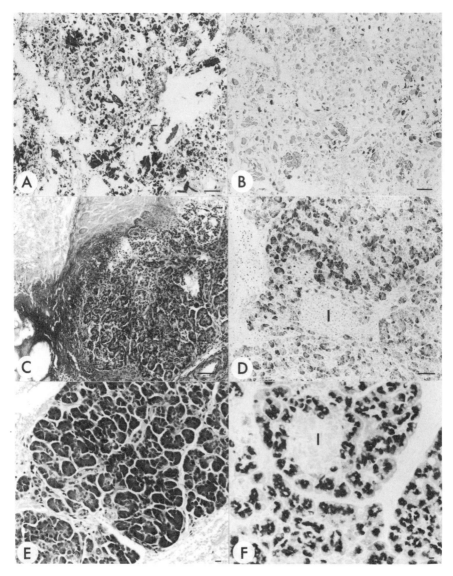

Figure 3. Pancreatic tissue stained with the van Gieson technique (A, C, E) and immunohistochemical demonstration of pancreatic PLA2 by a monoclonal anti-pancreatic PLA2 antibody 2E1 [16] (B, D, F). A and B: x80, bar 50 μm. Patient 2, see text to Figure 2 for details. Tissue is totally necrotic and structureless (A) and there is no immunoreactive PLA2 in the tissue (B).
C and D: x80, bar 50 μm. A 36-year-old man. Pancreas was resected because of severe necrotizing acute pancreatitis. Most of the pancreatic tissue is necrotic and infiltrated with polymorphs. There are a few non-necrotic acinar cells in the center of the tissue specimen (C). There are scattered acinar cells that contain immunoreactive PLA2 (D). I, an islet of Langerhans.
E and F: x160, bar 10 μm. Normal pancreatic tissue resected because of pancreatic cancer from a 65-year-old man. Acinar cells (E) and immunoreactive PLA2 in apical parts of acinar cells (F). I, an islet of Langerhans.

Acknowledgements

The authors thank Ms Maarit Kallio for secretarial assistance. Supported by grants from the Academy of Finland, The University of Turku Foundation, The University of Auckland Foundation and The Finnish Foundation for Alcohol Studies.

REFERENCES

1. Nevalainen T.J. The role of phospholipase A2 in acute pancreatitis. *Scand J Gastroenterol* 15;1980:641-650.
2. Nevalainen T.J. Phospholipase A2 in acute pancreatitis. *Scand J Gastroenterol* 23;1988:897-904.
3. Vadas P. and Pruzanski W. Role of secretory phospholipase A2 in the pathobiology of disease. *Lab Invest* 55;1986:391-404.
4. Uhl W., Büchler M., Nevalainen T.J., Deller A. and Beger H.G. Serum phospholipase A2 in patients with multiple injuries. *J Trauma* 30;1990:1285-1290.
5. Pruzanski W., and Vadas P. Phospholipase A2 — a mediator between proximal and distal effectors of inflammation. *Immunol Today* 12;1991:143-146.
6. Creutzfeldt W. and Schmidt H. Aetilogy and pathogenesis of pancreatitis. Current concepts. *Scand J Gastroenterol* (Suppl) 6;1970:47-62.
7. Heinrikson R.L., Krueger E.T. and Keim P.S. Amino acid sequence of phospholipase A2 from the venom of *Crotalus adamanteus*. A new classification of phospholipases A2 based upon structural determinants. *J Biol Chem* 252;1977:4913-4921.
8. van den Bosch H., Aarsman A.J., van Schaik R.H.N., Schalkwijk C.H., Nejs F.W. and Sturk A. Structural and enzymological properties of cellular phospholipase A2. *Biochem Soc Trans* 18;1990:781-786.
9. Reynolds L.J., Washburn W.N., Deems R.A. and Dennis E.A. Assay strategies and methods for phospholipases. *Methods Enzymol* 197;1991:3-23.
10. Patriarca P., Beckerdite S. and Elsbach P. Phospholipases and phospholipid turnover in *Escherichia coli* spheroplasts. *Biochim Biophys Acta* 260;1972:593-600.
11. Shakir K.M.M. Phospholipase A2 activity of post-heparin plasma; a rapid and sensitive assay and partial characterization. *Anal Biochem* 114;1981:64-70.
12. Schädlich H.R., Büchler M. and Beger H.G. Improved method for the determination of phospholipase A2 catalytic activity concentration in human serum and ascites. *J Clin Chem Clin Biochem* 25;1987:505-509.
13. Nishijima J., Okamoto M., Ogawa M., Kasaki G. and Yamano T. Purification and characterization of human pancreatic phospholipase A2 in development of a radioimmunoassay. *J Biochem* 94;1983:137-147.
14. Eskola J.U., Nevalainen T.J. and Lövgren T.N.-E. Time-resolved fluoroimmunoassay of human pancreatic phospholipase A2. *Clin Chem* 29;1983:1777-1780.
15. Oka Y., Ogawa M., Matsuda Y., Murata A., Nishijima J., Miyauchi K., Uda K., Yasuda T. and Mori T. Serum immunoreactive pancreatic phospholipase A2 in patients with various malignant tumours. *Enzyme* 43;1990:80-88.
16. Santavuori S.A., Kortesuo P.T., Eskola J.U. and Nevalainen T.J. Application of a new monoclonal antibody for time-resolved fluoroimmunoassay of human pancreatic phospholipase A2. *Eur J Clin Chem Clin Biochem* 29;1991:819-826.
17. Matsuda Y., Ogawa M., Sakamoto K., Yamashita S., Kanda A., Kohno M., Yoshida N., Nishijima J., Murata A. and Mori T. Development of a radioimmunoassay for human group II phospholipase A2 and demonstration of postoperative elevation. *Enzyme* 45;1991:200-208.
18. Nevalainen T.J., Kortesuo P.T., Rintala E. and Märki F. Immunochemical detection of group I and group II phospholipases A2 in human serum. *Clin Chem* 38;1992:1824-1829.

19. Ogawa M., Arakawa H., Yamashita S., Sakamoto K. and Ikei S. Postoperative elevations of serum interleukin 6 and group II phospholipase A2: group II phospholipase A2 in serum is an acute phase reactant. *Res Commun Chem Pathol Pharmacol* 75;1992:109-115.

20. Nevalainen T.J., Grönroos J.M. and Kortesuo P.T. Pancreatic and synovial-type phospholipase A2 in serum in severe acute pancreatitis. *Gut* 34;1993:1133-1136.

21. Rintala E.M. and Nevalainen T.J. Synovial-type (group II) phospholipase A2 in serum of febrile patients with a haematological malignancy. *Eur J Haematol* 50;1992:11-16.

22. Pulkkinen M.O., Kivikoski A.. ...d Nevalainen T.J. Group I and group II phospholipases A2 in serum during normal and pathological pregnancy. *Gynecol Obstet Invest*, in press 1993.

23. Schröder T., Kivilaakso E., Kinnunen P.K.J. and Lempinen M. Serum phospholipase A2 in human acute pancreatitis. *Scand J Gastroenterol* 15;1980:633-636.

24. Nevalainen T.J. and Haapanen T.J. Distribution of pancreatic (group I) and synovial-type (group II) phospholipases A2 in human tissues. *Inflammation* 17;1993:453-464.

25. Schröder T. The effect of early pancreatectomy and peritoneal lavage on the development of experimental hemorrhagic pancreatitis (EHP) in pigs. *Scand J Gastroenterol* 17;1982:167-171.

26. Eskola J.U., Nevalainen T.J. and Kortesuo P.T. Immunoreactive pancreatic phospholipase A2 and catalytically active phospholipase A2 in serum from patients with acute pancreatitis. *Clin Chem* 34;1988:1052-1054.

27. Büchler M., Malfertheiner P., Schädlich H., Nevalainen T.J., Friess H. and Beger H.G. Role of phospholipase A2 in human acute pancreatitis. *Gastroenterology* 97;1989:1521-1526.

28. Crowl R.M., Stoller T.J., Conroy R.R. and Stoner C.R. Induction of Phospholipase A2 gene expression in human hepatoma cells by mediators of acute phase response. *J Biol Chem* 266;1991:2647-2651.

29. Opie E.L. and Meakins J.C. Data concerning the etiology and pathology of hemorrhagic necrosis of the pancreas (acute hemorrhagic pancreatitis). *J Exp Med* 11;1909:561-578.

30. Schmidt H. and Creutzfeldt W. The possible role of phospholipase A in the pathogenesis of acute pancreatitis. *Scand J Gastroenterol* 4;1969:39-48.

31. Aho H.J., Nevalainen T.J., Lindberg R.L.P. and Aho A.J. Experimental pancreatitis in the rat. The role of phospholipase A in sodium taurocholate-induced acute haemorrhagic pancreatitis. *Scand J Gastroenterol* 15;1980:1027-1031.

32. Aho H.J. and Nevalainen T.J. Experimental pancreatitis in the rat. Light and electron microscopical observations on early pancreatic lesions induced by intraductal injection of trypsin, phospholipase A2, lysolecithin and non-ionic detergent. *Virchows Arch (Cell Pathol)* 40;1982:347-356.

33. Wilson H.A., Askari A.D., Neiderhiser D.H., Johnson A.M., Andrews B.S. and Hoskins L.C. Pancreatitis with arthropathy and subcutaneous fat necrosis. *Arthritis Rheum* 26;1983:121-126.

34. Goke B., Meyer T., Loth H., Adler G. and Arnold R. Characterization of phospholipase A2 activity in aspirates of human pancreatic pseudocysts after isolation by reversed-phase high performance liquid chromatography. *Klin Wochenschr* 67(3);1989:110-112.

35. Grönroos J.M. and Nevalainen T.J. Increased concentration of synovial-type phospholipase A2 in serum and pulmonary and renal complications in acute pancreatitis. *Digestion* 52;1992:232-236.

36. Yasuoka A. and Ogawa R. Changes in pancreatic lecithinase A and its significance during shock. *Jap J Anesthesiol* 28;1979:681-688.

37. Holm B.A., Keicher L., Liu M.Y., Sokolowski J. and Enhorning G. Inhibition of pulmonary surfactant function by phospholipases. *J Appl Physiol* 71;1991:317-321.

38. Bird N.C., Goodman A.J. and Johnsson A.G. Serum phospholipase A2 activity in acute pancreatitis: an early guide to severity. *Br J Surg* 76;1989:731-732.

39. Hietaranta A.J., Aho H.J., Grönroos J.M., Hua Z.Y. and Nevalainen T.J. Pancreatic phospholipase A2 in proximal tubules of rat kidney in experimental acute pancreatitis and after intravenous injection of the enzyme. *Pancreas* 7;1992:326-333.
40. Grönroos J.M., Hietaranta A.J. and Nevalainen T.J. Renal tubular cell injury and serum phospholipase A2 activity in human acute pancreatitis. *Br J Surg* 79;1992:800-801.
41. Matthys E., Yogendra P., Kreisberg J., Stewart J.H. and Venkatachalam M. Lipid alterations induced by renal ischaemia: pathogenic factor in membrane damage. *Kidney Int* 26;1984:153-161.
42. Kreutner W. and Siegal M.I. Biology of leukotrienes. *Ann Rep Med Chem* 19;1984:241-251.
43. Nguyen V.O.D., Cieslinski D.A. and Humes H.D. Importance of adenosine triphosphate in phospholipase A2-induced rabbit renal proximal tubular cell injury. *J Clin Invest* 82;1988:1098-1105.
44. Humes H.D., Nguyen V.O.D., Cieslinski D.A. and Messana J.M. The role of free fatty acids in hypoxia-induced injury to renal proximal tubule cells. *Am J Physiol* 256;1989:F688-F696.

METHODS FOR THE DETERMINATION OF THE ACTIVITY OF
PHOSPHOLIPASES AND THEIR CLINICAL APPLICATION

E.coli-based assay compared to a photometric micelle assay

Johannes Aufenanger[1], Michael Samman[1], and Michael Quintel[2]

[1]Institute for Clinical Chemistry
[2]Institute for Clinical Anaesthesiology and Surgical Intensive Care Medicine
Klinikum Mannheim
Faculty for Clinical Medicine of the University of Heidelberg
D-68135 Mannheim, Germany

INTRODUCTION

Many assays for the determination of phospholipase A_2 activity [for review: 1] have been developed. The choice of a detection method depends partly on the purpose of a particular experiment. For example, some assays can be used on purified enzymes but are incompatible with crude systems, some methods provide a continuous assay and generate a time course while others do not, and some methods are amenable to automation while others are not. However, the most important consideration in the choice of the detection method is the sensitivity required for the particular enzyme which depends on the quantity of enzyme available and on its specific activity. This point is especially important for the assay of phospholipases A_2 in human plasma, which are found in lower quantities and are, in general, less active than their counterparts from venom.

There are many assays based on artificial substrates such as liposomes or micelles which are possibly not the best substrates for the phospholipase A_2 in human plasma. Synovial phospholipase A_2, identical with serum phospholipase A_2, prefers membrane phospholipids as a substrate [2, 3]. There is some evidence that membranes of Gram-negative bacteria are the natural substrate of several extracellular phospholipases [4].

Other assays use membranes labeled only with one radioactive tracer , e.g. oleic acid. They do not monitor the "true activity" of the phospholipase A_2 because the released but unlabelled

Esterases, Lipases and Phospholipases, Edited by M.I. Mackness
and M. Clerc, Plenum Press, New York, 1994

fatty acids are not taken into account. Therefore, it is not possible to calculate the phospholipase A$_2$ activity concentration in "U/l".

The photometric micelle assay developed by Hoffmann [5] which has recently become commercially available suffers from lack of sensitivity for normal and moderately elevated phospholipase A$_2$ activity concentrations as shown later. But it monitors the true activity.

The *E.coli* membrane assay in the form described represents a method which combines the advantage of the sensitivity of a radiometric assay based on a natural substrate, with the determination of the true and total activity of the phospholipase A$_2$. For this purpose, the released radioactivity has been related to the sum of all liberated fatty acids. The principle of this method is the hydrolysis of radioactive oleate in the *sn-2* position of membrane phospholipids. For the quantitation of the enzyme activity, the liberated oleate has to be separated from the remaining radioactive phospholipids. We compared this method with the photometric assay [5].

We furthermore describe a modified application of the *E.coli*-based phospholipase A$_2$ assay, by which we are able to determine selectively the catalytically active pancreatic phospholipase A$_2$ without interference from of the inflammatory serum phospholipase A$_2$. This assay enables us to discriminate between edematous and necrotizing pancreatitis in the early stage of the illness. We are convinced that in this specific field the determination of the catalytic activity concentration of the pancreatic phospholipase A$_2$ will be superior to one determination of the enzyme concentration.

METHODS

E.coli-based phospholipase A$_2$ Assay

Phospholipase A$_2$ (EC 3.1.1.4) activity was determined by using [1-14C]oleate-labeled *Escherichiae coli* (K12) as substrate with modifications first described by Elsbach [6]. The essentials of the assay have been described earlier [7]. An *E.coli* culture was incubated with 7.4•10^6 Bq/l oleic acid bound to 0.4 g/l fatty acid free bovine serum albumin at 37°C for 75 min.. Thereafter the bacteria were sedimented, resuspended in fresh medium, and allowed to incorporate the rest of the labeled precursor. The labeled bacteria were washed once with 10 g/l albumin in saline followed by repeated washings with saline. After centrifugation the *E.coli* were autoclaved for 15 min at 120°C. Prior to use, the *E.coli* suspension was lyophilized in appropriate aliquots in extraction tubes. More than 95% of the incorporated label was found in phospholipids and, as demonstrated by bee venom phospholipase A$_2$ hydrolysis, more than 95% of the [1-14C]oleate was in the *sn-2* position as proven by thin-layer chromatography [7].

The lyophilized labeled *E.coli* membranes (18 nmol phospholipid, 180.000 dpm) were resuspended in an assay mixture containing 100 mM Tris-HCl (pH 8.0), 5 mM CaCl$_2$ and 20 g/l fatty acid free bovine serum albumin, and 25 µl of serum or other fluids in a final volume of 150 µl. Reaction mixtures were incubated at 37°C for 15 min while gently shaking to prevent the sedimention of the membranes. The reaction was stopped by addition with 100 µl of 30 mM EDTA, and the lipids extracted with 1250 µl of modified Dole [8] reagent (propane-2-ol/*n*-heptane/1 mol/l H$_2$SO$_4$, 40:10:1, v/v/v) followed by 200 µl of heptane and 750 µl of H$_2$O. After phase separation 0.5 ml of the organic phase was passed through a reusable column filled with 100 mg of aminopropyl solid phase (Macherey&Nagel, Düren, Germany). Free [1-^{14}C]oleic acid was quantitatively eluted with diethylether/acetic acid (2%) into a scintillation vial, while unhydrolyzed substrate remained on the column. Blank activity, resulting from

Table 1. Comparison of the procedures of the two presented PLA$_2$ assays radiometric *E.coli* assay and the photometric mixed micelle assay

	Radiometric *E.coli* assay	Photometric micelle assay
Substrate	Autoclaved membranes phosphatidylethanolamine/cardiolipin [80/20]	Mixed micelles, phosphatidylcholine/phosphatidylethanolamine [50/50], Triton-X, bile acids
Incubation	5 mmol/l Ca^2+ , 37°C, 15 min in the presence of albumin (2%)	4 mmol/l Ca^2+ , 37°C, 15 min
Reaction stop	EDTA (10 mmol/l)	EDTA (8 mmol/l)
Detection procedure	Radioactive free fatty acids: Dole´s extraction Solid phase extraction Scintillation	Increase in free fatty acids: Enzymic determination of free fatty acids Photometry

residual [1-^{14}C]oleic acid not incorporated into lipids, was substracted for each sample. Enzyme activity was calculated as released fatty acids based on the ratio between counted radioactivity and total liberated fatty acids as determined by HPLC [9, 10]. The fatty acid pattern and the relative concentrations for serum phospholipase A$_2$ were established with purified human serum phospholipase A$_2$ as standard. The phospholipase A$_2$ was isolated and purified from sera of intensive care patients [11]. 1 U/l was equivalent to 5.8 x 10^6 dpm/l per minute.

Photometric phospholipase A$_2$ assay

The activity of phospholipase A$_2$ was determined by the photometric assay recently described by Hoffmann et al. [5]. Briefly, a lyophilized phospholipid/detergent mixture containing phosphatidylcholine, phosphatidylethanolamine (50:50, w/w), Triton X-100 and sodium deoxycholate (Boehringer Mannheim, Germany) served as substrate. The total liberated free fatty acids were determined with an enzymic test kit (Boehringer Mannheim, Germany) based on a coenzyme A reaction. Fatty acids are incorporated into acyl-CoA using acyl-CoA synthetase then oxidized to 2,3-trans-enoyl-CoA with acyl-CoA oxidase. The peroxide generated in the latter step leads to the formation of a chromophore by oxidative coupling of an aminoantipyrine and a phenol derivative. For direct comparison, the *E.coli* assay and the photometric micelle assay are paralled in table 1.

Pancreatic phospholipase A$_2$ assay

Pancreatic phospholipase A$_2$ (type I) in human serum was determined by its catalytic activity after heat inactivation of type II phospholipase A$_2$ (60°C for 60 min) using the *E.coli* assay with reaction buffer of pH 6 [12]. The residual phospholipase A$_2$ activity was characterized as type I phospholipase A$_2$ by an inhibiting monoclonal antibody against type I phospholipase A$_2$.

RESULTS

Character of the substrate

We investigated the *E.coli* assay for interference from various lipases. Human serum contains several lipases and phospholipases, i.e. lipoprotein lipase, hepatic triacylglycerolipase, pancreatic lipase, which might influence the accuracy of the assay. Therefore, *E.coli* membranes were incubated with different (phospho)lipases (table 2).

The sera were diluted to approximately 20 U/l (332 nkat/l) as determined by the photometric micellar assay. Radioactive oleic acid was released from the *sn-2* position, as expected. The liberation of labeled palmitic acid by phospholipase A_2 activity can be explained by the co-labelling of the *sn-2* position. This effect reflects exactly the ratio of the label between the *sn-1* and *sn-2* position. Acute pancreatitis often causes an increase in phospholipase A_2, and it is important to discriminate this activity from that of the pancreatic lipase. Therefore, we incubated the labeled membranes with normal serum spiked with pancreatic lipase to 1000 U/l (16667 nkat/l). As shown in table 2 the release of labeled fatty acids is barely different from that of normal sera.

Linearity

Using sera from septic patients the linear range for serum phospholipase A_2 activity was established up to 7.2 U/l (120 nkat/l) (fig. 1) corresponding to a hydrolysis rate of 8%. Under assay conditions the curve deflected from linearity at $42 \cdot 10^6$ dpm/l per minute. To monitor the enzyme activity beyond this range, the sera of patients with septicemia and polytraumatic shock

Table 2. The specificity of the *E.coli* membranes as substrate for the serum phospholipase A_2 was verified by incubation with different (phospho)lipases. For this purpose, the phospholipids of *E.coli* membranes were labeled either with $[1-^{14}C]$palmitic acid in the *sn-1* position or with $[1-^{14}C]$oleic acid in the *sn-2* position to the ratio indicated in the table. To check the activity of serum lipases on the *E.coli* membrane healthy volunteers (n=3) were heparinized with 5000 i.U. of heparin and blood was drawn after 10 min. Such sera are rich in lipoprotein lipases und hepatic triacylglycerolipase. The latter exhibits mainly PLA_1 activity. Sera of intensive care patients (n=3) with nearly the same activity on the micellar substrate were chosen for comparison. Additionally, pancreatic lipase was added to a normal serum and tested for the release of radioactive fatty acids from the E.coli membranes. The sera of healthy volunteers (n=3) served as controls.

	Release of fatty acids from the phospholipids		
	Micelle assay	**E.coli based assay**	
	total fatty acids	$[1-^{14}C]$oleic acid	$[1-^{14}C]$palmitic acid
	µmol/min·l	nmol/min·l	nmol/min·l
Sera of healthy volunteers before heparin (n=3)	0.05 ± 0.01^1	5.1 ± 0.2	0.7 ± 0.2
Sera of healthy volunteers 10 min after heparin 5.000 i.E. (n=3)	20.4 ± 1.0	5.7 ± 0.3	3.6 ± 0.4
PLA_2 rich sera of intensive care patient (n=3)	20.0 ± 1.5	1360 ± 120	270 ± 21
Normal sera enriched with pancreatic lipase [1000 U/l] without colipase (n=3)		5.3 ± 0.4	6.4 ± 0.5
labeling ratio of *sn-2* and *sn-1* position		95:5	20:80

^{1} 24 hrs incubation

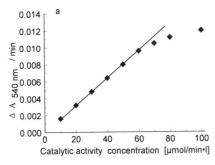 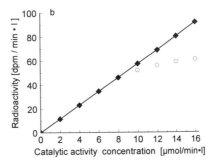

Figure 1. Linearity of the phospholipase A₂ assays.

a) The linear range for serum phospholipase A₂ was established by serial dilution ◆. The cut-off value for the photometric micelle assay was 58 U/l (967 nkat/l)

b) Using the the *E.coli* assay the linear range was established in the same manner ◯. The cut-off value for the radiometric *E.coli* assay of human serum was 7.2 U/l, (120 nkat/l)

Additionally, the linear range was determined with purified PLA₂ isolated from human sera with high phospholipase A₂ contents. The activity of purified human serum phospholipase A₂ was linear up to 14.4 U/l (240 nkat/l) ◆.

must be diluted many times. Surprisingly, the activity of purified human serum phospholipase A₂ was found to be linear up to a hydrolysis rate of 16% corresponding to 14.4 U/l (240 nkat/l). Spontaneous hydrolysis of membrane phospholipids was not observed. The range of the photometric micelle assay was linear up to 60 U/l (1000 nkat/l) [59].

Reliability

Quality control data for intra-assay and inter-assay imprecision for the *E. coli* and the photometric micelle assay, respectively, are given in table 3.

Sensitivity

The limit of detection for the photometric micelle assay was established using various concentrations of fatty acids in samples free of phospholipase A₂ activities. This was done in a different way to Hoffmann et al. [5] who verified the detection limit using water instead of serum. The detection limit depended on the fatty acid concentration and increased from 5.1 U/l (85 nkat/l) without fatty acids in the sample to 12.7 U/l (212 nkat/l) and for a sample with a fatty

Table 3. Intra-assay and inter-assay imprecision of the radiometric PLA₂ assay with membranes and the photometric micelle assay

Intra-assay imprecision[a]			Inter-assay imprecision[b]		
Mean [U/l]	SD [U/l]	CV (%)	Mean [U/l]	SD [U/l]	CV (%)
0.32	0.018	5.8	0.65	0.063	9.7
3.1	0.11	3.5	3.3	0.21	6.3
7.7^	0.22	2.9	7.1	0.38	5.4

[a] Intra-assay imprecision was determined with three different serum pools; (n=12 in each case)
[b] Inter-assay imprecision was determined with three other serum pools; (n=20 in each case)

acid concentration to 2150 µmol/l (fig. 2) The radiometric *E.coli* assay was almost independent of the fatty acid concentration, and the detection limit was 0.35 U/l (5.8 nkat/l) (fig 2). It should be mentioned that the activities of both assays are not strictly comparable because the substrates for the assays are different.

Reference range and patients:

Sera from healthy volunteers and intensive care patients were analyzed with the *E.coli* method (fig. 3a) and compared to the photometric assay. Normal ranges (2.5 to 97.5 percentile) were established in healthy women (n=45) as 0.088 - 2.22 U/l (1.46 - 36.8 nkat/l) and in healthy men (n=40) as lower than 0.88 U/l (14.66 nkat/l). This is the first observation of sex specific normal ranges for the serum phospholipase A_2, but at present, nothing can be said about this difference or its clinical implications. In addition, sera of patients with pathologically high phospholipase A_2 activities served to demonstrate the applicability of this assay. Sera of intensive care patients (n=36) suffering from septicemia, pancreatitis, acute lung failure and other diseases reached values up to 1080 U/l (18000 nkat/l).

For the photometric assay, we plotted the phospholipase A values of the same healthy volunteers (fig. 3b). The presumable normal range extented from 2 to 9 U/l (2.5 to 97.5 percentile). The free fatty acid concentrations of all volunteers were below 700 µmol/l. No sex specific differences could be observed. The sera of the patients of the intensive care unit reached values up to 290 U/l (4865 nkat/l). Actually, the normal range must be estimated within the sensitivity range corresponding to the actual fatty acid concentration in the specimen, i.e. probably below 7.9 U/l (132 nkat/l) for low fatty acid concentration and below 12.7 U/l (212 nkat/l) for high fatty acid concentration. Values below 7.9 U/l (132 nkat/l) have to be considered as accidental.

Correlation of the *E.coli* assay to the micelle assay

The photometric micelle assay assay lacks analytical sensitivity in the normal and moderately elevated phospholipase A_2 range. For this reason we compared both methods at higher phospholipase A_2 ranges only. The results obtained are plotted in fig. 4. They demonstrate an unexpected high correlation (r=0.97, p<0.001). However, the slope (4.1) indicates the obvious preference of the phospholipase A_2 for the membraneous substrate. This may partly be caused by the substrate composition, implying that phosphatidylethanolamine is possibly preferred over other phospholipids e.g. phosphatidylcholine. The micellar substrate contains phosphatidylcholi-

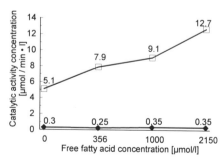

Figure 2. Detection limits for the *E.coli* assay ◆ and the photometric micelle assay ☐ using various concentrations of fatty acids in the sample without PLA₂ activities (abscissa). The detection limits of eachs point is given as mean + 3s (standard deviations)

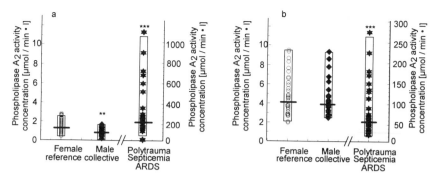

Figure 3. a) Phospholipase A$_2$ activities in sera from healthy male (n=43) and female volunteers (n=55) and intensive care patients (n=36) . The boxes mark the 2.5 and 97.5 percentile , and the horizontal bars the medians of the results. In the Mann-Whitney two sample test the differences between the groups were statistically significant (** p < 0.01, *** p < 0.001)

b) The normal ranges for the photometric assay were established with the same specimens. No statistical difference could be observed for the male and female group

ne and phosphatidylethanolamine in a ratio of 1:1. Since phosphatidylcholine was not hydrolized it is thus not a suitable substrate for the phospholipase A$_2$ [13]. The phospholipase A$_2$ activity, though, is largely dependent on the composition and conformation of the substrate [14].

Pancreatic phospholipase A$_2$

In cases of acute pancreatitis remarkably elevated levels of phospholipase A$_2$ are measurable in human serum samples (fig. 5a) . Two isoenzymes, both the pancreatic phospholipase A$_2$, type I, and the inflammatory serum phospholipase A$_2$ contribute to the elevated activities. In contrast to the type II phospholipases A$_2$ the pancreatic phospholipase A$_2$ is heat resistant (60°C for at least 60 min) as demonstrated with an inhibitory monoclonal antibody. Thus, the pancreatic phospholipase A$_2$ can easily be determined without interferences of the type II phospholipase A$_2$ [12]. Using the heat inactivation technique serum samples of 41 patients suffering from acute pancreatitis were analyzed for the catalytic activity of their pancreatic phospholipase A$_2$

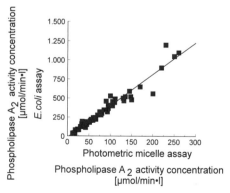

Figure 4. Comparison of the *E.coli* based radiometric assay and the micelle assay for the determination of phospholipase A$_2$ activities in human sera. The micelle assay showed markedly lower activities than the membrane assay. R = 0.97 ; Y = 4.1 X - 15.3; N = 55 . For comparison only phospholipase A$_2$ activities above 15 U/l were chosen because of the high detection limit of the photometric assay

on the first and second day of hospitalization. In 10 of the 41 cases clearly elevated values of catalytic heat resistant phospholipase A_2 (7.2 to 81.2 U/l; 120 to 1353 nkat/l) could be observed (fig. 5b). This group of patients was characterized by necrotizing pancreatitis proved by surgery. The other 31 patients with activities below 7 U/l (117 nkat/l) had almost no complications and recuperated rapidly.

DISCUSSION

The selection of a membrane fatty acid-labeled phospholipid substrate from *E.coli* allows an approximation of the physiological array of phospholipids and may avoid inherent problems associated with the physicochemical characteristics of synthetic substrates. The disadvantages are that the phospholipid composition is controlled by *E.coli* under highly sophisticated growth conditions. Secondly, the amount of phospholipid substrate present in the assay is extremely low. On the other hand, the characteristics of phospholipid micelles, i.e. packing density and lipid surface charge, can also have marked effects on phospholipase A_2 activation and hydrolysis [1, 14]. These circumstances make a standardization of a catalytic assay quite difficult.

This work indicates that phospholipase A_2 activity in human serum hydrolyzed phosphatidylcholine/phosphatidylethanolamine/deoxycholic acid/Triton X-100 mixed micelles poorly, and that autoclaved *E.coli* membranes containing primarily radiolabelled phosphatidylethanolamine were readily hydrolyzed by the enzyme, and thus provided the basis for a sensitive assay system. Similar results were described by Parks et al. [15]. There is no preference of any acyl chain in the *sn-2* position [16], in contrast to phospholipases A_2 from platelets and rheumatoid synovial fluid [17].

Another problem is inherent to the micellar photometric assay which is based on the determination of total fatty acid including those present in the serum of the patient as socalled free fatty acids before the phospholipase A_2 assay. The sera of intensive care patients often contain high free fatty acids concentrations and the additionally liberated fatty acids from the phospholi-

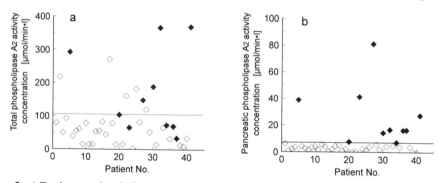

Figure 5. a) Total serum phospholipase A_2 was determined in 41 patients with acute pancreatitis on day 1 and 2 of the hospitalization, of which 31 patients had a mild course of the disease <> and 10 patients a necrotizing form. ◆. The data points represent the higher value of the two days. The difference of the two groups were highly significant (p = 0.0061, Whitney-Mann test). For a cut-off value of 100 U/l (1667 nkat/l) sensitivity was 0.6 and the specificity was 0.84. The total serum phospholipase A_2 does not allow a reliable assessment of the individual prognosis.

b) Heat resistant pancreatic phospholipase A_2 was determined in the same patients: <> mild course of the pancreatitis, ◆ necrotizing form. Each data point reflects the same patient no. from a). The difference of the two groups were very highly significant (p = 0.0001, Whitney-Mann test). For a cut-off value of 7 U/l (117 nkat/l) sensitivity was 1.0 and the specificity was 1.0 .

pase A_2 action contribute little to an increase in the total free fatty acids [7]. The multiple steps of the assay [5], impeded often by a lack of automation, imply that the increase in the phospholipase activity can not reliably be discriminated from the basal free fatty acid concentration, since low differences of absorbance lie within the sensitivity range. This explains why it is not possible to determine phospholipase A_2 activities below 8 U/l (133 nkat/l) by the photometric micelle assay with an acceptable accuracy. In comparison, the upper limit of the linear range of the *E.coli* assay was 7.2 U/l (120 nkat/l) and the assay shows good spreading in this range.

The *E.coli* membrane assay could be demonstrated to be very specific for the phospholipase A_2 of human serum and relatively insensitive to other lipases in human sera , i.e. lipoprotein lipases, hepatic triacylglycerolipase or pancreatic lipase. Although the activity of phospholipase A_1 (hepatic triacylglycerolipase) from human serum on *E.coli* membrane phospholipids appears to be very low in terms of cleaved labeled fatty acids the hydrolysis of the *sn-1* fatty acids resulted in a slight increase in radioactive lyso-phosphatidylethanolamine as assessed by the release of [^{14}C]palmitic acid (see table 2: 0.7 vs. 3.6 nmol/min* l). Only 5% to 10% of the fatty acid label is incorporated into the sn-1 position, and lyso-phosphatidylethanolamine is retained by the solid phase column technique for the separation of fatty acids. Consequently, it does not contribute to the result. In comparison, the photometric micelle assay is affected by the phospholipase A_1 activity of the specimen as we have recently demonstrated [16].

Activity was expressed accounting for the sum of the released fatty acids. In previous publications [13, 14] the phospholipase A_2 activity of the radiometric *E.coli* assay has been expressed as hydrolysis rate based only on the rate of release of the labeled fatty acids. This renders comparison and longitudinal studies questionable because of the possible variability in the degree of specific labelling from lot to lot. We could demonstrate an obvious preference of the enzyme to the membranous substrate (fig. 4) resulting in activities four times as high. This, in turn, contributes additionally to the higher sensitivity of the biomembrane assay.

The concordant kinetics of phospholipase A_2 activity and the other acute phase proteins (serum amyloid A and C-reactive protein) in the time course study are compatible with the conclusion that phospholipase A_2 in human plasma has its origin in hepatocytes, and can be considered as an acute phase protein itself. This hypothesis has been supported by Crowl et al [2] who could demonstrate a common genetic regulation of phospholipase A_2 and of the mentioned acute phase proteins in HepG2 cells. And indeed, we very recently proved the expression and secretion of phospholipase A_2 in primary human hepatocytes (data not yet published).

One of the most important questions concerning acute pancreatitis is the early differentiation between mild (odematous) and severe (necrotizing) forms of the disease. Most laboratory parameters such as lipase and amylase, and even the total serum phospholipase A_2 are not able to assess the individual prognosis in the early stage of the disease (fig. 5). In contrast, the catalytically active pancreatic phospholipase A_2 (type I) determined in human serum enables us to discriminate between the two major forms of acute pancreatitis. The demand for sensitivity and reliability is only fullfilled by the *E.coli* based assay or a comparably sensitive assay, but not by the commercially available photometric assay. We are convinced that in this specific field the determination of the catalytic activity concentration of the pancreatic phospholipase A_2 will be superior to the determination of the enzyme concentration by an antibody, and can help to estimate the high risk of an individual patient in the early stage of the pancreatitis [12].

In conclusion, the *E.coli* based assay is more reliable and more sensitive than the photometric micelle assay for the determination of the specific activity of phospholipase A_2 in human serum. The membrane substrate is easily prepared. The normal range is below 2.22 U/l

(37 nkat/l). With the method described the activities of phospholipase A_2 of human serum samples can be reliably determined. The inter- and intraassay imprecisions were below 5% and 10%, resp. Pathologically elevated phospholipase A_2 activities were increased up to 500 fold above the normal range which indicates a high diagnostic potential. An important diagnostic field has been disclosed for the use of the *E.coli* based assay in terms of the determination of the heat resistant phospholipase A_2 in patients with acute pancreatitis.

REFERENCES

1. Reynolds, L.J., Washburn, W.N., Deems, R.A., and Dennis, E.A., Assay strategies and methods for phospholipases. *Methods Enzym.* **197**: 1991; 3-23

2. Crowl, R.M., Stoller, T.J., Conroy, R.R., and Stoner, C.R. Induction of phospholipase A_2 gene expression in human hepatoma cells by mediators of the acute phase response. *J. Biol. Chem.* **266**: 1991; 2647-2651

3. Stefanski, E., Pruzanski, W., Sternby, B., and Vadas, P. Purification of a soluble phospholipase A_2 from synovial fluid in rheumatoid arthritis. *J. Biochem. Tokyo* **100**: 1986; 1297-1303

4. Elsbach, P., Weiss, J., Wright, G.W.; Forst, S., van den Bergh, C.J., and Verheij, H.M. Regulation and role of phospholipases in host-bacteria interaction. *Prog. Clin. Biol. Res.* **349**: 1990; 1-9

5. Hoffmann, G.E., and Neumann, U. Modified photometric method for the determination of phospholipase A activities. *Klin. Wochenschr.* **67**: 1989; 106-109

6. Elsbach, P., Weiss, J., Franson, R.C., Beckerdite-Quagliata, S., Scheider, A., and Harris, L. Separation and purification of a potent bactericidal/permeability-increasing protein and a closely associated phospholipase A_2 from rabbit polymorphonuclear leukocytes. *J. Biol. Chem.* **254**: 1979; 11000-11009

7. Aufenanger, J., Zimmer, W., and Kattermann, R. Characteristics of a radiometric E.coli-based assay - Modified method for the determination of human serum phospholipase A_2 and its clinical application. *Clin. Chem.* **39**: 1993; 605-613

8. Dole, V.P. and Meinertz, H. Microdetermination of long-chain fatty acids in plasma and tissues. *J. Biol. Chem.* **235**: 1969; 2595-2599

9. Durst, H.D., Milano, M., Kikta, E.J., Connelly, S.A., and Grushka, E. Phenylacyl esters of fatty acids via ether catalysts for enhanced ultraviolett detection in liquid chromatography. *Anal. Chem.* **41**: 1975; 1797-1801

10. Püttmann, M., Krug, H., v. Ochsenstein, E., and Kattermann, R. Determination of serum free fatty acids in the picomol range by an easy, fast and reliable high-performance liquid chromatography method. *Clin. Chem.* **39**: 1993; 825-832

11. Aufenanger, J. and Zimmer, W. 1993, Rapid isolation and purification of phospholipase A_2 from human serum *submitted*

12. Samman, M, Aufenanger, J. Püttmann, M., Zimmer, W., M. Quintel, and Katterman, R. Catalytically active pancreatic Phospholipase A2 as a prognostic marker in acute pancreatitis *Gastroenterology*: 1993 *submitted*

13. Märki, F., and Franson, R. Endogenous suppression of neutral-active and calcium-dependent phospholipase A_2 in human polymorphonuclear leukocytes. *Biochim. Biophys. Acta* **879**: 1986; 149-156

14. Waite, M. The Phospholipases. in: *Handbook of Lipid Research*, D.J. Hanahan (editor); Plenum Press, New York, Vol 5: 1987; 79-133

15. Parks, T.P., Lukas, S., and Hoffman, A.F. Purification and characterization of a phospholipase A2 from human osteoarthritic synovial fluid. in: *Phospholipase A_2*, P.Y.K. Wong and E.A. Dennis (eds.), Plenum Press, New York, 1990; 55-81,

16. Püttmann, M., Aufenanger, J., v. Ochsenstein, E., Dürholt, S., v. Ackern, K., Harenberg, J., and Hoffmann, G.E. Increased phospholipase A activities in sera of intensive-care patients show sn-2 specificity but no acyl-chain selectivity *Clin. Chem.* **39**: 1993; 782-788

17. Schalkwijk, C.G., Märki, F., and Van den Bosch, H. Studies on the acyl-chain selectivty of cellular phospholipases A_2. *Biochim. Biophys. Acta* **1044**: 1990; 139-146

18. Elsbach, P., Weiss, J., and Kao, L. The role of intramembrane Ca^{2+} in the hydrolysis of the phospholipids of Escherichia coli by Ca^{2+}-dependent phospholipases. *J. Biol. Chem.* **260**: 1985; 1618-1622

PHOSPHOLIPASES AS CLINICAL TOOLS: MEASUREMENT

Hugues Chap

INSERM, U326
Hopital Purpan
Place du Docteur Baylac
31059 Toulouse
France

The round table discussion of the use of phospholipases with particular reference to phospholipases A_2 as clinical tools in the diagnosis of septicaemia and other inflammatory diseases began with a comparison of the methods available for determining the enzyme in biological fluids particularly serum or plasma. Three methods are currently in routine use in specialist laboratories, two of which measure enzyme activity either by the use of a fluorescent substrate or by determining released fatty acids using a commercially available enzymic reagent. The third assay is a monoclonal antibody based immunoassay for the enzyme protein. Data presented by Professor Nevalainen showing that acute pancreatitis is accompanied by an elevation in the serum concentration of type II-phospholipase A_2 protein provided convincing evidence of the usefulness of the latter method. It was strongly enforcised that the measurement of plasma phospholipase A_2 for diagnostic purposes is a highly specialised task and should only be undertaken in competent laboratories as no quality control is currently available.

The discussion turned to the possible physiological function of phospholipase A_2. A clear understanding of the role of phospholipase A_2 in the pathophysiology of septic shock and several other inflammatory processes is not possible at the moment as the function of these enzymes has yet to be elucidated. Currently, the only clear hypothesis as to the physiological role of the enzymes concerns a possible bactericidal activity of type II- phospholipase A_2 due to its ability to hydrolyse membrane phospholipids. This may also be the reason for the high activity found in human seminal fluid, the prostatic origin of the enzyme caused considerable interest and it was noted that this is also the case for the natural inhibitors, the annexins.

Finally, two interesting presentations were given by Dr Hamosh on the pathological consequences of a paediatric deficiency of lecithin:cholesterol acyltransferase activity and by Dr Moss on a plasma phospholipase D which can hydrolyse the phospholipid anchor of membrane attached proteins, which ended the discussion.

LIST OF CONTRIBUTORS

D. Ameis (University Hospital Eppendorf, Hamburg, Germany)

J. Aufenanger (Institute for Clinical Chemistry, Mannheim, Germany)

L. Chan (Baylor College of Medicine, Houston, USA)

H. Chap (INSERM-U326, Purpan Hospital, Toulouse, France)

M. Clerc (University Hospital Saint-Andre, Bordeaux, France)

P. Durrington (Royal Infirmary, Manchester, UK)

C. Erlanson-Albertsson (University of Lund, Lund, Sweden)

G. Férard (University Louis Pasteur Strasbourg, Illkirch, France)

M. Hamosh (Georgetown University Medical Centre, Washington D.C. USA)

E. Heymann (University of Osnabruck, Osnabruck, Germany)

A. Iron (University of Bordeaux 2, Bordeaux, France)

R. James (University Hospital Geneva, Geneva, Switzerland)

W. Junge (Friedrich-Ebert Hospital, Neumunster, Germany)

K. Kutty (Memorial University of Newfoundland, St. Johns, Canada)

D. Lombardo (INSERM-U260, Marseille, France)

A. Lykidis (Aristotelian University, Thessaloniki, Greece)

M. Mackness (Royal Infirmary, Manchester UK)

T. Nevalainen (University of Turku, Turku, Finland)

J. Olivier (URA CNRS 1283, CHU Saint Antoine, Paris, France)

S. Parthasarathy (Emory University, Atlanta, USA)

M. Reddy (Children's Hospital, Milwaukee, USA)

E. Reiner (University of Zagreb, Zagreb, Croatia)

M. Robbi (Catholic University of Louvain, Brussels, Belgium)

M. Sanchez-Crespo (University of Valladolid, Valladolid, Spain)

E. Schmidt (Isernhagen, Germany)

W. Uhl (University of Berne (Inselspital) Berne, Switzerland)

P. Vadas (Wellesley Hospital, Toronto, Canada)

H. Van den Bosch (University of Utrecht, Utrecht, Netherlands)

H. Van Lith (University of Utrecht, Utrecht, Netherlands)

R. Verger (CNRS-ERS26, Marseille, France)

C. Walker (University of Reading, Reading, UK)

F. Zschunke (Georg-August University Gottingen, Gottingen, Germany)

INDEX

Lecithin-cholesterol acyltransferase *(cont)*
  functional regions, 170-171
  gene structure, 170-171
  heterogeneity, 171-174
  mutations, 172, 173
  Tangier disease, 170, 174
Lipase
  activity, 115-119, 121-128, 179-180, 183-185
  bile salt stimulated, 107-111, 149-157
  carboxylester, 107-111, 149-157; *see also* cholesteryl esterase
  catalytic properties, 188
  classification, 159-160
  diagnostic use, 179-182
  digestive, 139-147
  gastric
    amino acid sequence, 140
    in disease, 143-144
    human, 3-9, 139-147
    in infants, 141-143
    monoclonal antibody, 3-9
    quantitation, 3-9
    species differences, 141
    structure, 139-14
  hepatic triacylglycerol
    activity, 115-119, 122, 129
    in disease, 115-119, 122, 129
    expression, 116-117
    N-glycosylation, 115-117
    lipid metabolism, 115-119
    structure, 122
  lipoprotein
    activity, 115-119, 121-123, 129-138
    chylomicron remnant, 133
    deficiency, 122, 129
    in disease, 121-128, 129-138
    expression, 117
    lipid metabolism, 121-128, 129-138
  lysosomal acid
    activity, 123-125
    cholesteryl-ester storage disease, 121
    deficiency, 125-126
    lipid metabolism, 121-128
    physiological role, 124
    sequence, 124
    Wolman disease, 121
  pancreatic, 107-111, 149-157, 159-168, 179-182, 187-190
  preduodenal, 139-141
  purification, 187
  rabbit, 187-190
  reference method, 181
  in renal failure, 183-185

Lipase *(cont)*
  structural domain, 2-4
  X-ray crystallography, 1-2
Lipoprotein
  high-density, 65, 67-69, 103-105, 134, 169, 183
  low-density, 67, 69-71, 129-138, 169, 183, 245-251
    phospholipase, 245-251
  lyso-phosphatidylcholine, 245-251
  metabolism, 115-119, 121-123, 129-138, 169
  modification, 69, 245-251
  pathology, 129-138
  triglyceride, 129-138, 183-185
  very-low density, 115, 122, 129-139, 169, 183-185, 249

Parabene
  hydrolysis, 81-83
Paraoxonase, *see also* Phosphoric Triester Hydrolase
  catalytic properties, 58-60
  classification, 57, 64, 92, 99
  in disease, 59, 67
  human serum, 57-64, 65-73, 103-105
  and LDL-oxidation, 69-71
  monoclonal antibody, 103-105
  rat, 103-105
  pesticide hydrolysis, 91-98
  polymorphism, 66
Pesticide
  detoxication, 91-94
Phospholipase
  action, 245-251
  C, 134
  D, 274
  guinea-pig, 2-4
Phospholipase A$_2$
  activity, 235-244, 249
  in acute pancreatitis, 253-261
  assay comparison, 263-272, 273
  astrocytes, 213-222
  cellular, 193, 199, 203-204
  characterisation, 203-204
  chondrocytes, 213-222
  clinical application, 263-272
  cytokines, 193, 196, 216-217
  detection, 215-216, 225, 228, 236, 254, 263-272, 273
  in endotoxin shock, 203-211, 223, 235
  German multicentre trial, 223-232
  Hep G2 cells, 205, 207, 213-222
  high levels, 235